International Conference on Functional Materials and Nanotechnologies 2013

(FM&NT2-13)

IOP Conference Series: Materials Science and Engineering
Volume 49

Tartu, Estonia
21 - 24 April 2013

ISBN: 978-1-63439-860-2
ISSN: 1757-8981

Printed from e-media with permission by:

Curran Associates, Inc.
57 Morehouse Lane
Red Hook, NY 12571

Some format issues inherent in the e-media version may also appear in this print version.

Printed by Curran Associates, Inc. (2015)

For permission requests, please contact the Institute of Physics
at the address below.

Institute of Physics
Dirac House, Temple Back
Bristol BS1 6BE UK

Phone: 44 1 17 929 7481
Fax: 44 1 17 920 0979

techtracking@iop.org

Additional copies of this publication are available from:

Curran Associates, Inc.
57 Morehouse Lane
Red Hook, NY 12571 USA
Phone: 845-758-0400
Fax: 845-758-2634
Email: curran@proceedings.com
Web: www.proceedings.com

Table of contents

Volume 49

International Conference on Functional Materials and Nanotechnologies 2013 (FM&NT2-13)
21–24 April 2013, Institute of Physics, University of Tartu, Estonia

Accepted papers received: 4 November 2013
Published online: 13 December 2013

IOP Conf. Series: Materials Science and Engineering **49** (2013) 011001 doi:10.1088/1757-899X/49/1/011001

Preface

The International Conference Functional Materials and Nanotechnologies (FM&NT – 2013) was held in Tartu, 21–24 April 2013 at the Dorpat Conference Centre. The conference was organised by Institute of Physics University of Tartu.

The FM&NT conference series was started in 2006 by scientists from the Institute of Solid State Physics, University of Latvia. It is yearly conference bringing together researchers from whole world. The warm and open atmosphere of this scientific conference has turned it into event where people from rather different fields meet under the common name functional materials and nanotechnology. It is particularly important for early stage scientists who are looking for new knowledge and contacts with people from various fields. Our Latvian colleagues with their success in internationalization made us neighbouring Estonians so jealous that we could not withstand proposing organise the conference in every second year in Estonia. Actually this is in a way the continuation of idea of famous Baltic seminars which took place in last century over several decades. Due to political constraints these seminars were only opened to scientist of former Eastern Europe countries, but had a great popularity and attendance over whole Soviet Union. Much fruitful cooperation started from the initial personal contacts of scientists at these seminars hold twice per year once in Latvia and second time in Estonia. In last FM&NT 2012 conference, the decision was made that Institute of Physics, University of Tartu will organise the event in Tartu in 2013.

Along with traditional topics as multifunctional materials, nanomaterials, materials for sustainable energy applications and theory, this conference focused on studies with the help of synchrotron radiation and other novel light sources.

The number of registered participants from 21 countries was nearly 300. During the three days of the conference 14 invited, 45 oral, 5 commercial talks were delivered and more than 200 posters were presented. In framework of Conference more than 20 papers were submitted to *Topical issue of Physica Scripta*. 62 proceedings, based on the work presented, were included in this volume of *IOP Conference Series: Materials Science and Engineering*.

Additional information about FM&NT-2013 is available at its homepage http://fmnt.ut.ee/

The Organizing Committee would like to thank all the speakers, contributors, session chairs, referees and other involved staff for their efforts in making the FM&NT-2013 successful.

Special thanks for the hard work made with the conference papers goes to Guest Editors:
Prof. Mikhail Brik, Dr. Svetlana Zazubovitš, Dr. Arvo Kikas, Dr. Rainer Pärna.

The Organizing Committee hopes that Conference gave participants good insight into recent development in nanotechnology, sustainable energetics, processing and modelling of multifunctional materials and research made with synchrotron radiation or other novel light sources. We sincerely hope that Conference has provided support for information circulation and brought together many young and experienced scientists from different fields, leading to fruitful discussions. We hope that all participants had good and memorable time in Tartu.

On the behalf of organising committee
Prof. Ergo Nõmmiste
Dr. Marco Kirm
Dr. Toomas Plank

IOP Conf. Series: Materials Science and Engineering **49** (2013) 011001 doi:10.1088/1757-899X/49/1/011001

We would like to acknowledge financial support by Tartu University Development Fund, Estonian national projects for Materials Science, Science Internationalisation and Centre of Excellence programs based on contribution of European Structural Funds and Estonian national funding. Also graduate school on "Functional Materials and technologies" supported by European Social Fund has given generous contribution to help participation of our early stage researchers. This support is gratefully acknowledged. We also would like to acknowledge and thank all the sponsors. We hope that continuation of the tradition introduced with this conference will strengthen international cooperation of scientists from Baltic state with researchers of other countries.

UNIVERSITY OF TARTU 1632

European Union
European Social Fund

Investing in your future

EUROPEAN UNION
STRUCTURAL FUNDS

IOP Conf. Series: Materials Science and Engineering **49** (2013) 011001 doi:10.1088/1757-899X/49/1/011001

Sponsors

Raith

BRUKER

Vistec
Lithography

S.A.I.N.T
Tech
Security Analytics Industry Nano Technics

Doma
Group

ARMGATE

AMERICAN
ELEMENTS®
America's Leading Manufacturer of Engineered and Advanced Material Products

IOP Conf. Series: Materials Science and Engineering **49** (2013) 011001 doi:10.1088/1757-899X/49/1/011001

International Steering Committee

Chairperson: Andris Sternbergs, Institute of Solid State Physics, University of Latvia, Latvia

Members:

1. Juras Banys, Vilnius University, Lithuania
2. Gunnar Borstel, University of Osnabrueck, Germany
3. Niels E. Christensen, University of Aarhus, Denmark
4. Robert A. Evarestov, St. Petersburg State University, Russia
5. Claes-Goran Granqvist, Uppsala University, Sweden
6. Dag Høvik, The Research Council of Norway, Norway
7. Marco Kirm, University of Tartu, Estonia
8. Witold Łojkowski, Institute of High Pressure Physics, Poland
9. Ergo Nõmmiste, Institute of Physics, University of Tartu, Estonia
10. Helmut Schober, Institut Laue-Langevin, France
11. Ingólfur Thorbjörnsson, Icelandic Centre for Research, Iceland
12. Marcel H. Van de Voorde, University of Technology Delft, The Netherlands

Program Committee

Chairperson: Marco Kirm, University of Tartu, Estonia

Members

1. Ergo Nõmmiste, Institute of Physics, University of Tartu, Estonia
1. Enn Lust, Institute of Chemistry, University of Tartu, Estonia
2. Jaak Kikas, Institute of Physics, University of Tartu, Estonia
3. Enn Mellikov, Tallinn Technical University, Estonia
4. Alexey Romanov, Institute of Physics, University of Tartu, Estonia
5. Mikhail Brik, Institute of Physics, University of Tartu, Estonia

Local Organizing Committee

Chairperson: Ergo Nõmmiste, Institute of Physics, University of Tartu, Estonia

Members

1. Marco Kirm, University of Tartu, Estonia
2. Jaak Kikas, Institute of Physics, University of Tartu, Estonia
3. Toomas Plank, Institute of Physics, University of Tartu, Estonia
4. Väino Sammelselg, Institute of Physics, University of Tartu, Estonia
5. Vambola Kisand, Institute of Physics, University of Tartu, Estonia
6. Vladimir Hižnjakov, Institute of Physics, University of Tartu, Estonia
7. Aleksandr Luštšik, Institute of Physics, University of Tartu, Estonia
8. Rainer Pärna, Institute of Physics, University of Tartu, Estonia
9. Mikhail Brik, Institute of Physics, University of Tartu, Estonia
10. Enn Lust, Institute of Chemistry, University of Tartu, Estonia
11. Alexey Romanov, Institute of Physics, University of Tartu, Estonia

IOP Conf. Series: Materials Science and Engineering **49** (2013) 012001 doi:10.1088/1757-899X/49/1/012001

Growth of ZnO rods on FTO electrodes by spray pyrolysis

Tatjana Dedova[1], Olga Volobujeva[1], Malle Krunks[1], Valdek Mikli[2], Inga Gromyko[1], Atanas Katerski[1], Arvo Mere[1]

[1]Department of Materials Science, Tallinn University of Technology, Estonia,

[2]Centre for Materials Research, Tallinn University of Technology, Estonia

E-mail: tatjana.dedova@ttu.ee

Abstract. ZnO layers comprizing rods were deposited by chemical spray pyrolysis (CSP) method at 520 °C on different FTO (SnO2:F)/glass substrates using zinc chloride aqueous solutions. Substrates were purchased from different manufactures and differ by morphology, grain size and roughness of FTO electrode. FTO/glass substrates and ZnO layers grown on them were characterised with the help of XRD, AFM, high resolution SEM, EDX methods. The relationship between nanorod formation and substrate properties was studied. It was found that substrate roughness and grain size influence the ZnO rods formation. Deposition of rods (d=300 nm, L=1.4 μm) was successful on the FTO layers with grain sizes around 30-50 nm and roughness below 10 nm, whereas large-grained FTO (grain size > 130 nm) resulted in thick, low-aspect ratio crystals with diameter around 400 nm and length of about 400 nm.

Keywords: FTO, substrate, morphology, ZnO rod, spray pyrolysis, nanostructured layers, haze factor,

1. Introduction

ZnO nanostructured layers have been the research focus in recent years because of their potential applications in various nanodevices [1-5]. Key challenges in ZnO nanorod growth for device application are controlled growth using simple and inexpensive growth methods. ZnO nanorods can be produced by diverse methods, as summarized in numerous review articles [1, 2]. Among them, spray pyrolysis, presented in this study, is simple and low-cost method to produce large area thin films and nanostructured layers including ZnO nanorods [6-8].

Particular significance in nanorods growth has a substrate and its preparation. In many studies to contribute the rods growth or improve their vertical alignment a seed layer (ZnO nanoparticles) is required prior to the deposition of the ZnO nanorods [1, 9, 10]. However, for some applications (emitting diodes, solar cells), it is required to grow ZnO nanorod directly on a transparent conductive oxide [11, 12]. There are some reports on the direct deposition of ZnO rods onto Indium Tin Oxide (ITO) substrates [11, 13, 14]. However, due to the scarcity of indium search for alternative transparent conductive oxide (TCO) materials has been a topic of an active research. Such a good alternative could be fluorine-doped tin dioxide FTO, however, very little studies were reported on growth of ZnO rods on FTO substrates up to now [12, 15-17]. As FTO might have different morphologies depending on the manufacturer and preparation method, therefore it is important to find out a correlation between initial morphology of the substrate and final morphology of the grown layers.

In this study, we present a systematic study on ZnO rods on FTO substrates with different morphology. The morphology, structural and optical properties of the substrates are studied by means of high resolution SEM, AFM and UV-VIS techniques.

2. Experimental

ZnO layers were deposited using pneumatic spray pyrolysis method, described in details earlier [6, 7, 13, 16, 18]. To grow ZnO layers, 50 ml of 0.05 mol/l aqueous solution of $ZnCl_2$ (Fluka) (pH=5) was sprayed over the FTO (SnO_2:F) substrate surface at Tsub=520 °C. The solution spraying rate and carrier gas (compressed air) flow rate were 2.4 ml/min and 8 l/min, respectively. The variety of FTO substrates (further referred in a text as FTO-1, FTO-2, FTO-3, FTO-4, FTO-5) purchased from different manufactures was used in this study. All substrates undergone standard cleaning procedure: firstly in soap solution, then in isopropyl alcohol cleaning in ultrasonic bath for 1 minute and finally, treatment during 1 minute in concentrated H_2SO_4 (Tca ~40 °C). The samples were rinsed with deionised water after each cleaning step.

The morphology of the FTO substrates and ZnO layers were studied by a high resolution SEM on a Zeiss HR FESEM Ultra 55 apparatus at an accelerating voltage of 1.7 kV. Structural characterization of the FTO substrates was carried out on a Rigaku Ultima IV diffractometer using Cu K α radiation (λ=1.5406Å,40 kV at 40 mA). Surface topographies of the FTO substrates were studied by AFM using the NT-DT Solver47 Prosystem operated in a"semi-contact"(tapping) mode. The root mean square (RMS) and average roughness (Sa) values were taken on AFM from area of 2x2μm^2. The diffuse ($T_{diff.}$) and total (T_{tot}) transmittance spectra were measured in the wavelength range of 200 -2500 nm on a Jasco 670-V UV-VIS-NIR spectrophotometer equipped with an integrating sphere. Haze factor (HF) values were determined as HF=T_{diff}/T_{tot}.

3. Results and discussion

Figure 1 presents SEM and AFM images of FTO surface and of corresponding ZnO layers grown on them at similar growth conditions. The morphological parameters obtained from SEM and AFM studies for FTO and dimensions of ZnO crystals are summarized in Table 1.

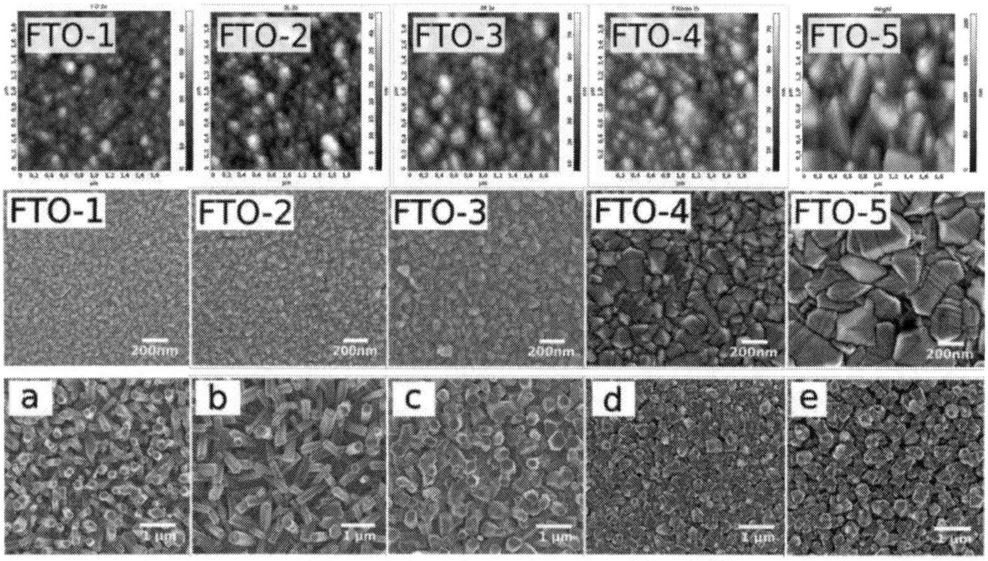

Figure 1. Influence of FTO substrate morphology on the formation of ZnO rods grown by spray pyrolysis. Upper row: AFM images of the substrates FTO-1 up to FTO-5; middle row: SEM images of the same substrates; bottom raw: SEM images of ZnO rods grown on these substrates (a, b, c, d, e correspond to FTO-1, 2, 3, 4 and 5, respectively).

As it can be seen from Fig.1 and Table 1, the substrates mainly differ by grain and crystallite size, roughness and grain shape. Elongated ZnO rods (L=800-1400 nm) were formed only on FTO -1 and 2 which have low roughness (RMS=7-9 nm) and smaller grains (ca. 30-50 nm). The FTOs with higher roughness ≥ 12 nm (FTO-3, 4 and 5) resulted in low aspect ratio ca. 1, fat ZnO crystals. FTO-5 and 6 differ from others by large and pyramidal shaped grains. It should be mentioned here, that ZnO crystals on FTO-4 and 5 are tightly standing next to each other. This could be explained by high number of nucleation centres on FTO-5 and 6 substrates. The step-like texture of FTO facets (clearly seen on SEM of FTO-5) might be a good platform for preferable sites for ZnO nucleation. High amount of ZnO nuclei lead to further coalescence of the crystals and formation of closely standing to each other thick low-aspect ratio rods.

Table 1. Root mean square roughness (RMS), average roughness (Sa), average grain and crystallite size of FTO substrates and dimensions of ZnO crystals grown on them.

FTO No.	Properties of FTO substrates				Dimensions of ZnO crystals		
	RMS, nm (AFM)	**Sa, nm (AFM)**	**Average grain size, nm (SEM)**	**Crystallite size, nm (XRD)**	**L, nm**	**D, nm**	**Aspect Ratio, L/D**
FTO-1	7	6	30	12	800	250	3
FTO-2	9	7	50	13	1400	300	4.5
FTO-3	12	10	80	15	600	500	1
FTO-4	15	12	130	17	300	200	1
FTO-5	36	30	250	40	300	300	1

Haze factor values presented in Figure 2 are in a good correspondence with AFM and SEM studies. The rougher FTO substrates is, the higher is light scattering and haze factor in the visible spectral region. Thus, haze factor measurements might be a handy and fast tool to estimate the surface roughness of the substrate and its suitability for growth of the ZnO rods. According to EDS study, the elemental composition of the FTO substrates (not presented here) is similar and, therefore impurity coming out from the substrate does not affect the rods formation in our case. According to the XRD patterns presented in Fig. 3, SnO_2 crystallites in FTO-5 are orientated along (110) plane parallel to the substrate, while in FTO-1 – FTO-4 the crystallites are preferably grown along (200) plane parallel to the substrate. Thus, the orientation of the crystallites in the substrate doesn't have strong influence on the formation of the rods. Thus, the main factor influencing the ZnO rods formation in our case is FTO roughness. Obtained result is in a good correlation with that reported for ZnO rods growth on GaN substrate by MOCVD [19] and ZnO seed layers by solution growth [20], where ZnO rod growth was favoured using relatively smooth substrates.

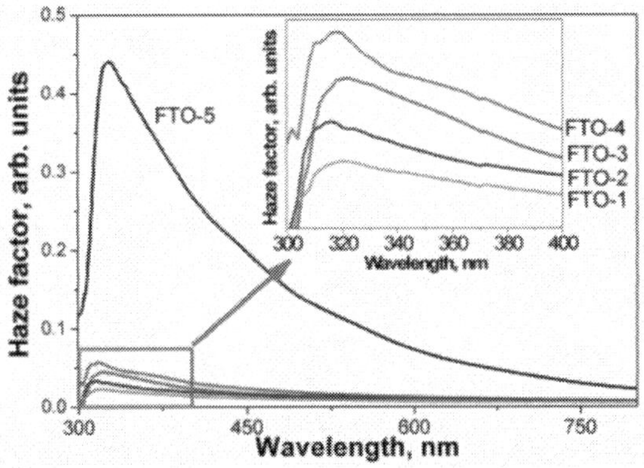

Figure 2. Haze factor spectra of FTO substrates.

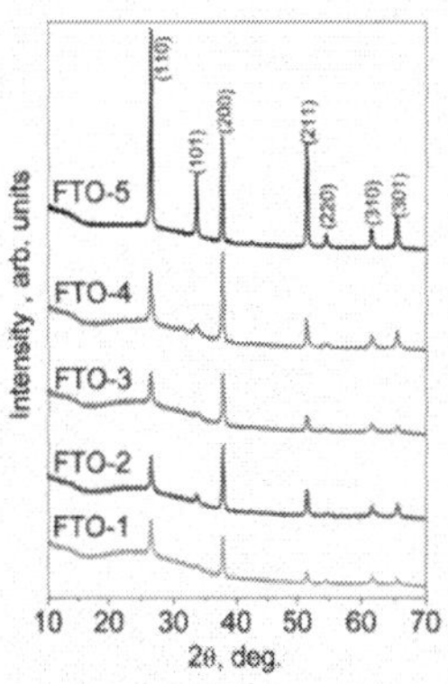

Figure 3. XRD patterns of FTO substrates.

4. Conclusions.

Herein we present a systematic study on ZnO rods growth by spray pyrolysis depending on the FTO substrates morphology. Depending on the manufacture, FTO surface morphology differs by roughness, grain size and shape, and crystallite orientation. Haze factor calculated from optical transmittance spectra measurements is a handy and fast tool for rough estimation of the surface roughness of the substrate and its suitability for growth of the ZnO rods. It was shown that rods formation is strongly affected by roughness and grain sizes of the FTO substrates. It was found that elongated ZnO rods (L=1400 nm, d=300 nm) can be produced on FTO with small roughness (ca. 8 nm) and grain sizes ca. 30-50 nm. FTO RMS roughness values higher that 12 resulted in thick ZnO crystals with low aspect ratio.

Acknowledgement.

This work is financially supported by the Estonian Ministry of Education and Research (SF0140092s08 and IUT19-4), Estonian Science Foundation (ETF8509) and European Regional Development Fund (Centre of Excellence "Mesosystems: Theory and Applications" TK114, 3.2.0101.11-0029 and project "Efficient plasmonic absorbers for solar cells" 3.2.1101.12-0023).

References:

[1] Yi G-C, Wang C, Park W I 2005 *Semicond. Sci. Technol.* **20** 22
[2] Weintraub B, Zhou Z, Li Y, Deng Y 2010 *Nanoscale* **2** 1573
[3] Yin Z, Wu S, Zhou X, Huang X, Zhang Q, Boey F, Zhang H 2010 *Small* **6** 307
[4] Acik Oja I, Katerski A, Mere A, Aarik J, Aidla A, Dedova T, Krunks M 2009 *Thin Solid Films* **517** 2443
[5] Belaidi A, Dittrich Th, Kieven D, Tornow J, Schwarzburg K, Kunst M, Allsop N, Lux-Steiner M-Ch, Gavrilov S 2009 *Sol. Energy Mater. Sol. Cells* **93** 1033
[6] Dedova T, Krunks M, Volobujeva O, Oja I 2005 *Phys. Status Solidi C* **2** 1161
[7] Dedova T, Klauson J, Badre C, Pauporté Th, Nisumaa R, Mere A, Volobujeva O, Krunks M 2008 *Phys. Status Solidi A* **205** 2355
[8] Dedova T, Krunks M, Oja Acik I, Klauson D, Volobujeva O and Mere A 2013 *Mater. Chem. Phys.* **141** 69
[9] Kim D C, Hyun Kong B, Cho H K 2008 *J. Mater. Sci.–Mater. Electron.* **19** 760
[10] Yin Y T, Que W X, Kam C H 2010 *J. Sol–gel Sci. Technol.* **53** 605
[11] Lee H K, Kim M S, Yu J S 2011 *Nanotechnology* **22** 445602
[12] Masuda Y, Kato K 2009 *Cryst. Growth Des.* **9** 3083
[13] Dedova T, Krunks M, Grossberg M, Volobujeva O, Oja Acik I 2007 *Superlattice Microstr.* **42** 444
[14] Krunks M, Dedova T, Kärber E, Mikli V, Oja Acik I, Grossberg M, Mere A 2009 *Phys. B Condens. Matter* **404** 4422
[15] Pauporté Th, Bataille G, Joulaud L, Vermersch F J 2010 *J. Phys. Chem. C* **114** 194
[16] Dedova T, Volobujeva O, Klauson J, Mere A, Krunks M 2007 *Nanoscale Res. Lett.* **2** 391
[17] Lee J H 2010 *Adv. Mat.Res.* **123-125** 715
[18] Dedova T, Oja Acik I, Krunks M, Mikli V, Volobujeva O, Mere A 2012 *Thin Solid Films* **520** 4650
[19] Park S-H, Seo S-Y, Kim S-H, Han S-W 2006 *Appl. Phys. Lett.* **88** 251903
[20] Lee Y-J, Sounart T L, Scrymgeour D A, Voigt J A, Hsu J W P 2007 *J Cryst. Growth* **304** 80

IOP Conf. Series: Materials Science and Engineering **49** (2013) 012002 doi:10.1088/1757-899X/49/1/012002

Novel fine-disperse bimetallic Pt-Pd/Al_2O_3 catalysts for glycerol oxidation with molecular oxygen

K Dubencovs[1], S Chornaja[1], E Sproge[1], V Kampars[1], D Markova[1], L Kulikova[2], V Serga[2], A Cvetkovs[2]
[1]Institute of Applied Chemistry, Riga Technical University, Latvia
[2]Institute of Inorganic Chemistry, Riga Technical University, Latvia

E-mail: Konstantins.Dubencovs@rtu.lv

Abstract. Using extractive-pyrolytic method several Pt-Pd bimetallic catalysts supported on plasma-processed alumina nanopowder were synthesized. Pt-Pd loading and glycerol oxidation process parameter influence on catalyst activity and selectivity was determined oxidizing glycerol in mild conditions. Novel bimetallic catalysts in neutral water solutions were practically inactive (glycerol conversion was only 3%) whereas in alkaline solutions they were active and selective to glyceric acid. Using 1.2%Pt-1.2%Pd/Al_2O_3 catalyst glyceric acid was obtained with 65% selectivity (glycerol conversion was 96%). It was shown that novel fine-disperse bimetallic Pt-Pd/Al_2O_3 catalysts were more active compared to analogous monometallic Pt/Al_2O_3 and Pd/Al_2O_3 catalysts.

1. Introduction

Glycerol is the main by-product in the transesterification process of biodiesel production from vegetable oil. Selective glycerol oxidation in presence of supported catalysts is one of the preferable glycerol conversion methods. Many valuable products are obtained in this process – glyceric acid, lactic acid, glycolic acid, tartronic acid etc. [1–3]. Method is environmentally friendly and catalysts can be reused. Generally glyceric acid is selectively produced in glycerol oxidation over monometallic catalysts [3, 4]. Usage of bimetallic catalysts gives opportunity to increase selectivity to glyceric acid [5] and obtain other important glycerol oxidation products like lactic acid or glycolic acid with a good yield [3, 6]. Bimetallic catalysts have some more advantages like higher activity and prolonged catalyst life [7]. In literature only Au-Pt and Au-Pd catalysts are known.

In this work we have synthesized novel fine-disperse Pt-Pd bimetallic catalysts supported on Al_2O_3 prepared by extractive-pyrolytic method described in our earlier papers [8–9]. We have investigated Pt-Pd loading and glycerol oxidation process parameter influence on catalyst activity and selectivity.

2. Experimental

All solutions used were prepared with Millipore superQ water (18 MΩ/cm) and analytical grade reagents. For preparation of catalysts (Pt-Pd nanoparticles on the surface of support) a mixture of platinum and palladium containing organic extracts were used as precursor. The method for the production of metal-containing organic extract is described in [8, 9]. Amounts of extracts were calculated based on each metal's loading in the composite (0.6, 1.2 and 2.5 wt.%). The support was impregnated with organic extracts. Nanopowder of Al_2O_3 obtained in plasma (SSA 46 m^2/g) was used as the support. The mixture was dried at room temperature and calcinated at 300 °C in air for 5 min. As synthesized nanopowders of catalysts were characterized by X-ray diffraction (XRD) using diffractometer D-8 Advance (Bruker AXS) with CuK_α radiation (λ = 1.5418 Å).The average size of Pt-Pd alloy crystallites was calculated by line broadening peak (111) using the Scherrer equation. Specific surface area (S.S.A.) of the powders was measured using HROM-3 chromatograph by Brunauer–Emmett–Teller (BET) method at temperature of liquid nitrogen.

Powder bimetallic catalyst activity and selectivity in the oxidation of glycerol were tested in thermostatted slurry bubble reactor (50 ml capacity) equipped with gas supply system. Glycerol oxidation process parameters, like NaOH initial concentration c_0(NaOH) = 0–1.5 M, glycerol and platinum molar ratio n(glyc.)/n(Pt-Pd) = 300–500 mol/mol and oxygen partial pressure from 0.2 to 1 atm were varied. Glycerol initial concentration c_0(glyc.) was 0.3 M, reaction temperature was 60 °C. Analysis of the reaction mixture was performed by high-performance chromatograph WATERS 2487.

3. Results

Bimetallic catalysts with Pt-Pd loading from 0.6 to 2.5% for each metal were prepared and tested. The XRD phase analysis of the obtained composites permits to detect formation of Pt-Pd alloy phase on the support (figure 1). As diffraction peaks of the support overlaps with peaks of the catalyst, the average size of Pt-Pd alloy crystallites was determined only for the catalyst with high metal loading (2.5%-Pt 2.5%Pd/Al_2O_3) and was 28 nm (figure 1).

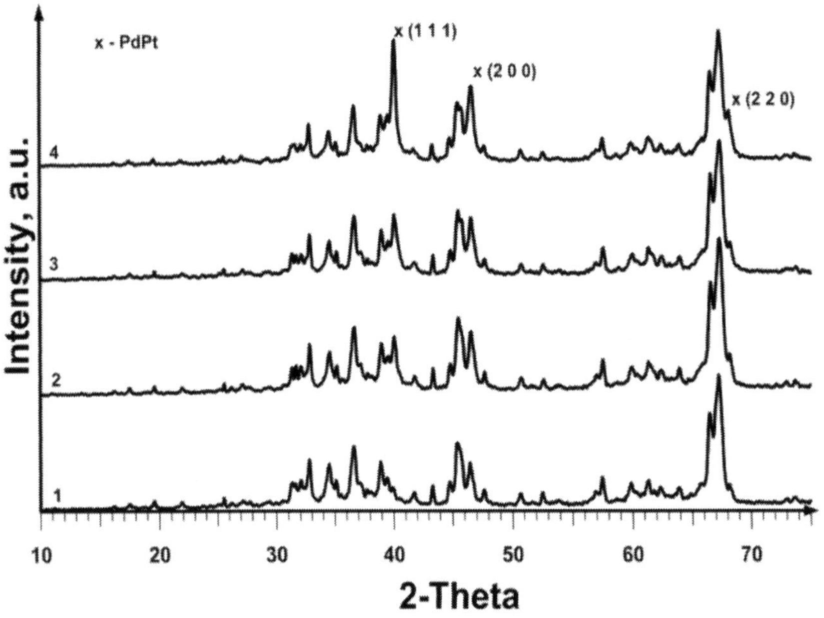

Figure 1. XRD patterns: 1 – Al_2O_3; 2 – 0.6%Pt-0.6%Pd/Al_2O_3; 3 – 1.2%Pt-1.2%Pd/Al_2O_3; 4 – 2.5%Pt-2.5%Pd/Al_2O_3.

The prepared catalysts were tested in glycerol oxidation processes. Catalyst activity was estimated by glycerol conversion. Experimentally it was found that catalyst activity is negligible in water solution without NaOH addition. Glycerol conversion was less than 3%. It is known that a basic pH increases the oxidation rate because it favors the deprotonation of the primary hydroxyl group of glycerol or because it favors the product desorption from the catalyst [10]. From table 1 it is visible that all catalysts were very active in the alkaline water solutions. Catalyst activity did not depend on Pt-Pd loading and was similar. Glycerol conversion ranged from 91 to 100%. Catalysts with loading 1.2-2.5% for each metal presented better selectivity to glyceric acid. It was concluded that 1.2%Pt-

1.2%Pd/Al_2O_3 catalyst is the best for glycerol oxidation to glyceric acid (selectivity to glyceric acid reached 42% with full glycerol conversion).

Also, table 1 data shows that in alkaline solutions novel bimetallic 1.2%Pt-1.2%Pd/Al_2O_3 catalyst was more active than analogous monometallic 2.5%Pt/Al_2O_3 and 2.5%Pd/Al_2O_3 catalysts. At similar glycerol conversion (66–74%) 1.2%Pt-1.2%Pd/Al_2O_3 catalyst selectivity to glyceric acid was equal to 2.5%Pd/Al_2O_3 catalyst selectivity and higher than 2.5%Pt/Al_2O_3 catalyst selectivity.

Table 1. Effect of Pt-Pd loading on the catalyst on glycerol conversion and glycerol oxidation product yield and selectivity.

Catalyst	Oxidation time, h	Glycerol conv., mol%	Selectivity, mol%				
			GLYA	TART	LACT	GLYC	OXAL
0.6%Pt-0.6%Pd/Al_2O_3	7	97	33	37	15	6	9
1.2%Pt-1.2%Pd/Al_2O_3	3	66	65	9	20	4	2
	7	100	42	26	20	6	6
2.5%Pt-2.5%Pd/Al_2O_3	7	91	41	26	23	5	5
2.5%Pt/Al_2O_3	7	69	49	9	24	13	2
2.5%Pd/Al_2O_3	7	74	66	8	17	5	4

Reaction conditions: c_o(NaOH) = 1.5 M, n(glyc.)/n(Me) = 300 mol/mol, $P(O_2)$ = 1 atm, 60 °C. GLYA – glyceric acid, TART – tartronic acid, LACT – lactic acid, GLYC – glycolic acid, OXAL – oxalic acid

Effects of NaOH initial concentration and glycerol/metal molar ratio n(glyc.)/n(Me) are summarized in table 2. Comparing glycerol oxidation in 0.7 M and 1.5 M NaOH solutions, from table 2 data it can be concluded that 1.2%Pt-1.2%Pd/Al_2O_3 catalyst in the 0.7 M NaOH solution was more active and selective to glyceric acid. While using 1.5 M NaOH water solution full glycerol conversion was reached only after 7 h, in 0.7 M NaOH solution similar glycerol conversion (96%) was reached already after 4 h. Selectivity to glyceric acid in the 0.7 M NaOH was higher by 23% compared to 1.5 M NaOH solution at the same glycerol conversion (96–100%). From table 2 data it is visible that in the 0.7 M NaOH water solutions the selectivity of glycerol oxidation products was not dependent on n(glyc.)/n(Me).

Table 2. Glycerol oxidation over 1.2%Pt-1.2%Pd/Al_2O_3 catalyst. Effect of NaOH initial concentration and n(glyc.)/n(Me).

c_o(NaOH), M	Oxidation time, h	Glycerol conv., mol%	Selectivity, %				
			GLYA	TART	LACT	GLYC	OXAL
0.7	4	96	65	11	14	6	4
	7^a	91	68	11	12	6	3
1.5	7	100	42	26	20	6	6

Reaction conditions: n(glyc.)/n(Me) = 300 mol/mol, $P(O_2)$ = 1 atm, 60 °C. [a]n(glyc.)/n(Me) = 500 mol/mol

From Fig. 2 it can be seen that oxygen partial pressure significantly influences rate of glycerol oxidation. In oxidation with molecular oxygen ($P(O_2)$ = 1 atm) after 7 h glycerol conversion was 90%, but reduction of oxygen partial pressure to 0.2 atm resulted in glycerol conversion of only 35%. At the same time selectivity to glyceric acid also dependent on $P(O_2)$. At similar glycerol conversion (till 35%) selectivity to glyceric acid at $P(O_2)$ = 1 atm was higher by 10% in comparison with selectivity at $P(O_2)$ = 0.2 atm. Fig. 2 show that the process selectivity to the main oxidation product, glyceric acid, slightly reduces. Selectivity to glyceric acid reduces due to glyceric acid oxidation to tartronic acid when the glycerol conversion degree increases.

Summarizing all the data it can be concluded that the best result (65% selectivity to glyceric acid with 96% glycerol conversion) was achieved using 1.2%Pt-1.2%Pd/Al_2O_3 catalyst when glycerol oxidation conditions were as follows: c_o(glyc.) = 0.3 M, c_o(NaOH) = 0.7 M, n(glyc.)/n(Me) = 300 mol/mol, t = 60 °C, $P(O_2)$ = 1 atm, oxidation time 4 hours.

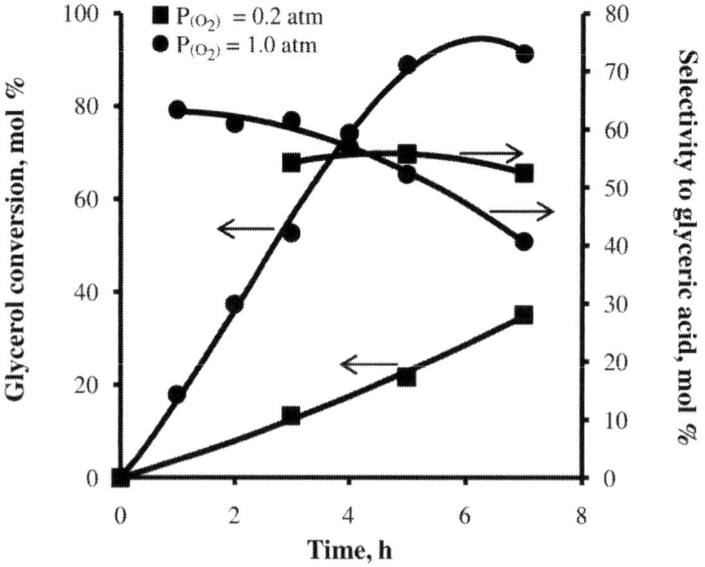

Figure 2. Influence of oxygen partial pressure on 2.5%Pt-2.5%Pd/Al_2O_3 catalyst activity and selectivity to glyceric acid.

4. Conclusions

Novel bimetallic Pt-Pd/Al_2O_3 catalysts with nanosized Pt-Pd alloy particles prepared by exctractive-pyrolytic method were very active in the glycerol oxidation processes when alkaline solutions were used. Optimal conditions for obtaining glyceric acid from glycerol were determined. Fine-disperse bimetallic Pt-Pd/Al_2O_3 catalysts were more active compared to analogous monometallic Pt/Al_2O_3 and Pd/Al_2O_3 catalysts.

Acknowledgment. This work was supported by the European Regional Development Fund – project No. 2010/0304/2DP/2.1.1.1.0/10/APIA/VIAA/087.

References

[1] Zhang M, Liang D, Nie R, Lu X, Chen P and Hou Z 2012 *Chin. J. Catal.* **33** 1340–46
[2] Royker M, Case J and Steen E 2012 *J S Afr I Min Metall* **7A** 577–581
[3] Bianchi C-L, Canton P, Dimitratos N, Porta F and Prati L 2005 *Catal. Today* 2005 **102–103** 203–212
[4] Liang D, Gao J, Sun H, Chen P, Hou Z and Zheng X 2011 *Appl. Catal. B* **106** 423–432
[5] Ketchie W-C, Murayama M and Davis R-J 2007 *J. Catal.* **250** 264–273
[6] Shen Y, Zhang S, Li H, Ren Y and Liu H 2010 *Chem. Eur. J.* **16** 7368–71
[7] Prati L, Villa A, Porta F, Wang D and Su D-S 2007 *Catal. Today* **122** 386–390
[8] Cornaja S, Kulikova L, Serga V, Kampars V, Dubencovs K, Zizkuna S and Muravjova O. Pat. EP 2 332 899 B1 (20.03.2013)
[9] Palcevskis E, Kulikova L, Serga V, Cvetkov A, Chornaja S, Sproge E and Dubencovs K 2012 *J. Serb. Chem. Soc.* **77** 1799–1806
[10] Chornaja S, Dubencov K, Kampars V, Stepanova O, Zhizhkun S, Serga V, Kulikova L 2013 *React. Kinet. Mech. Cat.* **108**(2) 341–357.

Characteristics of zirconia nanoparticles prepared by molten salts and microwave synthesis

J Grabis, Dz Jankovica, E Sokolova, I Steins

Institute of Inorganic Chemistry RTU, Miera 34, Salaspils, LV-2169, Latvia
E-mail: grabis@nki.lv

Abstract. Zirconia and yttria stabilized zirconia (3YSZ) nanoparticles were prepared from zirconia and yttria salts using molten salts (MS) and microwave (MW) synthesis. The crystalline ZrO_2 and 3YSZ nanoparticles with crystallite size in the range of 3–27 nm were obtained by MS synthesis in NaCl–$NaNO_3$ salts. The zirconia and 3YSZ powders with close characteristics were obtained by combining MW synthesis with calcination of products at 400–800 °C. The crystallite size depends upon synthesis or calcination temperature, and the precursors used. The powders prepared by MS and MW synthesis ensured manufacturing of bulk materials with relative density of 98.6% and 97.2% respectively by using spark plasma sintering at 1300 °C.

Introduction

Stabilized zirconia nanoparticles due to their unique physical, chemical and mechanical properties find application as solid electrolytes, gas sensors, catalysts, high-temperature and bio- materials [1]. Owing to the wide application of zirconia, several preparation methods with different particle size and phase composition have been elaborated [1, 2, 3]. At room temperature the high-temperature phases of ZrO_2 have been stabilizied by introducing trivalent oxides as solid solution in the ZrO_2 lattice [4]. These synthesis methods allow to prepare nanoparticles containing t-, m-ZrO_2 or pure monoclinic, tetragonal as well as cubic phase depending on the precursors used, calcination temperature and conditions of synthesis. Each method has its characteristic advantages and disadvantages which mainly relate to complexity of the process, expensive precursors, production rate, cost, and use of several solvents.

The present trends in technology are directed to development of highly effective, economic and environment friendly preparation methods of nanoparticles. From this point of view, fast microwave (MW) [1] and solvent free molten salts (MS) [2, 3] synthesis could be very perspective preparation methods. These methods can be used together with the fast spark plasma sintering (SPS) method for processing the powders produced. However, influence of different salts, precipitation agents, and synthesis temperature on characteristics of zirconia nanoparticles prepared by MS and MW synthesis and their sinterability by using the SPS method is not fully known.

In the present paper the characteristics and sinterability of the zirconia nanoparticles prepared by MS and MW synthesis are studied and compared.

2. Experimental

Experiments were performed by using following reagent grade chemicals $ZrO(NO_3)_2 \cdot 2H_2O$, $ZrOCl_2 \cdot 8H_2O$, $Y_2(NO_3)_3 \cdot 6H_2O$, NaCl, $NaNO_3$, NH_4OH, urea. For MS synthesis the Zr and Y salts were mixed with definite amounts of NaCl and $NaNO_3$ (ratio of Zr and sodium salts was varying in the range of 0.1–0.5). The melting temperature of the salts and therefore also synthesis temperature in the range of 400–800 °C was changed by varying the ratio of sodium salts. The ZrO_2 nanoparticles were extracted from the products by dissolving sodium salts in distilled water and washing with ethanol.

The MW synthesis of the zirconia nanoparticles was conducted using a commercial reactor Masterwave BTR (Anton Paar). The Zr salts or their mixture with $Y_2(NO_3)_3 \cdot 6H_2O$ (3 mol%) were dissolved in distilled water, stirred and then NH_4OH or urea was added. The synthesis of the zirconia was performed at 180 °C during 20 min at pressure about 26 bar with continuous stirring. The obtained particles were extracted by filtration and washing with distilled water and ethanol followed by drying at 100 °C for 20 h. ZrO_2 was calcinated at 400–800 °C.

The samples prepared by both methods were densified in vacuum at temperature 1200–1400 °C and a pressure of 30 MPa using the spark plasma sintering technique (SPS-825.CE).

The phase composition of powders and bulk materials was determined by X-ray diffraction (XRD) analysis at the room temperature (AdvanceD8, Bruker AXS). Presence of c-ZrO_2 phase was detected by splitting maxima of c-ZrO_2 and t-ZrO_2 at 2Θ in the range of 73–75 °C. The crystallite size was calculated using X-ray diffractometer software EVA12 based on the Scherrer formula. The specific surface area (SSA) of powders was determined by the argon absorption-desorption method. Density of bulk samples was determined by using the Archimede's method

3. Results and discussion

Typical parameters of the molten salts and microwave synthesis and characteristics of the prepared pure zirconia and stabilized with 3 mol% yttria (3YSZ) are shown in Table 1.

Table 1. Parameters of synthesis and characteristics of as-prepared zirconia particles

Sample	Precursors	SSA, m^2/g	Crystallite size, nm	Phase composition, %
	Molten salts synthesis, temperature 400 °C, time 120 min			
M1	$ZrO(NO_3)_2 \cdot 2H_2O$, $NaNO_3$, NaCl	104.0	8	m-ZrO_2 20%; t-ZrO_2 80%
M2	$ZrOCl_2 \cdot 8H_2O$, $NaNO_3$, NaCl	116.2	7	m-ZrO_2 15%; t-ZrO_2 85%
M3	$ZrO(NO_3)_2 \cdot 2H_2O$, $Y_2(NO_3)_3 \cdot 6H_2O$, $NaNO_3$, NaCl	107.1	5	t-, c-ZrO_2 100%
M4	$ZrOCl_2 \cdot 8H_2O$, $Y_2(NO_3)_3 \cdot 6H_2O$, $NaNO_3$, NaCl	120.9	3	t-, c-ZrO_2 100%
	Microwave synthesis, temperature 180 °C, time 20 min			
MW1	$ZrOCl_2 \cdot 8H_2O$, $Y_2(NO_3)_3 \cdot 6H_2O$, NH_4OH	140.2	...	X-ray amorphous
MW2	$ZrOCl_2 \cdot 8H_2O$, $Y_2(NO_3)_3 \cdot 6H2O$, CH_4N_2O	140.5	...	X-ray amorphous

All as-prepared powders had the high SSA. Data showed small increase of SSA and decrease of crystallite size in the samples prepared by MS synthesis using $ZrOCl_2 \cdot 8H_2O$ or Zr salts with yttria additives as precursors, possibly, due to the change of phase composition of powders.

The higher SSA of the samples prepared by MW synthesis can be explained by lower synthesis temperature and formation X-ray amorphous products. Besides this, the SSA is determined by concentration of zirconium in reaction medium. The increase of the ratio Zr/Na salts from 0.1 up to 0.5 decreased the SSA of powders prepared by MS synthesis from 142 m^2/g to 102 m^2/g.

The pure ZrO_2 powders prepared by MS synthesis contained t-ZrO_2 and some amount of m-ZrO_2 despite very small crystallite size. The MS synthesis of 3YSZ powders resulted in obtaining fully stabilized zirconia containing t-, c-ZrO_2 phases. The similar phase composition was detected for yttria stabilized zirconia samples prepared by MS synthesis using $NaNO_3$–KNO_3, $ZrOCl_2$, YCl_3 salts [2].

The XRD patterns of as-prepared ZrO_2 by MW synthesis showed very broad weak maxima indicating formation of hydrous-zirconia particles due to low synthesis temperature compared to the MS process. Such hydrous-zirconia has been prepared by hydrolysis of $ZrO(NO_3)_2$ in a reverse microemulsion at 60–75 °C for 72 h [5]. Additional calcination of powders at the temperature in the range of 400–800 °C led to formation of crystalline particles and mass loss up to 15.2% at 600 °C.

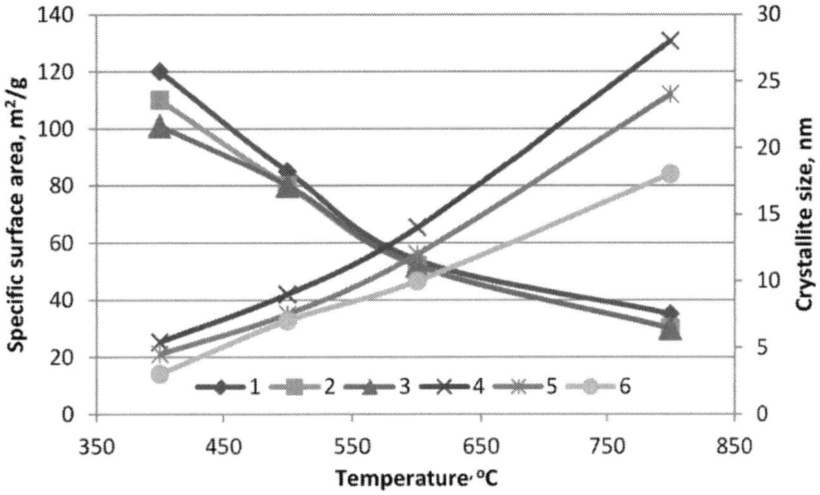

Figure 1. SSA (1, 2, 3) and crystallite size (4, 5, 6) of 3YSZ nanoparticles prepared by MS synthesis using $ZrOCl_2.8H_2O$ (1, 6) or $ZrO(NO_3)_2.2H_2O$ (3, 4) and by MW synthesis (2, 5) in dependence on synthesis or calcination temperature

The additional calcination of MW synthesis powders had siginificant impact on SSA and crystallite size (Figure 1). The SSA and crystallite size of the calcinated MW synthesis powders were very close to that of MS synthesis powders prepared at the same temperature. This indicated the important role of temperature in formation of zirconia nanoparticles.

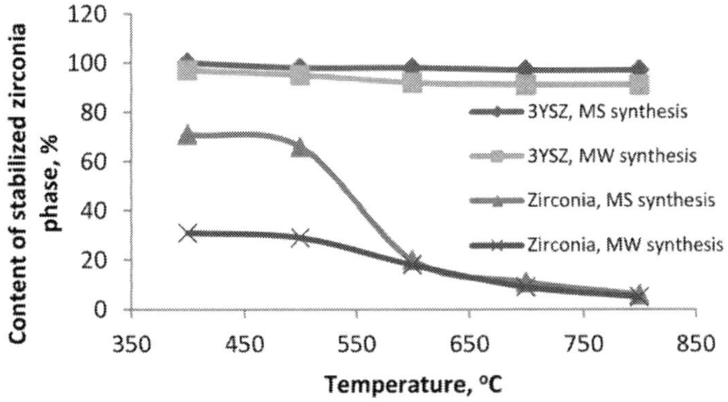

Figure 2. Content of stabilized zirconia phase in 3YSZ and zirconia powders prepared by MS and MW synthesis in dependence on synthesis or calcination temperature

The phase transition of powders depends upon the methods used in their preparation and preparation temperature (Figure 2). For unstabilized zirconia, calcination at 500–600 °C promoted formation of m-ZrO_2 phase although the crystallite size was below a critical size for tetragonal phase stabilization. The XRD patterns of 3YSZ powders prepared by MS synthesis at temperature higher as 500 °C showed only traces of m-ZrO_2 but MW synthesis powders calcinated at 400 °C and 800 °C contained about 3% and 9% of m-ZrO_2 respectively.

The densification of MS and MW synthesis powders during SPS process started at 820–840 °C depending on the SSA of the samples (Table 2) and was connected with their compacting as well as with release of absorbed gases. The final density was reached during 2 min at 1225–1250 °C.

Table 2. Spark plasma sintering process of powders.

Synthesis; temperature, °C	SSA, m^2/g	Temperature of shirinkage, °C		Relative density, %
		starting	final	
MS; 600	50.1	820	1225	98.3
MS; 800	33.4	850	1250	98.6
MW; 600	52.1	820	1220	97.0
MW; 800	36.9	840	1240	97.2

The different relative density of the bulk samples sintered from MS and MW synthesis powders could be explained by the higher content of monoclinic phase in MW powders. The increase of the sintering temperature up to 1400 °C did not improve relative density of the samples. The phase composition of the sintered bulk materials coresponded to that of the used powders.

4. Conclusions

1. Molten salts synthesis ensured preparation of crystalline zirconia and fully stabilized 3YSZ nanoparticles at temperature 400–800 °C with crystallite size in the range of 5–28 nm.
2. Microwave synthesis resulted in formation of hydrous zirconia which additional calcination at temperature 400–800 °C led to formation of crystalline zirconia nanoparticles with crystallite size close to that of molten salts process but increased content of monoclinic phase.
3. The 3YSZ nanoparticles prepared by simple one-stage molten salts process showed higher content of stabilized phase of zirconia and better sinterability with respect to microwave synthesis.

Acknowledgments

The research was supported by Latvian Science Council project 10.0032.

5. References

[1] Rizzuti A, Leonelli C, Corradi A, Caponetti E, Martino D C, Nasillo G, Saladino M L 2009 *J. Dispersion Sci. Technol.* **30** 151–16
[2] Descemond M, Brodhag C, Thevenot F, Durand B, Jebrouni M, Roubin M 1993 *J. Mater. Sci.* **28** 2283–88
[3] Du Yuansheng, Rogers P, Inman D 1996 *J. Mater. Sci.* **31** 3361–64
[4] Yanagida H, Koumoto K, Miyayama M 1996 *The Chemistry of Ceramics* (Chichester; New York: John Wiley & Sons)
[5] Clifford Y Tai, Bor-Yuan Hsio, Hsien-Yi Chiu 2004 *Colloids and Surfaces A:Physicochem. Eng. Aspects* **237** 105–111

IOP Conf. Series: Materials Science and Engineering **49** (2013) 012004 doi:10.1088/1757-899X/49/1/012004

Thermal stability of polyacetal/ethylene-octene copolymer/zinc oxide nanocomposites

A Grigalovica[1], R Merijs Meri[1], J Zicans[1], T Ivanova[1], J Grabis[2]
[1]Institute of Polymer Materials, Riga Technical University, Latvia
[2] Institute of Inorganic Chemistry of Riga Technical University, Riga, Latvia

E-mail: a.grigalovica@gmail.com

Abstract. In this work we investigate binary blends of polyoxymethylene and ethylene octene copolymer (EOC) and their composites with nanostructured zinc oxide (ZnO). EOC content in the composites varies from 0 to 50 wt. %. The amount of ZnO filler in the composites is changed in the interval from 0 to 5 wt. %. Thermal properties of composites are investigated with thermogravimetric analysis and differential scanning calorimetry. It is observed that ZnO addition increases thermal stability of the investigated composites.

1. Introduction

Polyoxymethylene (POM), also known as polyacetal, is one of the engineering polymers used in many applications (industrial, transport, construction, house holding, medicine etc.) as an excellent alternative material for the replacement of metals due to its high strength and stiffness. However, range of POM application is limited due to some properties, for example, poor impact strength. To eliminate it many researchers have blended POM with elastomers such as styrene rubber [1], ethylene copolymer [2] and thermoplastic polymer TPU [3].

In recent years polymer- nanoparticle composites have attracted much interest because addition of small amount of nano-sized fillers can yield to considerable property increment, unachievable by using traditional micros-sized fillers. Many review reports testify about significant improvements of mechanical, barrier, electrical and magnetic properties of polymer nanocomposites [4-6].

In the current research the effect of nanosized ZnO on the thermal stability of ethylene-octene copolymer (EOC) modified POM is investigated.

2. Experimental

2.1. Materials

Polyoxymethylene copolymer with melt flow index of 42 g/10min, density of 1.41 g/cm^3, crystallinty of 60 % and melting peak temperature of 166 ^{0}C (POM: Kocetal 900, Kolon Plastics Inc.) was used as a matrix. Ethylene octene copolymer with α- octene content of 38 %, melt flow index of 5 g/10min, density of 0.87 g/cm^3, crystallinty of 14 % and maximum melting peak temperature of 60 ^{0}C (EOC: Engage 8200, The Dow Chemical Company) was used as a toughening agent.

The ZnO nanoparticles were prepared by evaporation of coarse grained commercially available zinc oxide (purity of 99.7 %) powder in a radio frequency discharge air plasma. The experimental apparatus consists of radio-frequency oscillator (5.28 MHz) with maximum plate power of 100 kW, water cooled cylindrical stainless steel reactor, heat exchanger, gas and powder supply systems, and bag filter for collection of products. The flow rate of plasma forming gas air was 8.2 m^3/h and feed rate of powder was 1.2 kg/h. The raw powders were introduced into the plasma through 8 tubes with diameters of 2 mm by carrier gas. The growth of particles was controlled by introducing cold air at a distance of 25 μm below the introducing plane of ZnO.

Polyoxymethylene composites with various EOC and ZnO contents (0, 10, 30, 50 wt. % of EOC and 0, 2 and 5 wt. % of ZnO) were prepared by roll milling. POM based blends were processed at 170 ^{0}C, Blending time was 7 min.

2.2. Characterization

Differential scanning calorimetry (DSC) was carried out by using DSC1/200W apparatus (Mettler Toledo). All measurements were made under N_2 flow. Samples of about 10 mg were heated from -100 up to 200 ^{0}C with a heating rate of 10 ^{0}C/min. Results of the second heating run were analyzed in order to earse thermal history of the sample.

Thermogravimetric analysis (TGA) was performed by using TGA1/SF thermogravimetric analyzer (Mettler/Toledo). Samples of about 10 mg were heated from 25 ^{0}C to 600 ^{0}C with a heating rate 10 of ^{0}C/min.

3. Results and discussion

Firstly synthesized ZnO particles were characterized. The obtained products were fine crystalline ZnO nanoparticles – plate-like and whisker-like particles (Fig.1. a) with length of 20–100 nm and diameter of ~20 nm. The specific surface of area of the powders was in the range of 24–32 m^2/g in dependence on the concentration of particles in the plasma flow and cooling rate of the formed ZnO particles. According to XRD analysis the prepared powders contained only ZnO phase (Fig.1. b).

a) 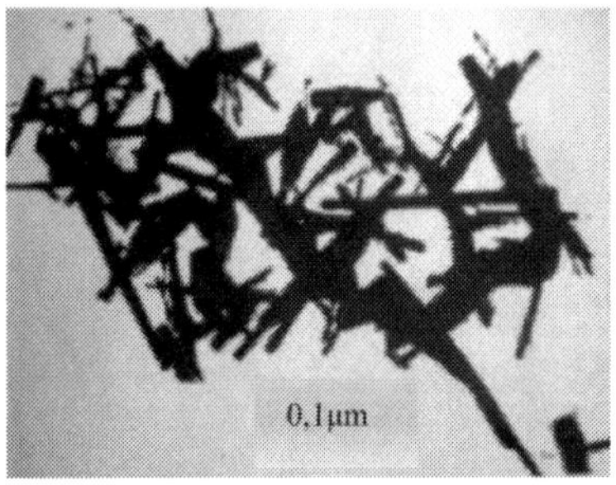

b) 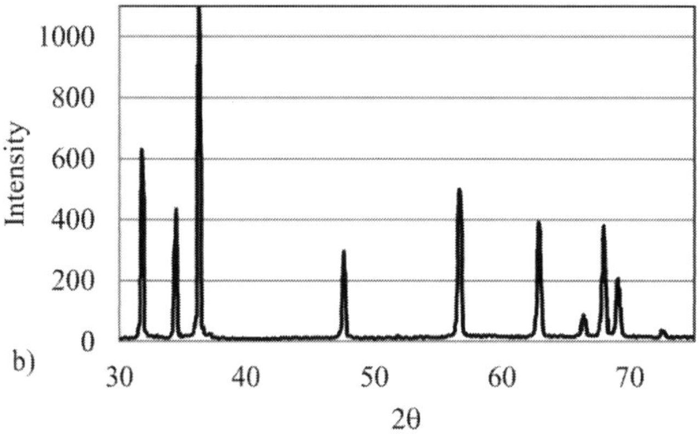

Figure1. Synthesized ZnO particles: a) TEM micrograph and b) XRD spectrum

Thermogravimetric curves of POM/EOC blends are shown in Fig.2. As one can see, POM and EOC, both degrade in one step; however, thermal degradation of EOC is observed 78 ^{0}C higher than POM testifying about higher thermal stability of the polymer, without any heteroatom in the hydrocarbon chain. Thermal degradation of POM macromolecule is initiated directly in the main chain by the scission of the -CH_2-O- linkages, which are the weakest points in the polymer macromolecule. In EOC macromolecule the weakest points are tertiary carbon atoms, at which breakage of the side-chains generally occurs. Consequently degradation process of EOC macromolecule is slower than that of POM macromolecule [8]. As we already expected, modification of POM with EOC, raised thermal resistance of the material. Besides it, two-step degradation behaviour is observed in the case of the investigated blends, the first degradation step being attributed to decomposition of POM, while the other one to degradation of EOC.

The effect of ZnO content on the degradation behaviour of all of the investigated blends is shown in Table1. The degradation temperatures are calculated at 5 %, 10 % and 20 % weight loss of the nanocomposites. It can be seen that thermal stability of all of the investigated nanocomposites is higher than that of pure polymer matrix, most probably because of adsorption of polymer chains on the surface of ZnO particles resulting in decreased segmental mobility of macromolecular chains due to polymer matrix- nanofiller interactions, as demonstrated earlier in the case of CdS and Fe_2O_3 nanoparticles modified polymer composites [9]. Especially effective in respect to thermal stability of the investigated nanocomposites is addition of 2 wt. % of ZnO, which is most probably connected with higher surface area of ZnO because of smaller agglomeration and better distribution of the nanofiller in the polymer blend matrix.

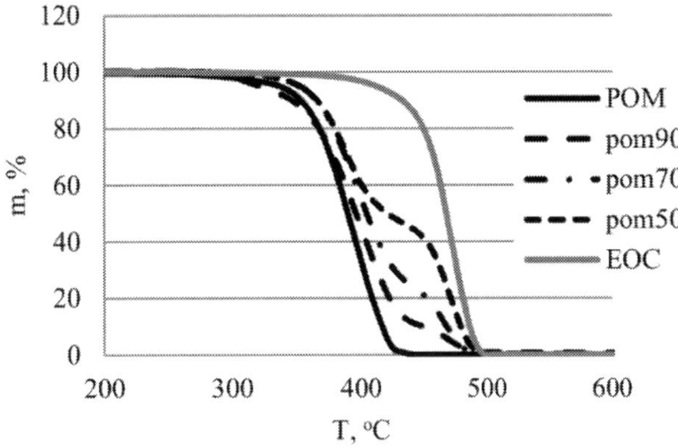

Figure 2. Thermogravimetric curves of POM, EOC and their binary blends.

Table1. Characteristic degradation temperatures of ZnO modified POM/EOC blends.

	5 % weight loss			10 % weight loss			20 % weight loss		
POM, wt.p.	0% ZnO	2% ZnO	5% ZnO	0% ZnO	2% ZnO	5% ZnO	0% ZnO	2% ZnO	5% ZnO
1	324	342	340	345	358	355	369	378	373
0.9	311	345	340	347	365	355	368	378	373
0.7	347	318	347	366	346	368	379	364	380
0.5	344	363	345	366	381	361	380	394	386
0	402	431	432	434	449	450	450	459	457

In order to evaluate the effect of ZnO and EOC on the structural characteristics of POM, DSC thermograms of the investigated composites are analyzed. After erasing of thermal history investigated POM/EOC blends exhibit 2 separate melting peak temperatures, characteristic for incompatible heterogeneous systems. As shown in Table 2, difference between the component melting temperatures in the blend is higher than that between neat POM and EOC. Absence of melting peak of the elastomeric component for the composition with 10 wt. % of EOC evidently is due to its small crystallinity. As one can see there is no considerable effect of either POM content on the maximum melting temperature of EOC or EOC on the maximum melting temperature of POM. In the Table 2 one can however see that both components hinder each other's crystallization.

Interaction of the components in the amorphous phases of the blend can be evaluated by following the changes of it glass transition temperatures T_{gt}. According to the DSC measurements T_{gt} of POM, could not be detected, evidently due to high crystallinity degree of the polymer; T_{gt} of EOC is -55 ^{0}C. T_{gt} of EOC in the investigated blends, however, is hardly noticeable due to high crystallinity of the material as well as because the rate of relaxation of the material evidently is faster than heating rate used in the experiment.

Table 2. Melting peak temperatures (T_M) and degree of crystallinity (K_M) of POM and EOC and their fractions in the blend.

	Melting peak temperature, °C		Difference between the component melting peak temperatures	Crystallinity degree, %	
POM, wt.p.	$T_{M\text{-}EOC}$	$T_{M\text{-}POM}$	$T_{M\text{-}POM}$ - $T_{M\text{-}EOC}$	$K_{M\text{-}EOK}$	$K_{M\text{-}POM}$
0	64	-	-	8	-
0.5	64	164	100	5	58
0.7	63	164	101	2	62
0.9	Not observed	164	-	Not observed	60
1	-	163	-	-	67

Addition of ZnO nanofiller to either POM or EOC or their binary blends does not unambiguously change its crystallization behavior as melting peak onset, maximum and offset temperatures fluctuate within the range of couple of degrees. It should however be mentioned that addition of ZnO nanofiller to EOC causes increase in its glass transition temperature by ~ 10 ^{0}C thus testifying about certain interaction of EOC macromolecules in the vicinity of ZnO nanofiller. It should be, however, mentioned that in the blends the effect of ZnO on the T_{gt} of EOC is less expressed, which is evidently caused because of lower concentration of EOC, causing difficulties in T_{gt} detection.

4. Conclusion

In the current research effects of nanostructured ZnO filler on the thermal and calorimetric properties of POM, EOC and their binary blends are investigated.

Results of DSC investigations show that in the blends POM and EOC hinder each other's crystallization. Addition of ZnO nanofiller either to POM or EOC or their binary blends does not unambiguously change crystallization behaviour of the material. Addition of ZnO nanofiller to EOC, however, causes increase in its glass transition temperature by ~10 ^{0}C thus testifying about certain interaction of EOC macromolecules in the vicinity of ZnO nanofiller.

Results of TGA reveal that addition of EOC improves thermal stability of POM/EOC blends. Addition of ZnO either to POM or EOC or their binary blends cause further increase in the thermal resistance of the material.

References

[1] Pan G O, Chen Y J, Li H L 2007 *Plast. Rubber Compos. Process.Appl* **36** 291-296
[2] Gao X, Qu C, Fu C, Peng Y, Zhang Q 2004 *Macromol. Mater. Eng* **289** 41-48
[3] Uthaman R N, Pandurangan A, Majeed S S M A 2007 *J. Polym. Res* **14** 441-447
[4] Sun L H, Yang Z G,Li X H 2008 *Polym. Eng. Sci* **48** 1824-1832
[5] Zhao X,Ye L 2009 *J.Appl.Polym.Sci* **111** 759-767
[6] Yu N, He L, Ren Y, Xu Q 2011 *Polymer* **52** 472-480
[7] Siengchin S, Psarras G C, Karger-Kocsis J 2010 *J. Appl. Polym. Sci* **117** 1804–1812
[8] Crompton T R 2010 *Thermo-oxydative degradation of polymers* (Smithers Rapra Technology)
[9] Salavati-Niasari M, Ghanbari D 2011 *Advances in Diverse Industrial Applications of Nanocomposites* (InTech)

Acknowledgments

This work has been supported by the European Social Fund within the project «Support for the implementation of doctoral studies at Riga Technical University».

IOP Conf. Series: Materials Science and Engineering **49** (2013) 012005 doi:10.1088/1757-899X/49/1/012005

Characterization of hydroxyapatite by time-resolved luminescence and FTIR spectroscopy

L.Grigorjeva, D.Millers, K.Smits, Dz.Jankovica, L.Pukina

Institute of solid State Physics, University of Latvia,

L.Grigorjeva:lgrig@latnet.lv

Abstract. Time-resolved luminescence and FTIR absorption spectra of undoped and Eu and Ce doped hydroxyapatite nanocrystalline powders prepared by sol-gel method were studied. The luminescence band at 350-400 nm was detected and two decay times (11 ns and 38 ns) was determinated for Ce doped samples. The luminescence spectra and decay kinetics were analized for Eu doped nanopowders. The Eu^{3+} ion was incorporated in different Ca sites. The process of energy transfer to Eu^{3+} excited state ($^5D_0$) was detected from luminescence decay kinetics.

1. Introduction

Hydroxiapatite $Ca_{10}(PO_4)_6(OH)_2$ (HAp) is a biomedical material – component of bone and teeth. During recent years HAp studies became topical in view of material non-toxicity, biocompatibility, important for different clinical and biological applications (drug delivery, implants, in *vivo* imaging etc.) [1]. The HAp doped with rare earth (RE) ions is fluorescent materials prospective for lightening [2]. The HAp can be synthesized a nanopowder by sol-gel, hydrothermal and others methods [3.4]. The luminescence properties of HAp were studied in limited number of papers and mostly contain the Eu^{3+} luminescence studies [1,3, 5]. The time-resolved luminescence is described in limited number of papers.

The HAp has hexagonal structure, space group $P6_3/m$ [6] and two non equivalent Ca^{2+} positions are – Ca(I) and Ca(II). Therefore the rare earth (RE) ions substitute for Ca^{2+} in HAp structure in two non-equal local positions and the RE ions luminescence was used as a local structure probe.

In this paper we will present the results of synthesis of undoped as well as doped with Eu and Ce HAp nanocrystalline powders. The XRD and FTIR studies were carried out for the controlling the structure and vibration spectra of prepared HAp. The luminescence properties - spectra and decay kinetics of HAp nanopowders were measured and analyzed.

2. Experimental

FTIR absorption spectra were recorded using spectrometer Bruker EQUINOX55 with KBr pellet method. The crystalline phases were identified by X-ray difractometer (Bruker AXS D8 Advance).
The luminescence spectra and decay kinetics were measured under YAG:Nd laser (266 nm, 2 ns) excitation.The luminescence was recorded through monohromator with a phonon counting head (HAMAMATSU H8259) and photon counting board FastComTech module P 7887 with 500 channels and minimal time bins 250 ps. The radioluminescence was excited by X-rays (30 kV, 10 mA) through a Be window; the monochromator SHAMROC303 coupled with CCD camera ANDOR iDUS DU401A-BV was used for the spectra registration. All experiments were carried out at 300K.

3. Results and discussion

3.1.1. Sample synthesis.
The sol-gel (co-precipitation) method was used for undoped and doped with Eu or Ce HAp nanocrystals synthesis. The synthesis was carried out by using the sheme similar to described in [7].

The starting materials was [$(NH_4)_2HPO_4$], $(CaNO_3)_2 \cdot 4H_2O$ and RE nitrates. The concentration of RE was calculated in at.%. During the reaction the NH_4OH was used to keep constant pH=10. After the reaction and washing in deionised water the powders were dried at 80 ^oC 72 h. All samples were prepared under the same conditions.

3.1.2. XRD difraction.
The XRD was measured to clear up the crystallinity of synthesized powders. Fig.1. shows the XRD patterns of undoped HAp, Ce and Eu doped HAp. The XRD pattern shows only HAp with well-crystallized hexagonal structure (JCPDS Nr.09-0432) and other phases were not found. The grain sizes calculated from XRD patern by Debay-Scherr's formula was 20-25nm.

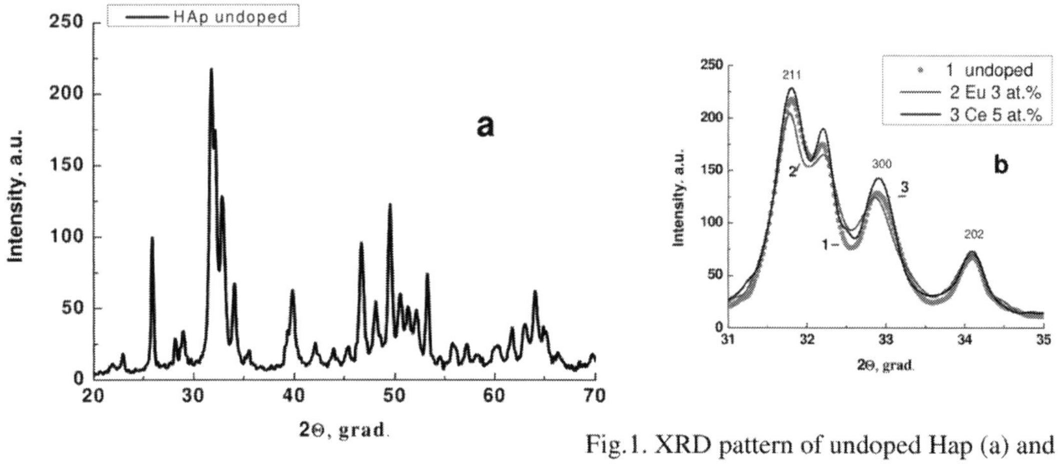

Fig.1. XRD pattern of undoped Hap (a) and doped with 3 at. % Eu and 5 at.% Ce (b)

3.1.3.FTIR
The FTIR absorption spectra of the obtained HAp nanopowders (Fig.2) show the absorption peaks concerned with stretching (ν) and bending modes (δ) of the hydroxyl, PO_4^{3-} and carbonate groups [4,8]. The narrow peak at 3570 cm^{-1} (O-H stretching mode for hydroxil group in regular HAp sites) confirms the synthesis HAp structure [9]. Along with PO_4^{3-} group vibration bands the traces of carbonate group bands [10] were identified (Fig.2).

3.1.4.Luminescence
In prepared undoped HAp sample the wide luminescence band peaking at 420 nm was detected at 300K. The time-resolved eksperiments showed that band consist of two peaks with different decay times (~6.5 ns and ~8.7 ns). The band must be due to two non-equal PO_4^{3-} groups.

In Ce doped samples the luminescence band at 350-400 nm was detected and two decay times (11 ns and 38 ns) was determinated (Fig.3). The decay time 38 ns are close to that observed for Ce^{3+} luminescence in other materials [11]. The two close bands are due to Ce ion ground state splitting to $^5F_{5/2}$ and $^2F_{7/2}$ (splitting 1800 cm^{-1}).

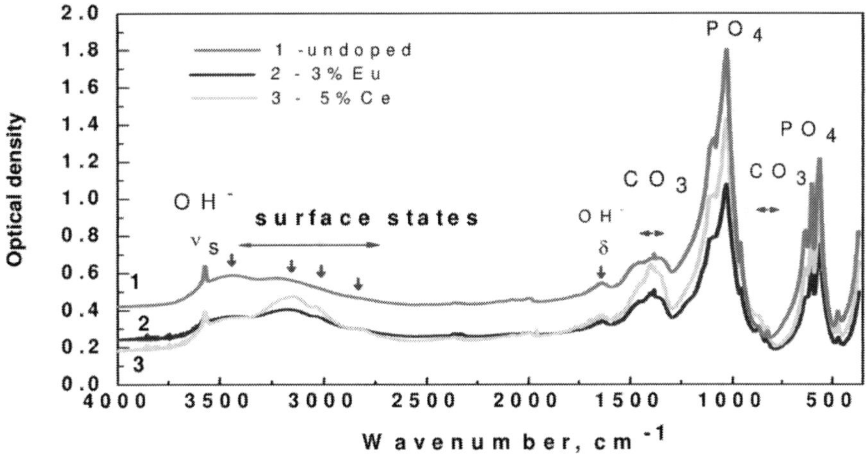

Fig.2.FTIR spectra of HAp nanopowders

Table 1. FTIR absorption peak positions and types of bonds of prepared HAP

Peak position, cm^{-1}	3570	3438 3151 3032	1468 1383 1346	1630	1110 1089 1030	960	873 821	634	602	573 563	472
Marking	ν	ν	ν_3	δ	ν_3	ν_1	δ	δ	ν_4	ν_4	δ
Type of bonds	OH^-	Surface OH^-	CO_3	H_2O	PO_4	PO_4	CO_3	OH^-	P-O	PO_4	O-P-O

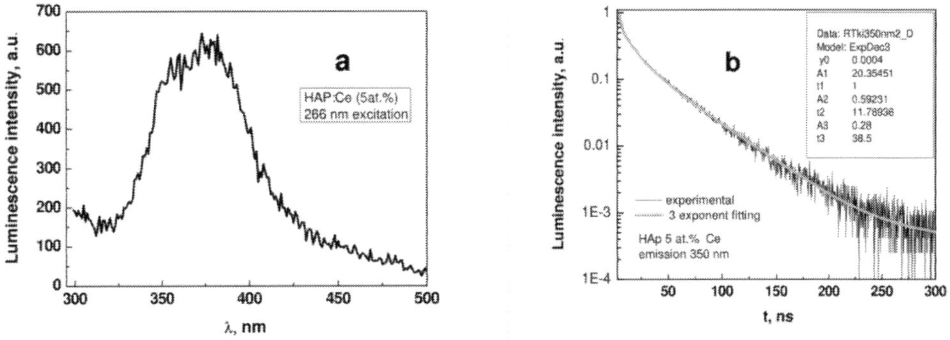

Fig.3. Luminescence spectrum (a) and luminiscence decay kinetic for HAp:Ce nanopowder.

The red luminescence of HAp: Eu is evidence of incorporation of Eu^{3+}. The spectra were recorded under 266 nm laser excitation and under X-ray excitation. The peak positions and the ratio of peak intensities are the same showing that both excitations band-to-band and charge transfer are efficient for Eu doped HAp. The luminescence spectra and luminescence decay kinetic of HAp:Eu nanopowder are presented in Fig.4. The band at 592 nm is characteristic of luminescence center with inversion center (magnetic dipole transition). The band at 619 nm is most intense and it is due to the Eu^{3+} center without inversion symmetry (electrical dipole). Hence the Eu^{3+} ion incorporates in two different sites in prepared HAp nanopowders. In [5] two sites C_3 and Cs was discussed for the Eu^{3+} incorporation in HAp.

The decay kinetics at 579 nm, 592 nm and 700 nm are close (~1.1 ms at 300K) (Fig.4b). The decay time of 619 nm band is ~540 μs that shows the different site symmetries of incorporatyed Eu^{3+} ions.

The rising front of the luminescence kinetics was detected for 592 nm, 619 nm and 700 nm bands. This behavior is related to energy transfer process during the excited state ($^5D_0$) formation.

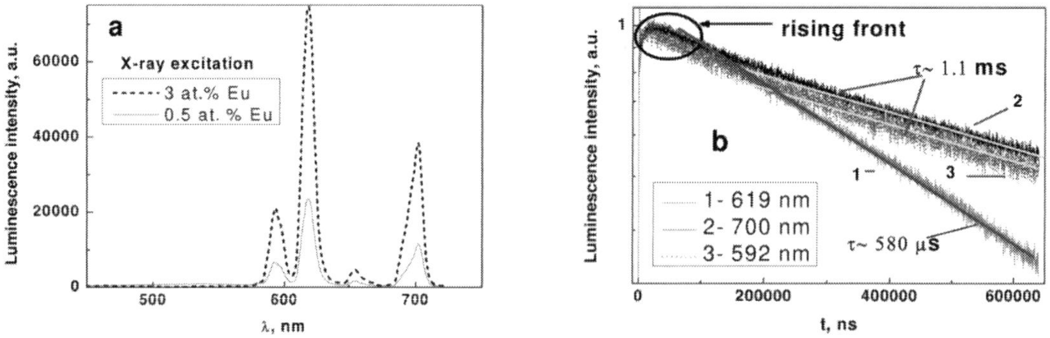

Fig.4. Luminescence spectra (a) and decay kinetics (b) of HAp:Eu nanopowders.

4. Conclusions

The analysis of XRD and FTIR absorption confirm the synthesis hexagonal HAp structure and the vibration spectrum did not showed traces of additiomal molecular impurities. The luminescence band peaking at 420 nm in undoped HAp nanopowders consist of two sub-bands with decay times τ_1 ~ 6.5 ns and τ_2~8.7 ns. We suggest that this luminescence is due to two different PO_4 groups in HAp structure. The Eu^{3+} luminescence spectra and decay kinetics were analyzed. It is shown that the mechanism of exited state ($^5D_0$) formation is complicate and includes the energy transfer process. The Eu^{3+} incorporates in Ca(I) position with C_3 site symmetry and in Ca(II) position with C_s symmetry.

Acknowledments. The work was supported by Latvian Material Science Program and LZP grant 2013.10-5/04.

References

[1] Chen F et al 2011 *Biomaterials* **32 (34)** 9031
[2] Shang M et al 2012 *J. of Physical Chemistry C* **116 (18)** 10222
[3] Han Y, Wang X, Dai H, and Li S 2013, *J. of Lumin.* **135** 281
[4] Bakan F, Lacin O and Sarac H. 2013 *Powder Technology* **233** 295
[5] Ternane R et al 2001 *Opt. Mater.* **16** 291
[6] Serret A, Cabanas M and Vallet-Regi M 2000 *Chem. Mater.* **12** 3836
[7] Ciobanu C S, Andronescu E and Predoi D 2011 *Digest J. of Nanomater.and Biostructure* **6 (2)** 1239
[8] Hayakawa C et al 2013 *Acta Biomaterialia* **9(1)** 4856
[9] Suchanek W L et al 2002 *Biomaterials* **23** 699
[10] Lacout J L, Mikou M and Bonel G 1990 *J. of Solid State Chem.* **86** 323
[11] de Queiroz T B, Ferrari C R, Ulbrich D, Doyle R and de Camargo A 2010 *Opt.Mater.* **32** 1480

IOP Conf. Series: Materials Science and Engineering **49** (2013) 012006 doi:10.1088/1757-899X/49/1/012006

Surface analysis of supercapacitor electrodes after long-lasting constant current tests

Alar Jänes, Jaanus Eskusson and Enn Lust

Institute of Chemistry, University of Tartu, 14 a Ravila Str., 50411, Tartu, Estonia

E-mail: alar.janes@ut.ee

Abstract. FIB-SEM, XPS, TOF-SIMS and electrochemical methods have been used for the characterisation of physical properties and chemical composition of microporous carbide derived carbon electrodes, prepared from TiC at 950 °C (noted as TiC-CDC) after 40000 charge/discharge cycles. Changes in surface chemical composition of TiC-CDC electrodes, includes partial contamination with reaction intermediates (F_2, CHO^-, CN^-, organic radicals), and Al current collectors, like partial dissolution of Al from positively charged electrode and deposition of Al onto the negatively charged TiC-CDC electrode surface, have been analysed.

1. Introduction

Mainly two types of electrochemical energy storage devices are available: supercapacitors (SC) and batteries. Supercapacitors (also called electrochemical capacitors), as energy storage devices with high power density and long cycle life, but moderate energy density, are considered as a promising high power density electrochemical energy sources in combination with batteries for hybrid electrical vehicles, combined solar cells and wind power systems etc., i.e., for development of combined energy recuperation complexes [1]. For the long-lasting exploitation and optimal specific energy — specific power regime, in addition to the optimized micro- and mesoporous structure of carbon material, the long-lasting electrochemical time stability of electrode material, electrical contacts and electrolyte is extremely important [2-5].

This paper reports the results of investigations of the electrochemical behaviour and surface chemical composition of the supercapacitor electrodes based on Al current collectors and deposited Al layers on microporous carbon TiC-CDC, prepared from TiC (Alfa Aesar, 99.5%, 2 μm powder) at T_{synt} = 950 °C and 1.5 M $(C_2H_5)_3CH_3NBF_4$ in acetonitrile (AN) as an electrolyte. Based on the N_2 sorption data, the multipoint BET surface area S_{BET} = 1698 $m^2\ g^{-1}$ and micropore area (obtained using *t-plot* method) S_{micro} = 1661 $m^2\ g^{-1}$ were calculated. Our previous SC studies indicate that TiC-CDC is an excellent supercapacitor electrode material containing micropores for adsorption of partially desolvated ions and a small quantity of mesopores for quick mass transfer of ions inside the porous TiC-CDC. However, from data available in literature it was recognized that, more detailed studies discussing the electrochemical behavior of Al-contact layers, deposited onto TiC-CDC carbon electrodes and Al current collectors at cell voltages $U \geq 3.2$ V, are inevitable [6-8].

2. Experimental

Before long-lasting electrochemical constant current charge/discharge (CC/CD) tests of SC the crystallinity and structure of the TiC-CDC electrodes were evaluated by focused ion beam scanning electron microscopy (FIB-SEM) and X-ray photoelectron spectroscopy (XPS) with a Cu anode to generate CuKα radiation (λ=1.5406 Å). The Raman spectra were recorded using a Renishaw micro-Raman spectrometer equipped with $50^{\times}$ objective and 488 nm Ar^+ ion laser with maximum radiation power of 15 mW on the sample. XRD and Raman data show that TiC-CDC carbon has mainly amorphous structure. The FIB-SEM data have been obtained using Helios™ Nanolab 600. The XPS experiments were carried out with a SCIEHTA SES-100 spectrometer by using an unmonochromated MgKα X-ray source with power of 300 W. SES-100 spectrometer was calibrated using Au 4f photolines. The pressure in the analysis chamber was varied within the range from 2×10^{-10} to 5×10^{-10} mbar. Time of flight secondary ion mass spectrometry (TOF-SIMS) spectra were measured using PHI TRIFT V *nanoTOF* surface analysis instrument (Physical Electronics, Inc., USA) with a 30 keV Au^+ ion gun. The acquisition time was 2.75 min and the TOF-SIMS spectra were collected in the range from 1 to 150 m/z in both the positive and negative polarities.

The SC electrode active material was prepared from TiC-CDC with 4 wt% addition of the binder polytetrafluoroethylene mixture (PTFE, 60 % solution in H_2O). The mixture received was laminated and roll-pressed to form a flexible layer of an active electrode material with thickness 100 ± 5 μm. The Al current collectors were deposited onto the one side of a TiC-CDC electrode using the plasma activated physical deposition method. Thereafter the two-electrode standard aluminium test cell with two identical electrodes (apparent area of ~1.53 cm^2) was completed and all electrochemical experiments were carried out at temperature 20 ± 1 °C. 25 μm thick TF4425 sheet was used for mechanical separation of the working TiC-CDC electrodes from each other and 1.5 M $(C_2H_5)_3CH_3NBF_4$ solution in acetonitrile (AN) was used as an electrolyte [2-14].

3. FIB-SEM, XPS and TOF-SIMS analysis

FIB-SEM combined energy dispersive X-ray (EDX) method was used for the quantitative analysis of the SC electrodes. It was found that before electrochemical tests the porous TiC-CDC electrodes consist of particles with variable linear dimension from 1 to 3 μm. Some oxygen has been observed on both TiC-CDC layers and Al current collectors being higher at Al containing ingredients. After 40000 charge/discharge cycles from 0.2 to 3.4 V the supercapacitor cells were opened and electrodes were washed and tried under vacuum. Thereafter TiC-CDC electrodes and Al current collectors were analyzed again by using XPS and FIB-SEM-EDX methods. Based on the analysis of data some changes in chemical composition of TiC-CDC electrode surface layers and Al current collectors have been detected after cycling. Weight % of TiC-CDC on carbon electrode was somewhat lower for negatively charged electrode (from 63 to 83 wt%) compared with that for positively charged electrode (from 87 to 92 wt%) or for the electrochemical untreated electrode. Noticeable increase of fluorine wt% for negatively charged electrode was detected, which is caused by the irreversible electroreduction of BF_4^- anions and adsorption of reaction intermediates or products (BF_3 and/or F_2) inside microporous TiC-CDC. Weak increase of N and O wt% at negatively charged electrode has been established caused by adsorption/absorption of $(C_2H_5)_3CH_3N^+$ and O_2 decomposition products (hydrogen peroxide) during complex electroreduction reactions. Differently from negatively charged TiC-CDC, only very small increase in N and O wt% has been established for positively charged TiC-CDC electrode.

After 40000 electrochemical charge/discharge cycles an increased amount of N as well as F wt% on Al-coated side of TiC-CDC was observed, being higher for negatively charged electrode. According to EDX data the oxygen amount at Al contacts for both electrodes was quite low (7-10 wt%) which indicates, that the irreversible oxidation of aluminium during long-lasting cycling at high cell voltage was very slow.

The XPS analysis (Fig. 1) shows that the intensity of XPS peaks is somewhat different for TiC-CDC electrodes before and after long-lasting constant current charging/discharging tests. The O(KLL, (Auger lines)), O1s and N1s peaks for positively and negatively charged TiC-CDC electrodes are more pronounced than for the electrochemically non-tested electrode indicating to the peroxide

IOP Conf. Series: Materials Science and Engineering **49** (2013) 012006 doi:10.1088/1757-899X/49/1/012006

formation (negatively charged electrode) and OH^- adsorption/absorption (positively charged electrode) in microporous carbon have taken place.

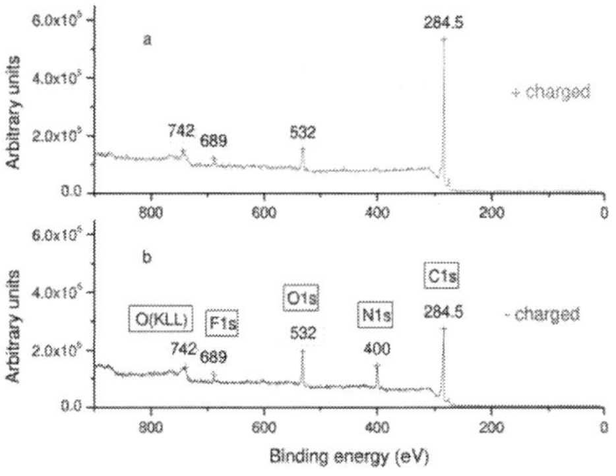

Fig.1. XPS spectra for positively (a) and negatively (b) charged electrodes after 40000 charge/discharge cycles.

Comparison of C1s peaks shows that after CC/CD tests the intensity of C1s peak is highest for positively charged TiC-CDC electrode, indicating to the formation of C-H bonds. However, this peak can be related to the modification of TiC-CDC surface as well as to the recrystallization of carbon particles from amorphous to more stable graphite like structure (so-called carbon electrode corrosion widely discussed in the case of H_2O medium). However, the intensity of F1s peaks is nearly independent of electrochemical testing time (i.e. polarization applied) indicating that there is no big degradation of Teflon binder under CC/CD cycling conditions. Existence of N1s peak in XPS spectrum indicates to the chemisorption of $(C_2H_5)_3CH_3N^+$ cations disproportionation products at negatively charge TiC-CDC electrode.

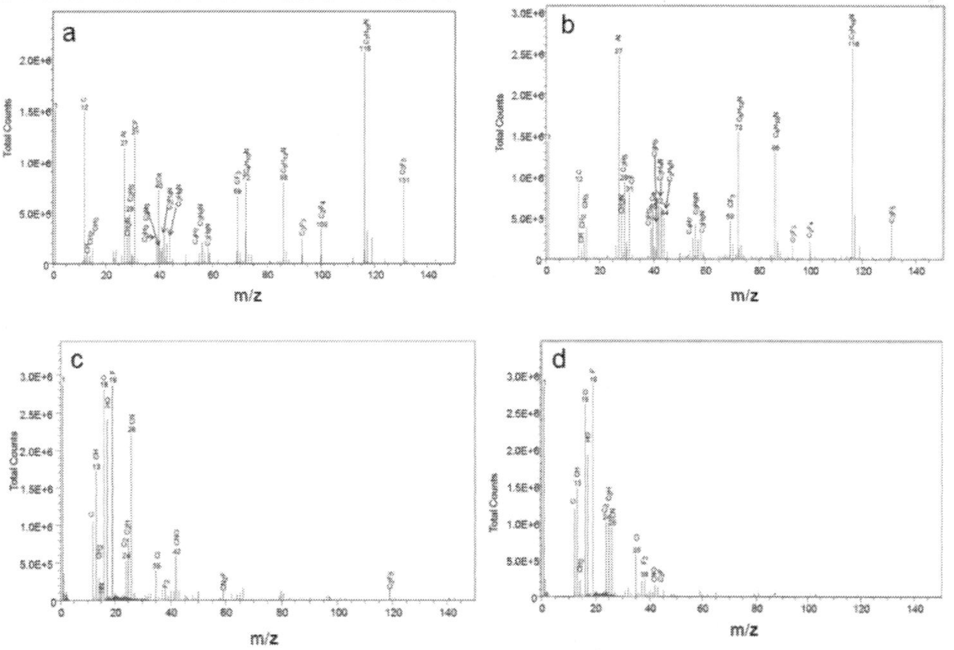

Fig.2. TOF-SIMS spectra for TiC-CDC supercapacitor for positively (a, c) and negatively (b, d) charged electrodes and for negatively (c, d) and positively (a, b) charged secondary ions after 40000 charge/discharge cycles.

After 40000 galvanostatic cycles from 0.2 to 3.4 V the supercapacitor electrodes were investigated by TOF-SIMS method up to ion masses m/z = 150. The mass spectra given in Figs. 2a – 2d demonstrate that the results for negatively and positively charged secondary ions obtained for positively and negatively charged electrodes are somewhat different. Mass spectrum for positively charged secondary ions shows mainly fragments or deposits formed from $(C_2H_5)_3CH_3N^+$ cations (13, 14, 15 m/z and 29 m/z for CH^+, CH_2^+, CH_3^+ and $C_2H_5^+$, respectively). Mass spectrum for negative charged secondary ions shows existence of CH^- (13 m/z), F^- (19 m/z), OH^- (17 m/z), CNO^- (42 m/z), O^- (16 m/z) and toxic CN^-(26 m/z) and other negatively charged fragments that are formed during $(C_2H_5)_3CH_3NBF_4$ and acetonitrile disproportionation and carbon electrode degradation processes. The positive characteristic fragment for PTFE ions and their nominal mass-to-charge ratios are: CF^+ (31 m/z), CF_3^+ (69 m/z), $C_3F_3^+$ (93 m/z), $C_2F_4^+$ (100 m/z), $C_3F_5^+$ (119 m/z) and $C_3F_5^+$ (131 m/z). In the negatively charged ion spectrum, the characteristic peaks of PTFE include $C_2F_5^-$ (119 m/z). In addition to the contaminants generated by electrochemically induced oxidation/reduction processes, common contaminants such as Na^+ (23m/z), K^+ (39 m/z), Ca^+ (40 m/z) and adventitious hydrocarbons in the positive ion spectra and Cl^- (from synthesis of $(C_2H_5)_3CH_3NBF_4$ salt) were observed.

4. Conclusions

FIB-SEM, XPS and TOF-SIMS methods have been used for the chemical analysis of supercapacitor electrodes prepared from microporous carbide derived carbon synthesised from TiC at 950 °C, noted as TiC-CDC. The FIB-SEM, XPS and TOF-SIMS data show that partial contamination of carbon electrode surfaces with reaction intermediates (F_2, CH^-, CHO^-, CN^-, organic radicals etc.) is a reality during galvanostatic charging and discharging. Al redeposition onto negatively charged carbon and restructuring of carbon particles was observed for positively charged TiC-CDC electrode.

Acknowledgements

This work was supported by the Estonian Science Foundation under Projects Nos. 8172 and 9184, Estonian Ministry of Education and Research project SF0180002s08 and European Regional Development Fund Project SLOKT10209T, Estonian Centres of Excellence in Research Project TK117T "High-technology Materials for Sustainable Development". Dr. L. Matisen and Dr. A. Kikas (Institute of Physics, University of Tartu) are thanked for the help with the XPS measurements. Dr. G.L. Fisher from Physical Electronics, Inc. (Chanhassen, MN, USA) is thanked for the assistance of TOF-SIMS measurements.

References

[1] Conway B E 1999 *Electrochemical Supercapacitors: Scientific Fundamentals and Technological Applications* (Kluwer Academic/Plenum Publishers, New York)
[2] Jänes A, Eskusson J, Kanarbik R, Saar A and Lust E 2012 *J. Electrochem. Soc.* **159** A1141
[3] Tallo I, Thomberg T, Kontturi K, Jänes A and Lust E 2011 *Carbon* **49** 4427
[4] Jänes A, Thomberg T and Lust E 2007 Carbon **45** 2717
[5] Eskusson J, Jänes A, Kikas A, Matisen L and Lust E 2011 *J Power Sources* **196** 4109
[6] Laheäär A, Jänes A and Lust E 2011 *Electrochim. Acta* **56** 9048
[7] Laheäär A, Jänes A and Lust E 2012 *Electrochim. Acta* **82** 309
[8] Jänes A, Permann L, Arulepp M and Lust E 2004 *J. Electroanal. Chem.* **569** 257
[9] Jänes A, Kurig H, Romann T and Lust E 2010 *Electrochem. Commun.* **12** 535
[10] Jänes A, Permann L, Nigu P and Lust E 2004 *Surf.Sci.* **560** 145
[11] Thomberg T, Jänes and Lust E 2009 *J. Electroanal. Chem.* **630** 55
[12] Thomberg T, Jänes A and Lust E 2010 *Electrochim. Acta* **55** 3138
[13] Laheäär A, Kurig H, Jänes A and Lust E 2009 *Electrochim. Acta* **54** 4587
[14] Palm R, Kurig H, Tõnurist K, Jänes A and Lust E 2012 *Electrochim. Acta* **85** 139

Impedance and admittance characteristics of Bi_2S_3 nanowire arrays

J Katkevics[1], G Kunakova[1], A Viksna[1], J D Holmes[2], D Erts[1].
[1]Institute of Chemical Physics, University of Latvia, Raina blv 19, Riga, LV-1586 Latvia
[2]Department of Chemistry, University College Cork, Dublin 2, Ireland

e-mail: juris.katkevics@lu.lv

Abstract. Current studies of the electrical impedance and admittance characteristics of the anodised aluminum oxide (AAO) nanoporous arrays and bismuth sulphide (Bi_2S_3) nanowire within AAO membranes are presented. The influence of potential and frequency scan rate effect produced on the real, imaginary and complex electrochemical impedance and double layer capacitance of the AAO nanopore and the Bi_2S_3 nanowire arrays were studied.

1. Introduction

Characterization of conductive properties of Bi_2S_3 nanowire within AAO membranes is important for perspective applications in nanoelectronics, photoelectrochemical solar cells and sensors [1]. Therefore it is important to determine the effects on electrical characteristics of B_2S_3 nanowire after repeated actions of current. It is shown that the resistance of Bi_2S_3 nanowire arrays is instable and varied between 1÷106 Ω cm [2, 3]. The resistance of nanowire arrays depends on method of preparation of nanowires, its size, and applied potential [2, 3].
Electrochemical impedance spectroscopy (EIS) is widely used for characterization of large area electrodes and the electrodes coated with porous layers [2, 5]. Recently EIS was applied for investigation of resistance and impedance of Bi_2S_3 nanowire arrays [4].
Here it is presented the studies of the electrical impedance and admittance characteristics of Bi_2S_3 nanowire arrays and AAO porous membranes. In particular it is addressed the influence of scan rate frequencies on the real, imaginary and complex electrical impedance and double layer capacitance of the AAO nanopores and the Bi_2S_3 nanowire arrays. The AAO membranes and Bi_2S_3 nanowire arrays were investigated applying two different modes of EIS - frequency scan at single potential and potential scan at single frequency.

2. Experimental

Bi_2S_3 nanowire arrays inside the pores of AAO membranes were fabricated by the thermolysis of the single source precursor bismuth bis(diethyldithiocarbamate) [$Bi(S_2CNEt_2)_3$] complex. This complex was prepared by the reaction of Bi_2O_3 with carbon disulfide and diethylamine in methanol and was thoroughly purified by repeated recrystallization as reported previously [6]. Scanning electron microscope Hitachi S4800 was used for visualization of samples (Figure 1 (a) and (b)). Samples containing Bi_2S_3 nanowires were polished mechanically on both sides to open the nanowires and remove reaction products from the surface. Then one side was coated by a 60 nm thick gold layer. For electrical characterization the specific measurement cell was designed in two electrode configuration (Figure 2 (a) and (b)). Deposited Au layer was used as a cathode and a Pt plate for AAO sample and drop of Hg for Bi_2S_3 nanowire arrays sample were used as an anode. Impedance spectra were obtained in potential range from: -0.4 to 0.4 V at scan rate of 10 ÷ 0.1 kHz, scanning number – 40, and measuring amplitude 0.01 V. The air humidity of laboratory was 75 % RH. All measurements were performed with potentiostat "Autolab PGSTA 30". Results were analyzed by the Frequency Response Analysis (FRA) program for

IOP Conf. Series: Materials Science and Engineering **49** (2013) 012007 doi:10.1088/1757-899X/49/1/012007

Windows version 4.9.007. In this report real (Z_r) and imaginary ($-Z_j$) component of impedance, real (Y_r) and imaginary ($-Y_j$) component of admittance is presented.

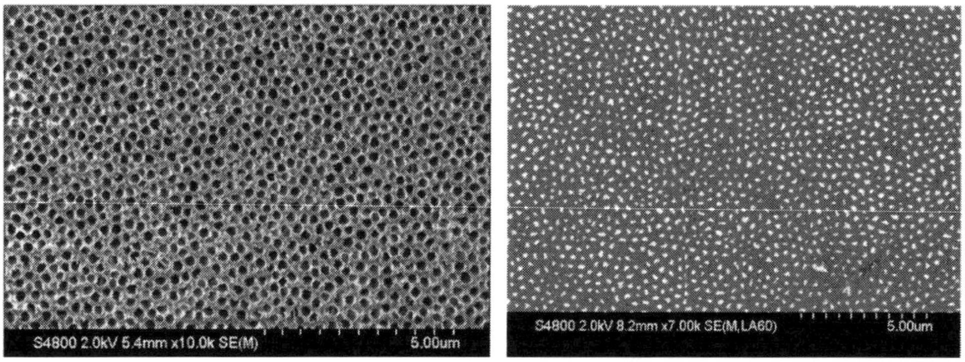

a b

Figure 1 SEM images of AAO nanopore arrays (a) and Bi_2S_3 nanowire arrays (b). All AAO membrane nanopores are packed with Bi_2S_3 substances (b).

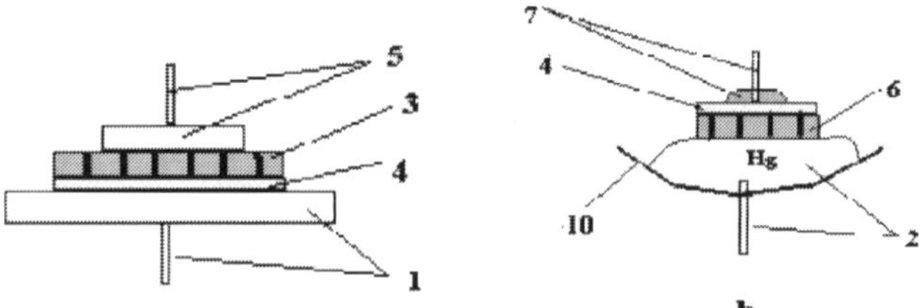

a b

Figure 2 Representation of electrochemical cells for EIS measurement, (a) – electrical impedance spectra measurements for AAO; (b) – electrical impedance measurements for Bi_2S_3 nanowire arrays: 1 – Pt anode; 2 – Hg anode; 3 – the AAO membrane; 4 – the Au layer; 5 – Pt plate (cathode); 6 – Bi_2S_3 nanowire arrays; 7 – the Pt wire on Au layer (cathode); 10 – the small glass cup with Hg droplet.

3. Results and Discussion

The real component of the impedance (Z_r) value for AAO membrane was analyzed at applied current frequency at 10 kHz is insignificant (2÷100 MΩ·cm^2) (Table 1, Figure 3a, 3e). When scanning rate of current frequency was increased from 10 kHz to 1 kHz then the real component of impedance AAO nanopore arrays increased from 2 to 100 MΩ·cm^2. The imaginary component of impedance ($-Z_j$) value at this scanning frequencies increase very fast (-Zj = 2.3 ÷ 7.5 GΩ·cm^2). The real and imaginary

Table 1 AAO nanopore arrays and Bi_2S_3 nanowire arrays impedance and admittance characteristic (in laboratory air, humidity - 75% RH)

Parameters	AAO nanopore arrays, air connect			Bi_2S_3 nanowire arrays, Hg connect, 1-10 kHz		
	-0.3 V	0.0 V	+0.3 V	-0.3 V	0.0 V	+0.3 V
Zr, MΩ cm^2	2-100	2-100	2-100	0.007-0.010	0.010-0.012	0.007-0.008
-Zj, MΩ cm^2	2300-4500	2300-4500	2300-4500	0.002-0.006	0.0034-0.010	0.0031-0.007
/Z/, MΩ cm^2	2300-4550	2300-4550	2300-4550	0.007	0.011	0.008
Yr, µS /cm^2	0.001	0.001	0.001	100	95	120
-Yj, µS /cm^2	0.0003	0.0004	0.0003	130	45	180
/Y/, µS /cm^2	0.001	0.0011	0.001	165	105	216

Table 2 The impedance and admittance characteristics of Al_2O_3 pressed powder and laboratory air at 75% RH humidity

Parameter	Z_r, MΩ cm^2	$-Z_j$, MΩ cm^2	Y_r, nS/cm^2	$-Y_j$, nS/cm^2
Al_2O_3 powder	5	40	200	25
Air layer of laboratory, 75% RH	9	32	10	4
Al_2O_3 powder, 70-80% relative humidity RH	Z = 5·10^6 - 2·10^8 Ω cm		Literature data [7]	
Air layer, absolute dray, 0% RH	Zeq = 1·10^{14} Ω cm			

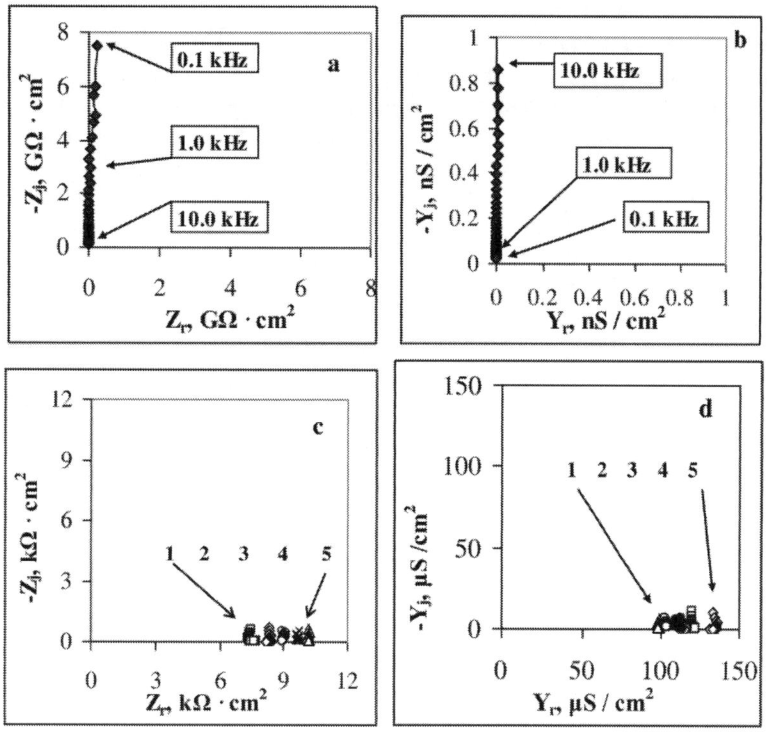

Figure 3 Complex impedance (a, c) and admittance (b, d) diagrams for AAO membrane (a, b) and for Bi_2S_3 nanowire arrays on the mercury droplet (c, d) for different applied potential: 1 -0.00 V; 2 -0.05 V; 3 -0.10 V; 4 -0.15V; 5 -0.20V

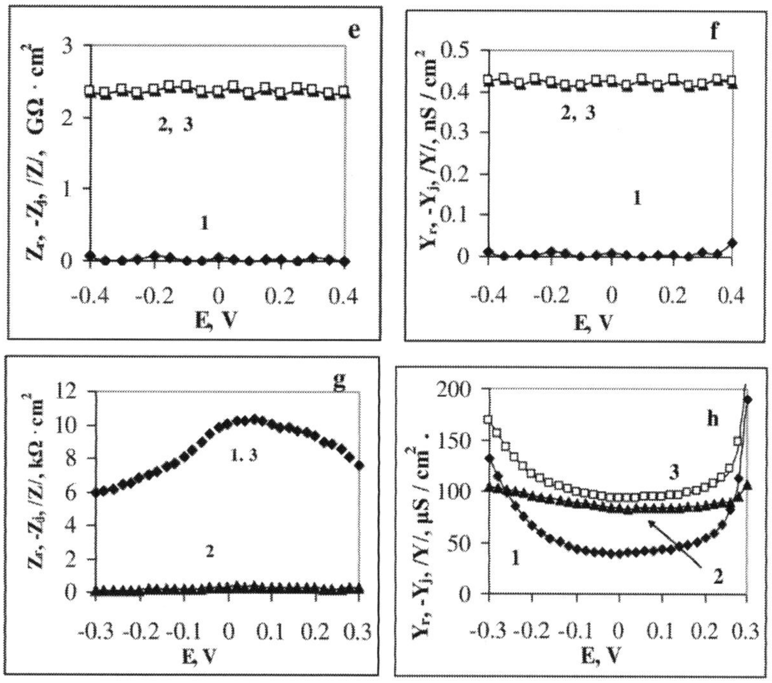

Figure 4 Impedance (e, g) and admittance (f, h) diagrams for AAO nanopore membrane (e, f), and for Bi_2S_3 nanowire arrays on the mercury drop (g, h) in applied potential range from -0.4 V to 0.4 V: 1 – real component of impedance (Z_r) and admittance (Y_r); 2 – imaginary component of impedance ($-Z_j$) and admittance ($-Y_j$); 3 – modulus of impedance (/Z/) and admittance (/Y/).

components of admittance (Table 1, Figure 3b, 4f,) are small – 1.0 nS/cm^2. It is easy to understand, because AAO (that is Al_2O_3) and air is characterized with high electrical resistance (Table 2.) The analysis of impedance spectrum using equivalent electrical circuit show that AAO characterized with one resistance elements (R) and capacitance component is very small, only some pF (Cs = 80 pF/cm^2). The experiments shows that Bi_2S_3 nanowire arrays real component of impedance (Z_r) and real component of admittance (Y_r) using applied potentials from 0.0 to -0.2 V is rather constant from 7.0 to 10.0 kΩ·cm^2 and from 95 to 120 μS/cm^2, respectively (Table 1, Figure 3c, 3d, 4g, 4h). The both, imaginary component of impedance ($-Z_j$) and imaginary component of admittance ($-Y_j$) are very small at scanning rate of current frequencies below 1 kHz. But applying scan rate at larger frequencies (1.0 - 10 kHz) these components are from 2 to 3.4 kΩ·cm^2 and from 45 to 180 μS/cm^2 respectively. These facts denote that characteristics of Bi_2S_3 nanowire arrays applied at higher frequencies are different than characteristics of Bi_2S_3 nanowire applied at low frequencies.

The equivalent circuit for Bi_2S_3 nanowire arrays is denoted as $R_1(R_2C_1)$, where R_1 is 7.2 kΩ·cm^2, R_2 - 3.0 kΩ·cm^2 and C_1 - 2 nF/cm^2. AAO membrane and Bi_2S_3 nanowire arrays are characterized of two and three parallel positioned impedance elements: $Z_{Al_2O_3}$, Z_{air} and $Z_{Al_2O_3}$, $Z_{Bi_2S_3,bulk}$, Z_{Cs} for AAO membrane and Bi_2S_3 nanowires arrays respectively. For the calculation of impedance components AAO membrane and Bi_2S_3 nanowire arrays formulas **a)** and **b)** are used, respectively. Formula **c)** is used for the calculation of capacitance component of impedance. Coefficient 0.8 in formula **a)** and **b)** is a fraction of surface area covered by nanopore wall cross section but coefficient 0.2 is relative air double cavity or Bi_2S_3 nanowire area. Z is impedance and Cs is capacitance.

a) $$\frac{1}{Z_{AAO}} = \frac{1}{Z_{Al2O3}} \times 0.8 + \frac{1}{Zair} \times 0.2;$$ **b)** $$\frac{1}{Z_{Bi2S3}} = \frac{1}{Z_{Al2O3}} \times 0.8 + \frac{1}{Z_{Bi2S3,bulk}} \times 0.2 + \frac{1}{Z_{Cs}};$$

c) $$Cs = \frac{1}{2\pi f Z_{Cs}}; \quad (1)$$

The pure real calculated component of impedance ($-Z_{r,Bi_2S_3,\ bulk}$) for Bi_2S_3 nanowire arrays is 50 Ω·cm^2, but imaginary component ($-Z_{j,Bi_2S_3}$) is 2 Ω·cm^2, that is smaller than previously calculated [4].

4. Conclusion

It is experimentally defined that for AAO nanopore membrane the imaginary component of impedance ($-Z_j$) is one order of magnitude higher than the real component of impedance (Z_r), but for Bi_2S_3 nanowire arrays the real component of impedance (Z_r) is order higher than the imaginary component of impedance ($-Z_j$). It is determined that for Bi_2S_3 nanowire array real component of impedance is smaller than it was calculated previously.

Acknowledgements

Work was granted by ERAF Project Nr. 2010/0251/2DP/2.1.1.1.0/10/APIA/VIAA/096

References

[1] A. A. Tahir, M. A. Ehsan, M. Mazhar, K. G. U. Wijayantha, M. Zeller, A. D. Hunter. Chem. Mater., 22: 5984-5092 (2010).
[2] K. A. Z. Syed Abuthahir, R. Jagannathan. Materials Chemistry and Physics, 121 ,184-192 (2010).
[3] P. Birjukovs, N. Petkov, J. Svirksts, J. J. Boland, J. D. Holmes, D. Erts. J. Phys. Chem. C., 112: 19680-19695 (2008).
[4] J. Katkevics, A. Viksna, G. Kunakova, A. Pastare, J. D. Holms, D. Erts. International conference "Functional materials and nanotechnologies 2011", Institute of Solid state physics University of Latvia , April 5 – 8, Riga, p. 186, (2011)
[5] M. E. Orazem, B. Tribollet. ECS-The Electrochemical Society, 159; 533 (2008)
[6] J. Xu, N. Petkov, X. Wu, D. Iacopino, A. J. Quinn, G. Redmond, T. Bein, M. A. Morris, J. D. Holmes. Chem. Phys. Chem., 8, 235-240 (2007)
[7] F. Faver, F. Villieras, Y. Duval, E. McRae, C. J. Rapin. Colloid Interface Sci., Jun 15; 286 (2):615-620 (2005).

Electrodeposition of Nanoporous Nickel Layers Using Inductive Voltage Pulses

Martins Vanags, Janis Kleperis, Gunars Bajars, Vladimirs Nemcevs

Institute of Solid State Physics, University of Latvia, Latvia

kleperis@latnet.lv

Abstract. We used inductive voltage pulses for electro-deposition of porous nickel thin film onto steel electrode. Short pulses (tp<1 μs) with variable amplitude (to ensure electro-deposition current density in region 124-224 A/dm^2) were applied to electrode and resulting coating were analyzed with electrochemical and microscopic methods. At lower current densities only smooth nickel coatings growth, while at higher current densities the bubbles appear and porous layer was formed. Electrochemical impedance spectra of smooth and porous layers are measured in deionized water and in 1M KOH solution. The capacity in equivalent scheme is proportional to electrode surface, and from impedance spectra it is calculated that porous layer has 20 times larger active surface comparing to smooth layer (in KOH solution). From electrochemical measurements it is estimated that more efficient hydrogen evolution reaction occurs on electrode with porous nickel layer obtained at 223 A/dm^2. It is shown in this work that inductive short pulse method can be used to obtain nano-porous nickel coatings on electrodes for efficient electrolysis cell.

1. Introduction

The pulse electrolysis in scientific literature is recognized as a powerful method to obtain electrochemical coatings [1]. Compared to constant voltage electrolysis, where only the value of applied potential can be changed, the more power parameters are changed in pulse electrolysis – the length of pulse, pulse tracking frequency, pulse polarity, pulse amplitude etc. By changing the characteristics of the pulse power, also changes the structure of obtained coatings – they are smoother and with better adhesion, as it is commonly recognized and explained by specific peculiarities of pulsed electric double layer[2]. Double layer is divided into two parts: the pulsating layer and diffusion layer. The concentration of active ions is periodically changing in the pulsating layer, reaching maximum at the momentum when pulse appears, and the minimum of the pulse ended. The layer thickness is proportional to the pulse duration, so the pulsating thin layer is duplicating the electrode surface very well, resulting in a smooth coating [3]. It is also observed that coating hardness increased by increasing the peak current [4], and if very high hardness internal stresses occur on the surface, it may be so high that the coating cracks. Generally, electro-coating crystallization process takes place in two steps. In the first step, the metal ion discharge and in the second step the metal atom is generated. The second step takes place in two scenarios: the metal atom is included in the bran of crystal and crystal growth, or new kernel is formed. The second scenario occurs when the crystal growth rate is so slow that generated metal atoms they can not add. It is shown that the radius of the surface of generated kernel is inversely proportional to the over-voltage. This suggests that increasing the over-voltage, the grain radius is decreasing [5]. When the pulse rate is high, Nernst diffusion layer non-stationary mass transfer is dominating, and higher is metal ion concentration gradient in the interface. The stationary mass transfer is dominating in the DC or pulse electrolysis at low frequencies [6]. Although pulsed electrolysis is widely used to obtain different coatings, there are not found many research works directed to pulse power applying to water electrolysis.

We have considered the inductive voltage electrolysis process [7,8], and raised the hypothesis that such a process is separating the charging current of cell geometric capacitance and double layer from the current f electrochemical reactions, and found that this type of pulse is limited by the energy before it is supplied to the electrochemical cell.

In this work, the nickel layer was grown up of stainless steel wire as working electrode in electrolysis cell with inductive voltage pulses. Changing the voltage pulse amplitude, different types of films are obtained smooth and porous. For each sample the specific surface area, exchange current, Tafel slope and the equilibrium potential are measured and compared with smooth wire.

2. Experimental

We used nickel plating solution (80g $NiSO_47H_2O$, 80g $NiCl_26H_2O$ and 45g H_3BO_3 in 1 litter water), stainless steel wire 316L as working electrode (diameter 1 mm, length 10 mm), nickel sheet with large surface area as counter electrode and standard calomel electrode as reference electrode. Thin nickel film was grown onto the working electrode in two ways: using direct current mode or inductive voltage pulse mode. Stainless steel wire was mechanically polished, washed with acetone and rinsed in deionised water before experiment. Then electrode was placed in 5M KOH solution at a temperature 60 °C for 5 minutes, again washed with deionised water and rinsed in nickel plating solution. Inductive voltage pulses were generated by the generator, which is detailed in our previous publications [7,8]. Different coatings were obtained adjusting the pulse amplitude. Typical coating time 2 min in all cases was used. After coating, the samples were rinsed with deionised water. The impedance spectrum of each sample in deionised water was measured with HP potentiostat- bridge within a frequency range from 100Hz to 40MHz, using modulation with pulse amplitude of 20mV and DC component 0V. This was done in order to judge the specific surface area of electrode in dependence from the type of plating. The impedance spectrum of all coatings in 0.1 M KOH solution also was measured with the same settings to measure the double layer capacitance of electrode/electrolyte interface for each coating. Potentiostat PGZ301 VoltaLab is used to measure VA curves in 1M KOH aqueous solution for each sample, using voltage scanning from -250 to -2500mV with scan rate 50 mV/s and the Tafel slope value and exchange current were determinate. To study the morphology of nickel coatings, scanning electron microscopy (SEM, Hitachi S-4800, Institute of Physical Chemistry, University of Latvia) was used.

3. Results

4. Voltage oscillograms taken during deposition of thin nickel layer using pulses of different amplitude conditions are shown in Figure 1. The current density values marked in figure 1 is the current pulse peak values during deposition. In the case of current densities 207, 213 and 223 A/dm^2, in addition to nickel loading reaction, gas evolution at the working electrode was noticed. As a result of gas evolution, obtained coating was porous and homogeneous over the entire deposition surface. Gas bubbles certainly contributed in additional recovering of the nickel loading process simply mixing the electrolyte around the electrode. Gas bubble encapsulation in deposited nickel layer is responsible for the formation of pores in the coating. Obtained electrode coatings appeared homogeneous over the entire surface, since the release of gas after visual inspection were relatively uniform during all coating time. In voltage oscillograms the rapidly growing voltage region appears followed with slow changing region (Figure 1). As already discussed in our previous works [7,8], fast growing front is purely capacitive charging of electrochemical cell, while the slow discharge is related to the electrochemical reaction. In other words, at starting very short-time region is only charging process and electrochemical processes are turned off. The energy supply to the cell ended up charging when voltage pulse front reach the maxima, and since energy is supplied to the capacity, small decrease of it

during the process may further increase the potential on it. Two reasons can be to decrease the geometric capacity of cell: gas bubbles are blocking the surface, and erosion occurring on nickel grains. Since both of these processes are electrochemical reactions, the flat voltage increases the beginning of the electrochemical reaction start time for Ni coating process (this time is set around 1 microsecond). In addition to the slow voltage growth region in oscillograms, previously debated fact is confirmed, that during the inductive voltage pulse electrolysis, the energy supplied to the interface metal - electrolyte is determined prior to the pulse inducing magnetic field in primary circuit coil. Hence it is easy to deal with power of pulse electrolysis system – firstly calculate the reactive power of coil in primary circuit which will determine the power of electrolysis system in secondary circuit.

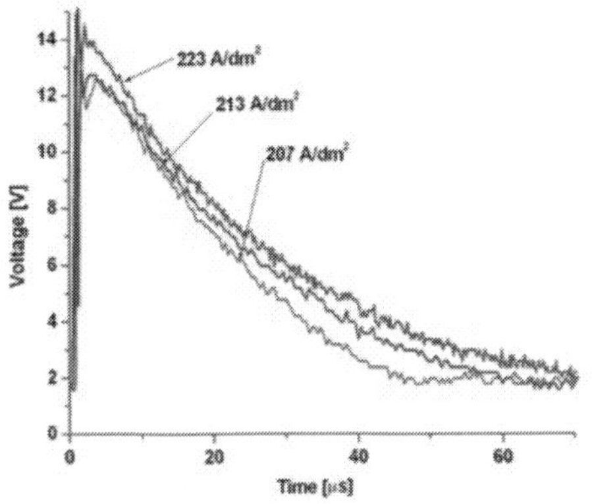

Figure 1. Voltage oscillograms taken during pulse plating with different peak currents.

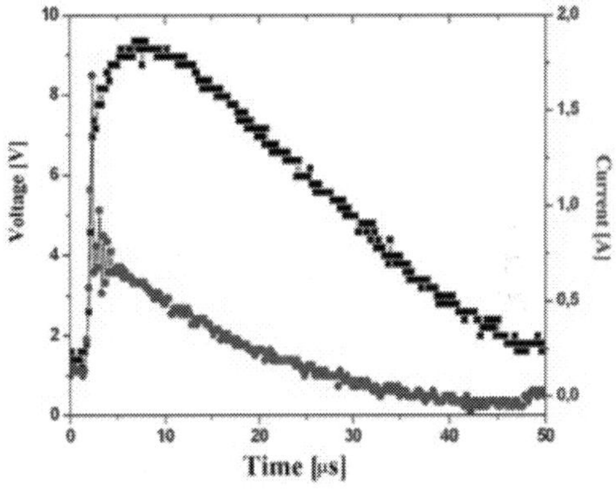

Figure 2. Voltage and current oscillograms taken during pulse plating with peak current 213 A/dm^2.

SEM images shows that gas bubbles during pulse coating are encapsulated and later remains as pores, while in DC mode the coating is smooth (Figure 3).

Figure 3. SEM images of Ni coatings: a), b) - a porous coating with inductive voltage pulses, c), d) - smooth coating with DC voltage mode

From impedance spectra the capacitance of electrode/electrolyte interface is calculated for porous pulse plated and smooth DC plated Ni coatings. In case of smooth steel electrode in water the value 5.08 pF is obtained what is very close to measured value 9.64 pF. In case of porous coating rough estimation gives value 2 times larger comparing with smooth layer. Impedance measurements of electroes without in with different coatings in an alkaline solution gives evidence that the double layer capacity of porous coating is six times larger than that of smooth coating. It means that the difference between surfaces of porous and smooth coatings is already about 20 times.

From electrochemical measurements it is estimated that more efficient hydrogen evolution reaction occurs on electrode with porous nickel layer obtained at 223 A/dm^2 (Table 1).

Table 1. Parameters of Tafel equation for porous nickel coating and smooth steel (accordingly SCE).

Material	E_0 [mV]	I_0 [μA/cm^2]	b [mV/dec]
Steel (smooth)	-998	50,2	158
Porous nickel coating (mode 213 A/dm^2)	-934	16,08	142
Porous nickel coating (mode 223 A/dm^2)	-926	18,17	135

5. Conclusions

A variety Ni layer structures on stainless steel wire is obtained using inductive voltage pulses. The porous layer is formed in pulse coating mode because of encapsulation of small gas bubbles in the volume of coating film. The main difference with DC mode is that in pulse mode the immediately after rapid voltage rise the interface is charging followed by a slow voltage decrease. The authors explain this by the surface of the capacity reduction as a result of the constant power supplied to the cell voltage increase. Capacity reduction on the surface can be attributed to two reasons: the erosion of nickel grains, which reduces the surface area and gas bubbles blocking the surface. Since both of these processes are electrochemical reactions, the flat voltage increases the beginning of the electrochemical reaction start time for Ni coating process (this time is set around 1 microsecond). The highest measured HER kinetics parameters has sample obtained at highest current 223 A/dm^2, observed for porous nickel coating obtained in the pulse mode and having the equilibrium potential of -926 mV (SCE), 18.17 μA/cm^2 – the exchange current value and 135 mV/dec - Tafel slope. As the great advantage of such a porous Ni layer on the metal surface in pulse mode plating the films are the high power density and high voltage amplitude obtained at the same time.

Acknowledgement

The authors acknowledge National Research program LATENERGI for financial support. J. Kleperis acknowledges European Regional Development Fund project No. 2010/0204/2DP/2.1.1.2.0./10/APIA/VIAA/010 for travel support..

References

[1] Yagi S, Murase K, Hirato T, Awakura Y 2007 J. Electrochem. Soc. **154(6)** D304
[2] Ibl N, Puippe JCl and Angerer H 1978 Surf. Tech **6** 287
[3] Chandrasekar MS, PushpavanamM 2008 Electrochimica Acta **53** 3313
[4] Qu NS, Zhu D, Chan KC, Lei WN 2003 Surface and Coatings Technology **168** 123
[5] Paunovic M, Schlesinger M 1998 Fundamentals of Electrochemical Deposition, Electrochemical Society Series, Wiley-Interscience 2 edition 373 pages
[6] Landolt D, Marlot A 2003 Surface and Coatings Technology **169 –170** 8
[7] VanagsM, Kleperis J and Bajars G 2011 Int. J. Hydrogen Energy **36** 1316
[8] Vanags M, Kleperis J and Bajars G 2011 Latvian J. Phys. Techn. Sci. **48** 34

IOP Conf. Series: Materials Science and Engineering **49** (2013) 012009 doi:10.1088/1757-899X/49/1/012009

Nanotubes folded from cubic and orthorhombic $SrZrO_3$: First-principles study

A V Bandura, R A Evarestov and D D Kuruch[1]

Department of Chemistry, Quantum Chemistry Division,
St. Petersburg State University, Universitetsky Prosp. 26, Petergof,
St. Petersburg, 198504, Russia

E-mail: di_ma_rex@front.ru

Abstract. Hybrid functional of the density functional theory has been applied for investigations of the electronic and atomic structure of nanotubes (NTs) based on cubic and orthorhombic strontium zirconate. The full optimization of all atomic positions in the considered systems has been performed to study the atomic relaxation and to determine the most favourable nanostructures. The stability of the single-walled nanotubes of chiralities (n, 0) and (n, n) for the cubic, and (n_1, 0) and (0, n_2) for the orthorhombic $SrZrO_3$ has been investigated in dependence on their diameter, the number of layers (2 or 4) in the slab, and the outer surface termination (SrO or ZrO_2). The spontaneous splitting of the SrO-terminated (n, n) 4-layer nanotubes folded from the cubic phase into the two separated NTs has been found. The NTs folded from the orthorhombic phase, generally, preserve the atomic distortions inherent to the bulk $SrZrO_3$ orthorhombic phases. However, the concrete NT surface structure depends on the chirality, the number of layers and termination type.

1. Introduction

Presently, ABO_3 type perovskites are widely studied objects that exhibit unique electronic, magnetic and optical properties. Perovskite based nanomaterials can significantly improve the performance of a variety of electronic, electro-optical, electrochemical, and electromechanical devices and systems. Nanoscale systems based on $SrZrO_3$ (hereinafter SZO) and $BaZrO_3$ exhibit great promise as a solid fuel cells and hydrogen sensors due to their high chemical and mechanical stability and high proton conductivity at high temperatures [1]. Additionally, SZO, having a high dielectric constant, is great for creating high-voltage capacitors [1].

In this work we consider the different types of stoichiometric nanotubes (NTs) folded from the cubic and orthorhombic SZO. SZO undergoes [2] three temperature-dependent phase transitions. It is orthorhombic (*Pbnm*) at room temperature and it becomes cubic ($Pm\bar{3}m$) above 1360 K. In one of our previous works [3] the surface relaxation of SZO has been investigated. The first goal of the present study is to provide the understanding of the relation between the bulk SZO atomic structure and the stability of SZO-based nanotubes. The second goal of this work is to compare properties of the NTs folded from the layers of SZO with those of the NTs folded from the layers of $SrTiO_3$ studied earlier [4]. For the rolling up of nanotubes we use the relaxed or reconstructed thin slabs obtained by cleaving the perovskite SZO crystals along the nonequivalent atomic planes.

[1] To whom any correspondence should be addressed.

2. Computational details

All the calculations have been performed within the periodic LCAO method using the CRYSTAL-2009 computer code [5]. A hybrid exchange-correlation functional PBE0 [6] has been applied. All-electron basis set for O atom has been taken from [7]. The CRENBL [8] small-core pseudopotentials and the corresponding basis set has been used for Sr and Zr atoms. To exclude the basis set linear dependence in LCAO crystal calculations, the diffuse *s*-, *p*-, *d*-, and *f*-Gaussian-type orbitals with exponents less than 0.1 have been removed from the corresponding basis sets. The method chosen provides the good quality of the calculated SZO bulk properties. Thus, optimized lattice constants of cubic and orthorhombic SZO differ from the experimental values by less than 1% (see Table 1).

Stoichiometric nanosheets composed of 2 and 4 alternating (001) SrO and ZrO_2 atomic planes from the cubic phase and (001) or (110) atomic layers from orthorhombic phase have been constructed and studied as nanotube precursors. The lattice parameters and atomic structure of initial slabs have been optimized before folding. It has been proved that the obtained lattice constants of two-dimensional square or rectangular lattices are only slightly decreased compared to the bulk values (see Table 1).

Table 1. Calculated lattice parameters, formation and surface energies of nanosheets cut from $SrZrO_3$

Value	System	Parent phase and surface index		
		Cubic, $Pm\bar{3}m$	Orthorhombic, *Pbnm*	
		001	001	110
Lattice constants, *a*, *b*, *c*, Å	Bulk[a]	4.123 (4.154)	5.750, 5.819, 8.173 (5.786, 5.815, 8.196)	
	2 layer slab	4.115	a=5.782, b=5.867	$a/\sqrt{2}$=5.802, $b/\sqrt{2}$=5.854
	4 layer slab	4.104	a=5.766, b=5.778	$a/\sqrt{2}$=5.704, $b/\sqrt{2}$=5.813
Formation energy, kJ/mol	2 layer slab	282.2	276.7	275.6
	4 layer slab	145.7	157.2	156.8
Surface energy, kJ/m^2	2 layer slab	1.221	1.355	1.347
	4 layer slab	1.437	1.568	1.571

[a] Bulk values of experimental lattice constants [2] are given in parenthesis.

3. Results and discussion

The stability of the single-walled nanotubes of chiralities (n, 0) and (n, n) for the cubic, and (n_1, 0) and (0, n_2) for the orthorhombic SZO has been investigated in dependence on their diameter, the number of layers in the slab, and the outer surface termination (SrO or ZrO_2). The first (I) type of NTs has been obtained by rolling up of (001) slabs of the cubic phase, the second (II) and the third (III) types were folded from (001) and (110) slabs of the orthorhombic phase, accordingly.

To estimate the NT stability we use the strain energy E_{str} per formula unit:

$$E_{str}(\mathrm{NT}) = E_l(\mathrm{NT})/l - E_m(\mathrm{slab})/m, \quad (1)$$

where E_l(NT) – total energy of NT unit cell consisting of l bulk formula units, and E_m(slab) – total energy of slab unit cell containing m bulk formula units.

The atomic positions on the boundary NT surfaces exhibit a noticeable relaxation. Type I NTs folded from 2-layer slabs, generally, keep the initial structure while maintaining the original symmetry during the optimization procedure. However, if the symmetry decreases due to removing of the screw rotations, the nanotube structure exhibits considerable reconstruction with splitting of ZrO_2 shells into two atomic (O and Zr) subshells, thus resembling the surface of SZO orthorhombic phases. Strain energy of those NTs is significantly less than that of the tubes with a full set of symmetry operations. The NTs generated from the layers of cubic $SrTiO_3$ in ref. [4] do not exhibit such behavior.

It has also been observed a spontaneous splitting of the SrO-terminated (n, n) 4-layer nanotubes into two separated NTs. The resulted double-walled nanotubes are energetically more favorable than the two isolated single-walled nanotubes (see Figure.1), consisting of ZrO_2 (L_1) and SrO–ZrO_2–SrO (L_2) shells, correspondingly. The atomic positions in every one of these two single-walled nanotubes were optimized (see Table 2).

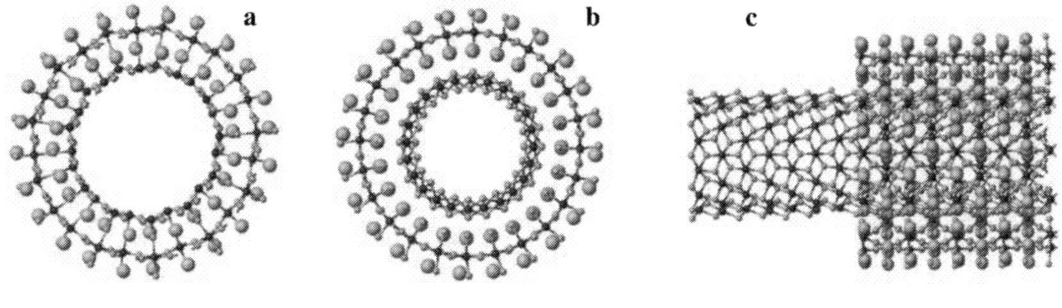

Figure 1. Spontaneous splitting of the type I 4-layer SrO-terminated (*n*, *n*) nanotubes into two separated NTs. Cross-section view before optimization (a), cross-section view after optimization (b), and side view after the optimization (c). Legend: Sr – large light balls, Zr – small dark balls, O – small light balls.

Table 2. Binding energy (ΔE) of two SWNTs into DWNT and corresponding NTs dimensions[a].

Chirality	ΔE, kJ/mol (per formula unit)	DWNT				After the optimization	
		Before the optimization		After the optimization		SWNT L_1	SWNT L_2
		R_{min}, Å	R_{max}, Å	R_{min}, Å	R_{max}, Å	R_{min}, Å	R_{max}, Å
(12, 12)	–58.7	8.10	14.21	5.65	13.42	5.35	13.42
(14, 14)	–49.3	9.95	16.06	6.89	14.88	6.37	15.18
(16, 16)	–33.2	11.80	17.90	8.20	16.33	7.39	16.96

[a] Radii R_{min} and R_{max} of NTs are estimated as the distances from NT axis to the nearest and the most remote atom, correspondingly.

The NTs folded from the orthorhombic SZO phase (of types II and III), generally, preserve the atomic distortions inherent to the orthorhombic structure. However, the concrete NT surface structure depends on the chirality, the number of layers and the termination type. The reconstruction of ZrO_2-terminated outmost surfaces due to break of the Zr–O bonds has been found in all types of NTs simulated from 4-layer slabs.

In Figure 2 we depict the dependence of the NT strain energy on its average diameter calculated as sum of R_{min} and R_{max}. Figure 2 shows that strain energy of SrO-terminated NTs decreases with increasing the size of NTs but that strain energy of ZrO_2-terminated NTs increases. In the latter case the negative strain energy is a consequence of the large positive formation energy of the initial slabs. On the other hand, the SrO-terminated 2-layer NTs have the positive or zero strain energies because of the unfavorable interaction between the O atoms in the internal ZrO_2 surface due to its shrinking. In both cases the strain energy tends to zero with increasing of the NT size, since the total energy of the NT converges to the total energy of the slab. The final relaxed structure of ZrO_2-terminated NTs folded from the two layers of both the cubic and orthorhombic phases is very similar and corresponds to the latter phase, which is more stable. Raising the diameter of the (*n*, *n*) SrO-terminated SZO NTs above the 50 Å promotes the transition from the cubic-like to the orthorhombic-like NT shell structure, which results in drastic fall of the strain energies (see Figure 2a). Figure 2b effectively demonstrates that the strain energy of (n_1, 0) and (0, n_2) 2-layer NTs coincides for the both terminations. It means that there is no difference between relaxed (n_1, 0) and (0, n_2) type II NTs. The same can be seen on Figure 2c for (n_1, 0) and (0, n_2) ZrO_2-terminated 2-layer type III NTs.

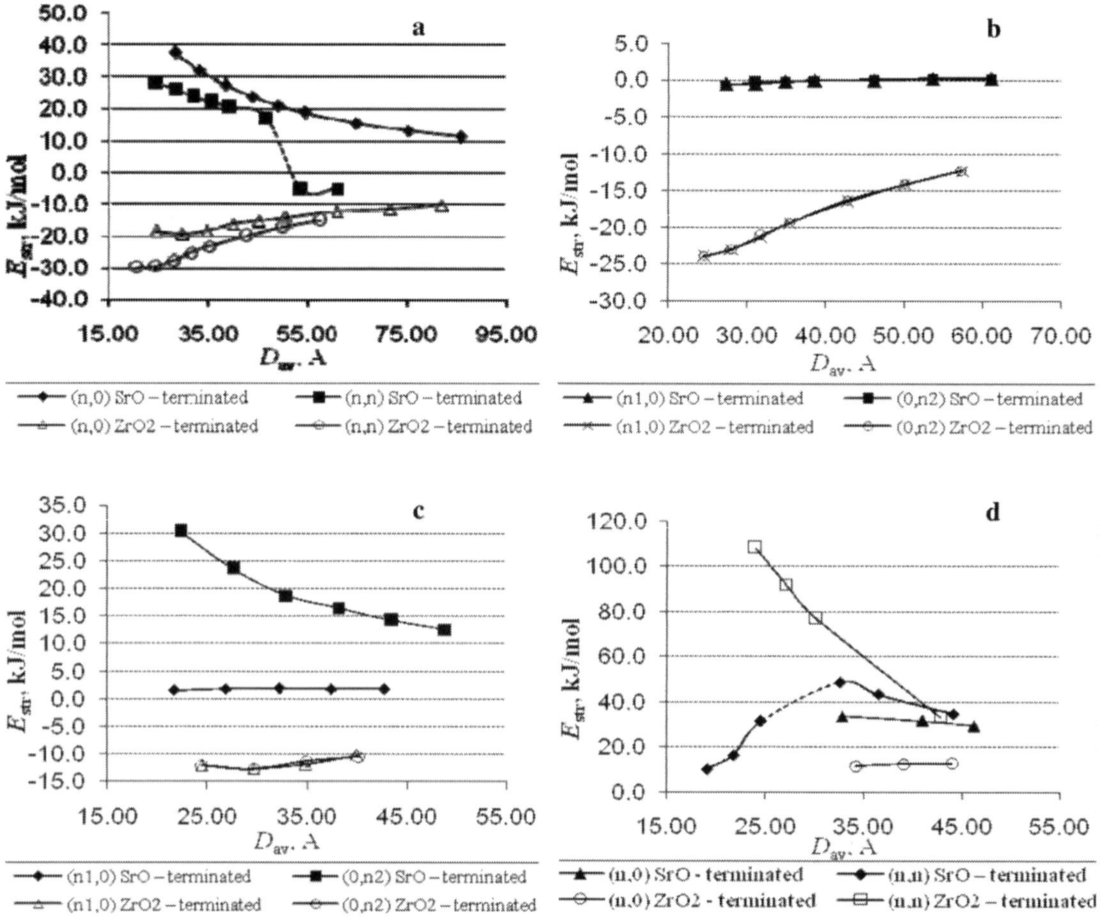

Figure 2. Dependence of strain energy E_{str} on average tube diameter D_{av} for 2-layer type I NTs (a), 2-layer type II NTs(b), 2-layer type III NTs (c), and 4-layer type I (d) NTs.

In the Figure 2d we can see the region where the E_{str} increases with increasing the diameter of SrO-terminated type I NTs. This just is the case for spontaneous splitting of 4 layer nanotubes into two separated NTs that was discussed above.

Acknowledgments
Author thanks St. Petersburg State University for the financial support (Grant 12.37.142.2011).

References

[1] Wang Y Xu, Arai M, Sasaki T and Wang C L 2006 *Appl. Phys. Lett.* **88** 091909
[2] Kennedy B J, Howard Ch J and Chakoumakos B C 1999 *Phys. Rev.* B **59** 4023
[3] Evarestov R A, Bandura A V and Alexandrov V E 2006 *Phys. Status Solidi* B **243** 2756
[4] Evarestov R and Bandura A 2011 *IOP Conf. Ser.: Mater. Sci. Eng.* **23** 012013
[5] Dovesi R, Orlando R, Civalleri B, Roetti C, Saunders V R and Zicovich-Wilson C M 2005 Z. *Kristallogr.* **220** 571
[6] Perdew J P, Ernzerhof M and Burke K 1996 *J. Chem. Phys.* **105** 9982
[7] Hariharan P C and Pople J A 1973 *Theoret. Chim. Acta* **28** 213
[8] LaJohn L A, Christiansen P A, Ross R B, Atashroo T and Ermler W C 1987 *J. Chem. Phys.* **87** 2812

IOP Conf. Series: Materials Science and Engineering **49** (2013) 012010 doi:10.1088/1757-899X/49/1/012010

Conoscopic study of strontium-barium niobate single crystals

Aleksandr Kolesnikov[1], Rostislav Grechishkin[1], Olga Malyshkina[1], Yury Malyshkin[1], Jan Dec[2], Tadeusz Łukasiewicz[3], Aleksandra Ivanova[1]

[1]Tver State University, Tver, Russia
[2]Institute of Materials Science, University of Silesia, Katowice, Poland
[3]Institute of Electronic Materials Technology, Warsaw, Poland

e-mail: Olga.Malyshkina@mail.ru

Optically transparent single crystals of strontium-barium niobate, $Sr_xBa_{1-x}Nb_2O_6$, of different compositions (x = 0.26…0.7) were examined with the aid of conoscopic light interference figures. A regular change of the isochrome concentric ring number and diameters consistent with the temperature variation of the value of birefringence is demonstrated by direct observations of polar cuts of optically uniaxial samples. Anomalous violations of the conventional (uniaxial) interference patterns occur occasionaly in some samples being indicative of the existence of biaxial trait in their behaviour even though no voltage is applied. These features may depend on annealing treatments at elevated temperatures. The results of the study show that conoscopic images may serve as a sensitive indicator of the structural state of SBN crystals related to the effects of stress–induced change of optical anisotropy and temperature dependent birefringence parameters.

1. Introduction

Strontium barium niobate $Sr_xBa_{1-x}Nb_2O_6$ (SBN:x) lead-free environmentally friendly system represents a specific group of materials in which ferroelectric–relaxor behaviour depends on the x ratio Sr/(Ba+Sr) [1–7]. This system is characterized by a number of useful properties including piezoelectric, electro-optic and pyroelectric ones. Of special interest are large electro-optic coefficients (420×10^{-12} m/V) and high photorefractive sensitivity, which can be effectively used in optoelectronic devices [1]. One more advantage of SBN is the relatively low temperature where the structural phase transitions take place [7]

In spite of technological importance the understanding of SBN crystal properties is far from completeness, because, in addition to chemical composition, they strongly depend on the processing parameters, such as crystal growth method, heat treatment regimes and atmosphere, structure of dislocations and other defects, optical nonuniformity, etc.

In the present work we applied the method of conoscopic light interference pattern to the study optical properties of SBN single crystals with different concentrations of strontium. Conoscopic observations are known to be a powerful tool for characterization of the optic dielectric tensor of anisotropic media [8] including materials with phase transitions and internal stresses (see, e.g. [9, 10]).

2. Experimental

Single crystals of SBN were pulled along [001] crystallographic direction from a melt by the Czochralski method. Details of the growing procedure are described elsewhere [11]. After X-ray

IOP Conf. Series: Materials Science and Engineering **49** (2013) 012010 doi:10.1088/1757-899X/49/1/012010

orientation and cutting into polar slices 5×5×1 mm the sample surface was carefully polished by diamond pastes. The conoscopic figures were observed in an optical polarizing microscope with Bertrand lens and narrow-band interference filters making use of objectives and condensers with a numerical aperture of 0.65 or higher. Following the suggestions of Steward [12] and to minimize the risk of damaging the microscope optics during heating the sample a very thin hot stage supplied with thin film thermocouple was constructed and installed on the microscope stage.

3. Results and Discussion

Figure 1 presents an excerpt from the videofilm illustrating the changes of the conoscopic figures of SBN70 polar cut as observed within the temperature range of 295...381 K. It is seen that at room temperature (RT) the interference pattern is presented by isochromes in the form of regular concentric rings having superimposed isogyres in the shape of a maltese cross. This pattern is typical of optically uniaxial crystals in a satisfactory agreement with the existing theoretical predictions [13, 14].

During heating above RT this pattern undergoes regular changes consisting firstly in an increase of the diameters of the isochromes which expand smoothly in radial directions. With this provision the external circles are one by one displaced from the field of view (FOV). As a result the total number of isochromes in the FOV reduces so that finally at T = 358 K one observes a uniform structureless dark background to be indicative of optical isotropy of the sample. Further heating above this particular isotropic point, however, results in a reappearance of the interference effects (Fig.1). This sequence of events is completely replicated in a reverse order with cooling back to the starting state.

Similar behaviour (radial expansion of isochromes – decreasing their number – transition to isotropy – reappearance of interference features with further heating – reversible return to the starting point) were observed in SBN35 samples (Fig.2), although the specific diameters of isochromes and their number are different.

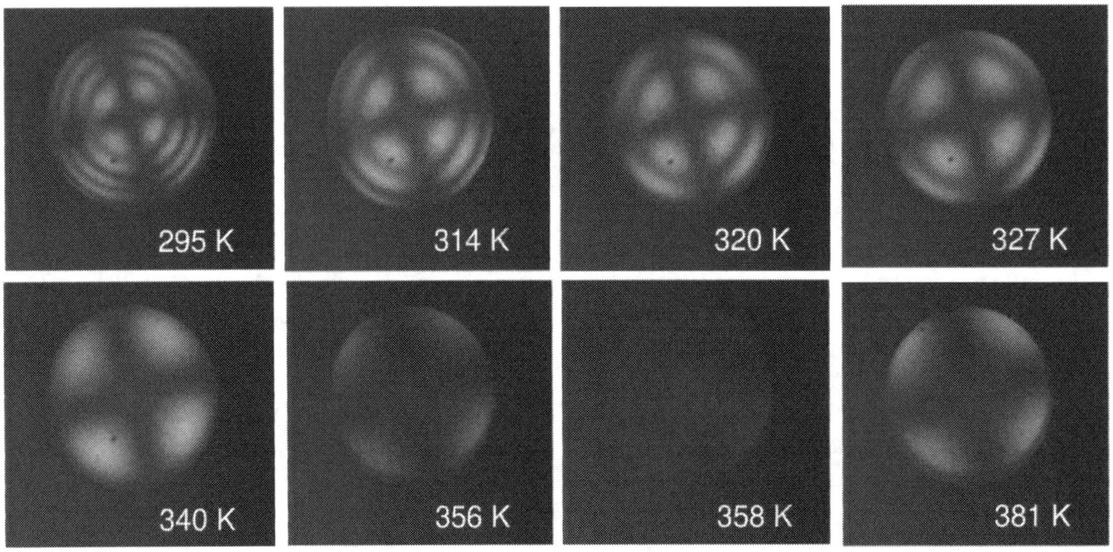

Figure 1. Effect of temperature on the conoscopic figures of SBN70 polar cut. Before observations the sample was annealed at 580 K for 2 hours.

The observed radial expansion of isochrome rings typical of uniaxial crystals and decrease of their number with temperature in SBN crystals may be understood as follows. The refractive indices n_0 of SBN are known to increase slightly with temperature [15]. At room temperature the refractive indices n_e are smaller than n_o. However, n_e increases with temperature much faster than n_o, so the difference $\Delta n = n_o - n_e$ decreases with heating giving rise to the isochrome diameters. At some point Δn becomes

IOP Conf. Series: Materials Science and Engineering **49** (2013) 012010 doi:10.1088/1757-899X/49/1/012010

zero (crystal becomes isotropic); further heating changes the optical sign of the crystal as demonstrated by the experiment (Fig. 1).

However, detailed examination of SBN samples has shown that the uniaxial type of conoscopic patterns is often violated by the appearance of interference effects peculiar to optically biaxial materials. Some examples of such patterns are shown in Figure 3,*a-c* for SBN26, SBN61 and SBN70 samples. For the sake of comparison Figure 3,*d* shows the conoscopic figure of gadolinium molybdate (GMO) single crystal, which is a well known typical representative of the biaxial medium. It is seen that the given examples of the SBN samples bear the peculiarities of both uniaxial and biaxial behaviour, although the latter are not as distinct as in purely biaxial crystals. It should be added that in some cases these anomalies may be removed by annealing. For example, the distorted conoscopic figure of SBN70 shown in Figure 3,*c* was transformed to a normal uniaxial appearance shown in Figure 1 by heat treatment at 580 K for 2 hours.

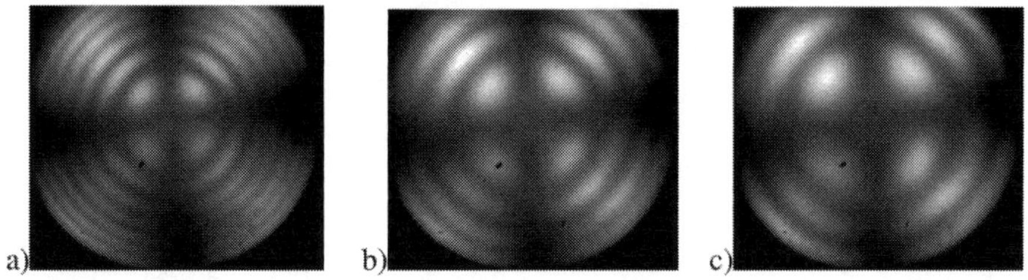

Figure 2. The conoscopic figures of SBN35. 320 K (*a*); 340 K (*b*); 350 K (*c*)

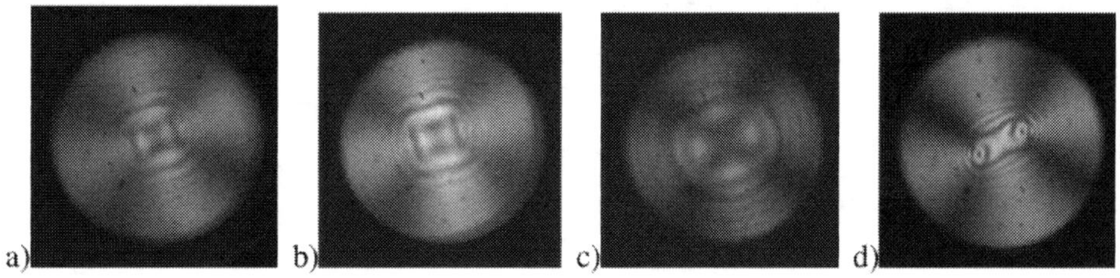

Figure 3. The conoscopic figures of SBN26 (*a*), SBN61 (*b*), SBN70 (*c*) and GMO (*d*)

The effect of anomalous biaxiality in uniaxial crystals may be understood in the following way. In general the biaxiality may be induced by electrooptic and piezooptic effects or induced by internal mechanical stresses [16]. Since the samples under study possess piezoelectric properties, the respective piezomodule d_{31} as determined at room temperature by the resonance–antiresonance method are given in Table 1.

Table 1. Piezoelectric modulus values of the SBN crystals.

Composition	SBN35	SBN50	SBN61
d_{31}, 10^{-12} C/N	−10	−17	−20

The existence of the piezoeffect permits us to estimate the mechanical stresses using the relation for the angle between optical axes (2*V*) [13]:

$$\tan V = \frac{\sqrt[4]{\left[(\pi_{1\mu} - \pi_{2\mu})\sigma_\mu\right]^2 + \left(2\pi_{6\mu}\sigma_\mu\right)^2}}{\sqrt{n_o^{-2} - n_e^{-2}}},$$

where n_o and n_e are the ordinary and extraordinary indices of refraction, $\pi_{i\mu}$ is the matrix of piezooptic coefficients, σ_μ are the mechanical stresses. The data for the piezooptic coefficients and refractive indices for SBN70 were taken from the references [1, 15, 17], while the value of $\tan V$ was obtained with the aid of a calibrating sample of gadolinium molybdate crystal as schematically shown in Figure 4. It follows that

$$V_{\mathrm{SBN}} = V_{\mathrm{GMO}} \frac{n_{\mathrm{GMO}}}{n_{\mathrm{SBN}}} \frac{L_{\mathrm{SBN}}}{L_{\mathrm{GMO}}} .$$

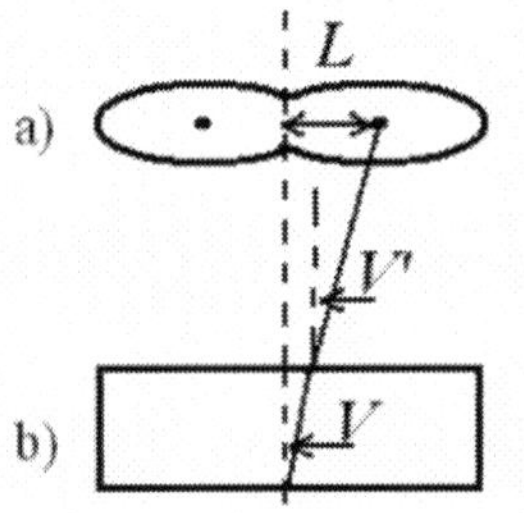

Figure 4. Diagram for determining the optical angle. (*a*) anomalous isochromes as observed from the top view, (*b*) transmission of the light through the sample

As a result we obtain an estimate of the order of reasonable mechanical stresses $2\cdot10^7\ N/C < \sigma < 3\cdot10^7\ N/C$ which may account for the observed anomalies in SBN70.

4. Conclusion

The performed study shows that the conoscopic method provides useful information on the characterization of SBN crystals. Experimental data on the effect of temperature on the conoscopic figures were presented for the first time and explained by the known temperature behaviour of SBN refractive indices. Specific biaxial optical anomalies in (uniaxial) SBN crystals not reported earlier were detected, thus demonstrating usefulness of the conoscopy technique in the studies of complex structure transformation in SBN system.

References

[1] Prokhorov A M, Kuz'minov Yu S 1990 *Ferroelectric Crystals for Laser Radiation Control* (Bristol: The Adam Hilger)
[2] Ivanov N R, Volk T R, Ivleva L I, Chumakova S P, and Ginzberg A V 2002 *Cryst. Reports* **47** 1092-9
[3] Dec J, Kleemann W, Pankrath R 2000 *Europ. Phys. J. B* **14** 627
[4] Niemann R, Buse K, Pankrath R, and Neumann M 1996 *Solid State Commun.* **98** 209
[5] Woike Th, Weckwerth G, Palme H, and Pankrath R 1997 *Solid State Commun* **102** 743
[6] Malyshkina O, Lisitsin V, Movchikova A, Dec J, Lukasiewicz T 2012 *Ferroelectrics* **426:1** 230
[7] Dorywalski K, Andriyevsky B, Cobet C, Piasecki M, Kityk I V, Esser N, Lukasiewicz T, Patryn A 2013 *Mater.* **35** 887
[8] Wahlstrom E E 1979 *Optical Crystallography* (New York: Wiley)
[9] Yamamoto N, Mamedov N, Shinohara T, Kunie A 2002 *J. Cryst. Growth* **237-239** 2023
[10] Rinaldi D, Pietroni P, Davi F 2009 *Nucl. Instr. Meth. in Phys. Res.* A **603** 294
[11] Lukasiewicz T, Swirkowicz M, Dec J, Hofman W, Szyrski W, 2008 *J. Cryst. Growth* **310** 1464-9
[12] Steward E G 1952 *J. Sci. Instr.* **29** 214
[13] Sirotin Yu I, Shaskolskaja M P 1979 *The basics of crystallophysics.* (Moscow: Nauka) p 639
[14] Shtukenberg A, Punin Y O 2007 *Optically Anomalous Crystals* (New York: Springer)
[15] Venturini E L, Spencer E G, Lenzo P V, Ballmann A A 1968 *J. Appl. Phys.* **39** 343
[16] Kolesnikov A I, Kaplunov I A, Ivanova A I, Tretiakov S A, Malyshkina O V, Grechishkin R M 2012 *Ferroelectrics* **441** 84
[17] Blistanov A A, Bondarenko V S, Perelomova N V 1982 *Acoustic Crystals* (Moscow: Nauka)

Modification of LiF structure by irradiation with swift heavy ions under oblique incidence

I. Manika[1], J. Maniks[1], R. Zabels[1], K. Schwartz[2], R. Grants[1], A. Dauletbekova[3], A.Rusakova[3], M. Zdorovets[4]
[1]Institute of Solid State Physics, University of Latvia, Riga, Latvia
[2]GSI Helmholtz Centre, Planckstr 1, 64291 Darmstadt, Germany
[3]L.N. Gumilyov Eurasian National University, Astana, Kazakhstan
[4]Institute of Nuclear Physics, Kazakhstan

E-mail: manik@latnet.lv

Abstract. The structural modifications of LiF irradiated with swift heavy ions under oblique angles have been investigated using AFM, SEM, chemical etching, nanoindentation and optical absorption spectroscopy. LiF crystals were irradiated under incidence angles of 30 and 70 degrees with 2.2 GeV Au (fluence 5×10^{11} ions·cm^{-2}) and 150 MeV Kr ions (fluence 10^{12}-10^{14} ions·cm^{-2}). Structural study on sample cross-sections shows that two damage regions – (1) nanostructured zone and (2) dislocation - rich zone, which are typical for irradiations at normal incidence, appear also in samples irradiated under oblique angles. However in the latter case a more complex structure is formed that leads to stronger ion-induced hardening.

1. Introduction

The modifications of the structure of materials on the micro- and nanometer scale and an improvement of their properties by irradiation with swift heavy ions (SHI) are of importance on both the fundamental and technological standpoint. High fluence irradiation of ionic crystals with SHI leads to severe structural modifications including formation of latent tracks and surface and bulk nanostructures [1-3]. A significant role in the self-organized nanostructuring play stress fields of extended defects (such as tracks and dislocations) as well as crystallographic orientation. The latter is of special importance for ionic crystals which maintain crystallinity even under high-fluence irradiation. In the majority of studies the ion beam has been oriented normal to the crystal surface and tracks propagate along particular crystallographic direction. Peculiarities of track damage were observed at irradiation of crystal surfaces under grazing incidence of ion beam [4].

In this study the structural modifications in the bulk of LiF crystals irradiated under oblique incidence of ions has been investigated using AFM, SEM, chemical etching, nanoindentation and optical absorption spectroscopy. Irradiation was performed with 2.2 GeV Au ions, which in LiF produce complex tracks consisting of core region containing small defect aggregates and a surrounding halo region where mostly single defects are created [1]. Such tracks similar to dislocations exhibit stress field and can be revealed by selective chemical etching. In this study also 150 MeV Kr ions have been used which are attractive in a way that the threshold of energy loss for the core damage is surpassed only in a certain part of the ion range. As a result, two distinctively different in structure regions on the etched profile surfaces are observed [3].

2. Experimental

Irradiations were performed on nominally pure LiF crystals (Korth Kristalle, Germany) at the UNILAC linear accelerator of the GSI Darmstadt with 2.2 GeV Au ions at dose of 5×10^{11} ions·cm^{-2} under normal and oblique (30 degrees) incidence to the (100) cleavage face and at the cyclotron DC-60 (Astana, Kazakhstan) with 150 MeV $^{84}Kr^{+14}$ ions in the fluence range 10^{11} - 10^{14} ions·cm^{-2} under incidence angles of 30 and 70 degrees. All irradiations were performed at room temperature. Ion tracks and dislocations were revealed using chemical etching

3. Results and discussion

3.1 Dependence of the thickness of irradiated layer on angle of incidence

The thickness of the irradiated layer in the case of normally and under oblique incidence irradiated samples was estimated from optical microscopy of etched profile surfaces. The irradiated layer can be observed as striated structure oriented along the direction of the ion beam. For the normally irradiated samples the thickness of the irradiated layer coincided with the ion range R, which was calculated using the SRIM code [5]. For the samples irradiated under oblique angles the damaged layer was thinner. Its thickness is in agreement with the estimates performed from the geometrical considerations as $h = R\sin\alpha$, where α is the angle of incidence (Fig. 1). The result confirms the independence of ion range of the ion incidence angle. Besides, samples which are inclined against ion beam receive correspondingly lower ion fluence due to projection of exposure over larger area.

The results of optical absorption spectroscopy for simultaneously irradiated samples under normal and oblique incidences are also in agreement with this model. Weaker coloration was found in samples irradiated under lower angles of incidence (Fig. 2). Spectra for all samples were similar and showed F, F_2 and F_n absorption bands typical for high fluence irradiations [1,3].

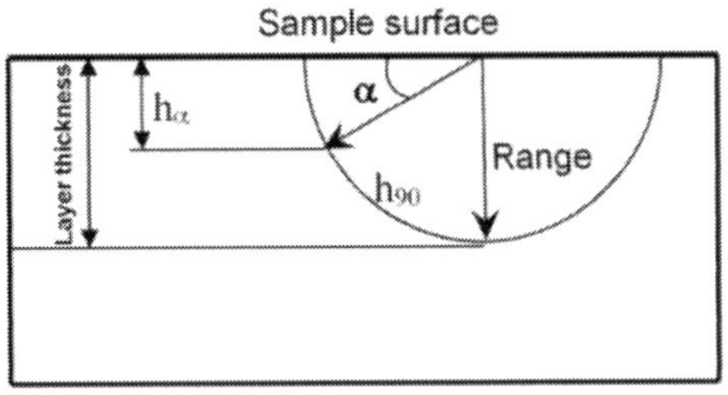

Figure 1. The thickness of the irradiated layer in dependence on the incidence angle of ions (scheme).

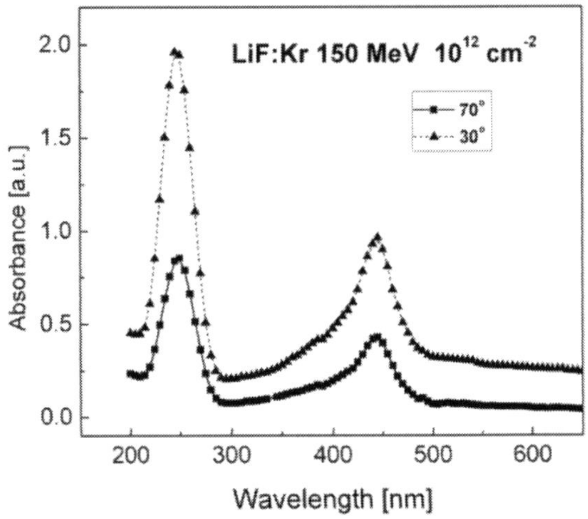

Figure 2. Absorption spectra for LiF crystals irradiated with150 MeV Kr ions at 10^{12} ions·cm^{-2} under70^0 and 30^0 incidence.

3.2 Structural study on the irradiated surface

On the surfaces irradiated with Au and Kr projectiles etchable tracks were observed (Fig. 3a). Such result was expected because in both cases the energy loss for the ions surpasses the critical threshold of 10 keV/nm for core damage and track etching [1]. For the samples irradiated under oblique angle also the etch pits of tracks were tilted (Fig. 3a, inset).

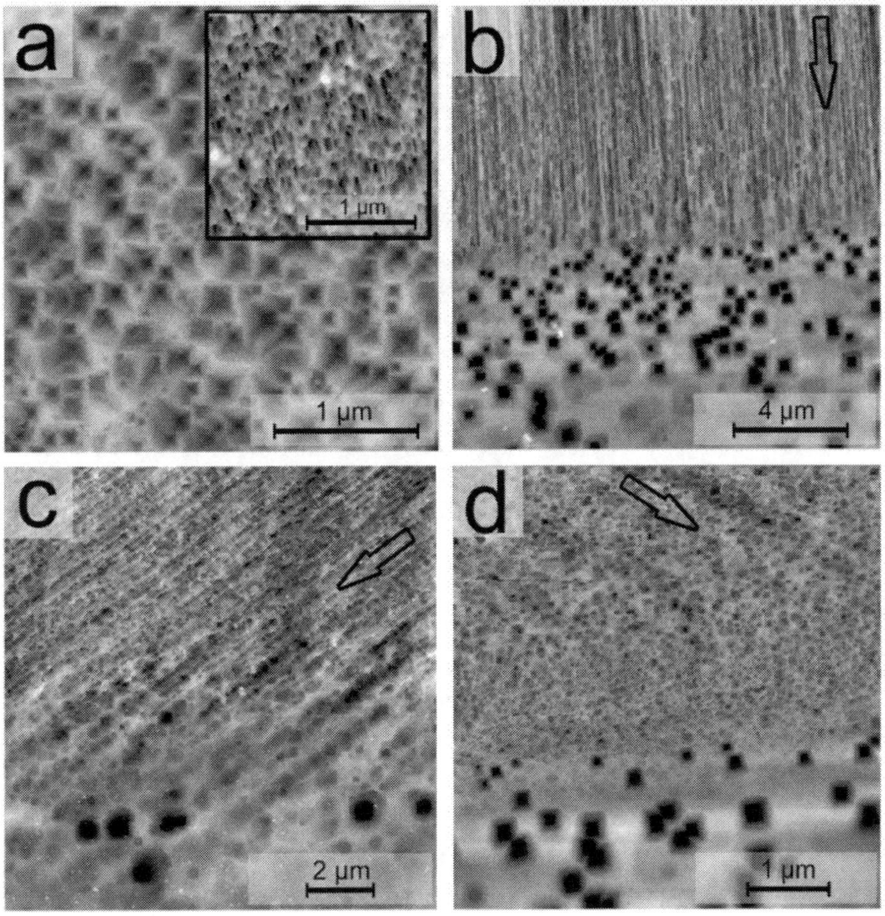

Figure 3. Damage structures created under normal and oblique incidence of ions (AFM and SEM images): (a) – etching of Au tracks on normally irradiated surface; fluence $5x10^{11}$ ions·cm^{-2}. Inset shows etched tracks under 30° incidence; (b) - profile surface, normal irradiation with Kr ions at 10^{12} ions·cm^{-2}; (c) profile surface, Au ions under 30° incidence , fluence $2.5x10^{11}$ ions·cm^{-2}; (d) profile surface, irradiation with Kr ions under 30° incidence at 10^{12} ions·cm^{-2}. Arrows show the direction of irradiation.

3.3 Structural study on the cross-sections of samples

The structural study was performed on cross sections prepared by cleaving perpendicular to irradiated (001) face. At all irradiations under oblique angles the total thickness of irradiated layer was lower than the ion range and varied with the incidence angle as described in section *3.1*.

Our previous studies showed that at high fluence irradiation with Kr ions under normal angle, two regions exhibiting different types of structural features appear: (1) the nanostructured zone which can be characterized as mosaic type structure with small subgrains and (2) the zone containing numerous ion-induced dislocations (Fig. 3b) [2,3]. The two structural types were obtained also for all investigated incidence angles of Kr ions. The proportion of their extension was approximately the same as for irradiations under normal angle. Figure 3d shows the structure formed on profile surface of sample irradiated with Kr ions at the angle of 30^0.

In the case of Au ions, the energy loss of which in LiF strongly exceeds that for Kr ions, the high fluence irradiation under normal and oblique angles in a dominant part of the irradiation zone creates a typical for nanostructures striated relief oriented in the direction of incoming ions (Fig. 3c)

3.4 Nanoindentation measurements

Nanoindentation testing on normally irradiated surface revealed substantial increase of hardness in agreement with our earlier results [2,3]. Nevertheless in samples irradiated under oblique angle the hardening effect was stronger despite of the lower received irradiation dose (Fig. 4). Taking into account that hardness is sensitive to defect aggregates, the result points to certain difference in their ordering in dependence of irradiation angle. In agreement with the scheme (Fig. 1), the thickness of the hardened zone at oblique irradiation (30^0) is about half of that for perpendicularly irradiated sample

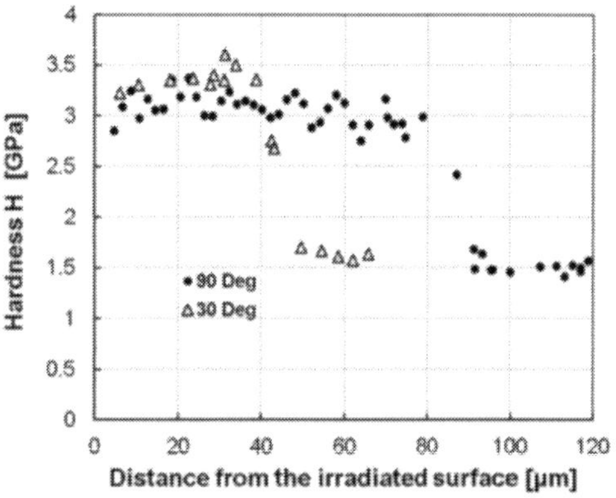

Figure 4. Hardness *vs.* indentation depth on profile surface of samples irradiated with Au ions under 90^0 (solid symbols) and 30^0 (open symbols) incidence angle. The fluence was $5x10^{11}$ and 2.5 $x10^{11}$ ions·cm^{-2}, correspondingly.

4. Conclusion

The thickness of irradiated layer varies with the incidence angle of ions. Maximum thickness which equals to the calculated ion range R is observed at irradiation perpendicularly to the (001) surface of a sample. The thickness of the irradiated layer decreases with decreasing of the incidence angle as h= R sinα in agreement with geometrical considerations. The results confirm that the ion range is independent of the incidence angle of ions.

Structural study shows that two zones of structural damage – nanostructured zone and dislocation rich zone, which are typical for irradiations at normal incidence [2,3], appear also in samples irradiated under oblique angles. The extension of these zones depends on energy loss of projectiles.

The irradiated layer for all investigated incidence angles exhibits increased hardness that gives evidence for formation of defect aggregates, such as dislocations and other extended defects.

Acknowledgments

The research is carried out within the framework of the ERAF project Nr. 2010/0253/2DP / 2.1.1.1.0/10/APIA/VIAA//079.

R.Zabels acknowledges the financial support from the European Social Fund within the project «Support for Doctoral Studies at University of Latvia».

References

[1] Trautmann C, Schwartz K, Costantini JM, Steckenreiter T, Toulemonde M. Nucl. Instr. and Meth. B **146** (1998)367-378.

[2] Maniks J, Manika I, Zabels R, Grants R, Tamanis E, Schwartz K. Nucl. Instr. and Meth. B **282** (2012) 81

[3] Dauletbekova A, Maniks J, Manika I, Zabels R, Aklibekov AT, Zdorovets MV, Bikert Y, Schwartz K. Nucl. Instr. and Meth. B**286** (2012) 56-60

[4] Vorobyova IV. Nucl. Instr. and Meth. B **198** (2002) 119-128.

[5] Ziegler JF, Biersack P, and Littmark U. The Stopping and Range of Ions in Matter, Pergamon Press, New York (1985).

IOP Conf. Series: Materials Science and Engineering **49** (2013) 012012 doi:10.1088/1757-899X/49/1/012012

Structure, micromechanical and magnetic properties of polycarbonate nanocomposites

J Maniks[1], R Zabels[1], R Merijs Meri[2], J Zicans[2]
[1]Institute of Solid State Physics, University of Latvia, Latvia
[2]Institute of Polymer Materials, Riga Technical University

E-mail: manik@latnet.lv

Abstract. The current study evaluates the applicability of polycarbonate (PC) for development of magnetic polymer nanocomposites with $CoFe_2O_4$ nanofiller, the amount of which was changed from 0 to 5 wt. %. Ethylene-vinyl acetate elastomer in the amount of 10 wt. % was added as toughener. Upon introduction of the magnetic filler a magnetic hysteresis loop was observed: at 5 wt.% of $CoFe_2O_4$ saturation magnetization of the nanocomposite was 2,2 emu/g, remanent magnetization was 0,8 emu/g and coercivity is 1200 G. Nanoindentation tests showed that nanofiller-reinforced samples maintain reasonable plasticity characterized by work of plastic indentation, while their modulus and hardness were improved by up to 14% and 20%, correspondingly. Structural study revealed clustering of the nanofiller at the contents above 2 wt.% that led to reduction of strengthening effect.

1. Introduction

Modification of polymers with nanoparticle fillers offers multiple advantages over traditional microcomposites. Polymer-matrix nanocomposites usually possess considerably improved properties already at small nanofiller content. Depending on the intrinsic nature of the nanofiller, some specific properties such as magnetism, electrical conductivity and others can be assigned to polymers. There is a lot of evidence about modification of primary aromatic thermoplastic polyesters, such as polycarbonate (PC), with inorganic anisometric fillers, particularly layered silicates [1], carbon nanotubes [2] and graphene [3]. Considerably less research has been devoted to modification of PC with magnetic nanofillers. Our research is aimed towards this problem and includes preparation of nanocomposites and characterization of their magnetic and mechanical properties.

2. Experimental

Polymer nanocomposites were manufactured by the method of twin screw extrusion. Temperature profile along the barrel length of the extruder was following: 230(feed)-245-265-275-265(die) °C. Both primary (Calibre 300 15, manufactured by The Dow Chemical Company) and recycled polycarbonate PC (obtained from used CD/DVDs) was used as a matrix for nanocomposites. Commercial nanostructured $CoFe_2O_4$ powder (manufactured by Nanostructured & Amorphous Materials, Inc.) with an average size of 35-55 nm and spherical morphology served as nanofiller in the

amount of 1, 2 and 5wt. % (0.22, 0.64 and 1.18 vol.%, correspondingly). Ethylene-vinyl acetate (EVA) elastomer (11507-375, manufactured by JSC NEFTEKHIM SEVILEN, vinyl acetate group content - 26%) in the amount of 10 wt. % was added as toughener for recycled PC.
Structure of nanocomposites was studied by Atomic Force Microscopy (AFM) on fracture surfaces. Magnetic properties were characterized by means of Lake Shore 7404 vibrating sample magnetometer. Nanoindentation tests were performed by MTS-Agilent G200 nanoindenter with a Berkovich diamond tip using the continuous stiffness measurement technique. The measurements were conducted at the strain rate 0.05 s^{-1}. The nanoindenter was calibrated using a reference sample of fused silica. The hardness, Young's modulus and standard deviation of the measurements were calculated from experimentally obtained loading-unloading curves.

3. Results and discussion

3.1 Magnetic properties

Fig.1. depicts the effect of $CoFe_2O_4$ on the magnetic behaviour of the polymer nanocomposites. As can be seen, upon addition of the ferrite nanoparticles to the polymer matrix paramagnetic properties are assigned to the composite. With increasing $CoFe_2O_4$ concentration magnetic properties of the composite improve. Consequently saturation magnetization of the composite containing 5 wt.% of ferrite nanofiller is 2,2 emu/g, remanent magnetization is 0,8 emu/g and coercivity is 1200 G. Respective values for the composite containing 2 wt.% of the ferrite nanofiller are 0,6 emu/g, 0,2 emu/g and 1200 G. It should be mentioned that the effect of $CoFe_2O_4$ on the magnetic behaviour of EVA toughened recycled PC was similar.

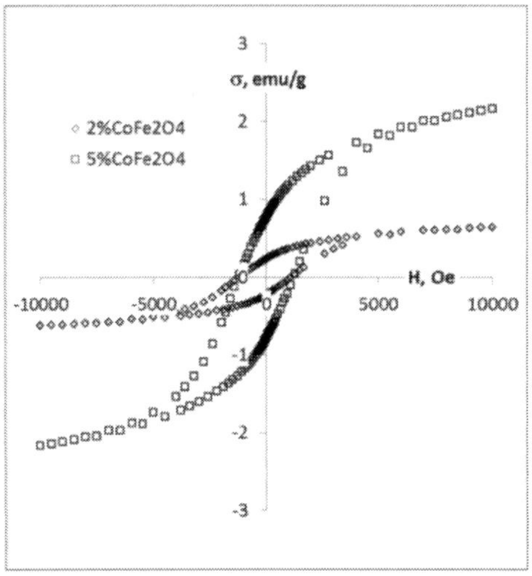

Fig.1 Magnetization curves of PC/$CoFe_2O_4$ nanocomposites with 2 and 5 wt.% filler content.

3.2 Nanoindentation study

The plasticity is an important parameter that characterizes mechanical properties of functional nanocomposites. The plasticity can be determined from the load-displacement curves obtained during nanoindentation [4]. The load- indentation depth curves of PC matrix and PC composite with 2% filler content are shown in Fig.2 a. The area between loading and unloading curves characterizes the work of plastic deformation but the area under the loading curve corresponds to the work of whole deformation (including also elastic). It can be seen that the course of loading and unloading curves for

the matrix and composite is very similar. The estimation shows that the plasticity characterized by the ratio of plastic deformation to whole deformation for the nanocomposite is even slightly higher (65%) in comparison to PC matrix (63%). It means that the cobalt ferrite nanocomposite maintains reasonable plasticity. This points to satisfactory adherence between ferrite particles and the matrix, which determines properties of interfaces and the behavior of the material during plastic deformation.
Fig.2 b shows the hardness dependence on the indenter penetration depth for PC matrix and cobalt ferrite nanocomposites. The stabilized values of hardness and modulus at indentation depth around 2000 nm are used to characterize nanocomposites in this study.

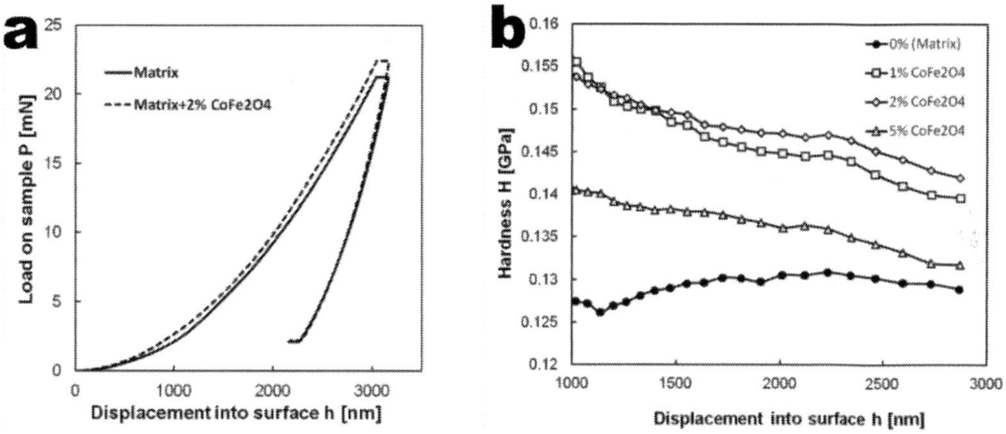

Fig.2. Loading-unloading curves for PC matrix (solid) and PC nanocomposite with 2wt% $CoFe_2O_4$ nanofiller (dashed) (a) and hardness *vs.* indentation depth for PC matrix and PC nanocomposites at 1%, 2% and 5 wt.% concentrations of $CoFe_2O_4$ nanofiller (b).

The results of nanoindentation affirm that the cobalt ferrite nanocomposites have higher hardness(Fig. 2b) and modulus in comparison to PC matrix. Such improvement of mechanical properties in the result of reinforcement effect has been observed in many polymer nanocomposites using different inorganic fillers of various size and shape [1-4].
The observed change of hardness and Young's modulus in dependence on $CoFe_2O_4$ content is not linear. The maximum effect in Young's modulus (16% increase) and hardness (20% increase) is achieved on samples containing 2 wt.% (0.64 vol.%) magnetic nanofiller. Similar effect of reinforcement by magnetic nanoparticles was observed in polypropylene matrix nanocomposites [6]. Appearance of the reinforcement effect at such low volume additions of nanoparticles is a characteristic feature of nanocomposites. However, a further increase of filler concentration resulted in the decrease of mechanical properties, which tend towards the level of PC matrix. It is well known that besides the size and content of nanofiller the strengthening effect depends also on the distribution of particles and particle/matrix interfacial adhesion which determine effective stress transfer in two-phase composite materials.
To investigate the causes of decrease in the reinforcement effect the structure of nanocomposites was studied by AFM. The imaging was performed on the fracture surfaces. To ensure a brittle fracture, the samples were broken at low (liquid nitrogen) temperature. Particles are invisible in topography image, however, the phase image from the same area clearly shows a non-uniform distribution of particles for samples with 5 wt.% concentrations of nanofiller (Fig.3). The size of particle-rich areas in nanocomposite reaches up to several micrometers. The non-uniformity of nanoparticle distribution in polymer matrix composites with inorganic nanofiller is a serious problem which is related mainly to compatibility problem of components in nanocomposite [7]. It follows from the obtained results that mechanical properties are more sensitive to the uniformity of nanofiller distribution in matrix

compared to magnetic measurements. The particle aggregates can act as stress concentrators which reduce the strength properties of the material during plastic deformation.

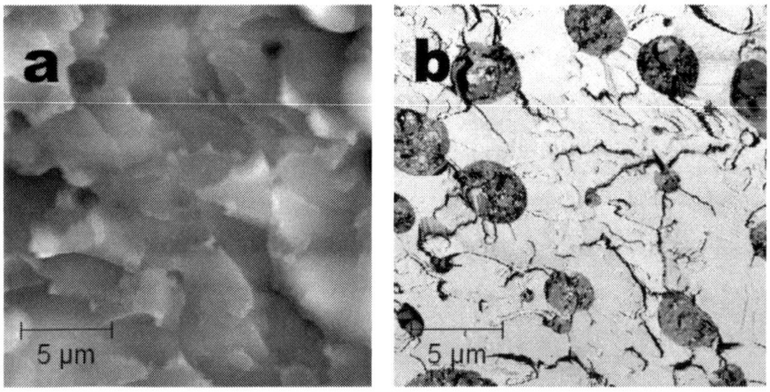

Fig.3. AFM images from fracture surfaces of PC composite with 5wt% $CoFe_2O_4$ nanofiller: (a)-topography, (b) –phase.

5. Conclusion

- Addition of $CoFe_2O_4$ nanoparticles either to commercial PC or EVA toughened recycled PC allows to assign magnetic characteristics to the material. Saturation magnetization, remanent magnetization and coercivity all increase along with raising the ferrite content.
- For the $CoFe_2O_4$ particle – reinforced nanocomposites the results show an improvement of mechanical properties. An increase in hardness and Young's modulus in comparison to PC matrix is observed. However, the improvement of mechanical properties in dependence on the concentration of filler particles is non-linear. When the filler concentration exceeds ~2 wt.% the reinforcement effect decreases. Structural investigation using AFM shows that this can be attributed to non-uniform distribution of filler particles.
- During indentation where compressive stresses are dominating cobalt ferrite nanocomposite maintains plasticity comparable with that of the matrix.

Acknowledgements:
The research is carried out within the framework of the ERDF project Nr.2010/0209/2DP/2.1.1.1.0/10 APIA/VIAA/028.

References

[1] Guduri B R, Luyt A S 2008 J. Nanosci. Nanotechnol. **8** (4) 1880-1885
[2] Wang M, Li B, Wang J and Bai P 2011 Polym. Adv. Technol. **22** 1738–1746
[3] Yoonessi M, Gaier J R 2010 ACS Nano, **4** (12) 7211-20
[4] Milman Yu V, 2008 J.Phys. D:Appl.Phys. **41** 074013-074022.
[5] Hussain F, Hojjati M, Okamoto H, Gorga R E 2006 J.Composite Materials **40** 1511-1575
[6] Reinholds I,. Kalkis V, Zicans J, Merijs Meri R, Grigalovica A, Maiorov M 2012 IOP conf. Ser.:Mater. Sci. Eng. **38** 012030
[7] Fisher H 2003 Mater. Sci.Eng.:C **23** 763-772

Deuterium desorption temperature of Mg-Ti composites prepared by the method of atom-by-atom component mixing

O.M. Morozov[*1], V.G. Kulish[1], V.I. Zhurba[1], I.M. Neklyudov[1], V.O. Progolaieva[1], A.S. Kuprin[1], N.S. Lomino[1], V.D. Ovcharenko[1], I.V. Kolodiy[1], O.G. Galitskiy[2]

[1]National Science Center "Kharkov Institute of Physics and Technology", 61108 Kharkov, Ukraine
[2]G.S. Skovoroda National Pedagogical University at Kharkov, 61029 Kharkov, Ukraine

*E-mail: morozov@kipt.kharkov.ua

Abstract. The plasma evaporation-sputtering method has been applied to obtain composite materials of the Mg-Ti system. Variations in the temperature of the ion-implanted deuterium desorption as a function of the component concentration are studied. It has been established, that introduction of titanium into magnesium leads to the significant decrease of deuterium desorption temperature, namely, to 400-450 K as compared to ~800 K in the case of deuterium release (desorption) from magnesium. The step-like shape of the deuterium desorption temperature curve evidences on the existence of two different structural states of the Mg-Ti composite depending on the ratio of components. The deuterium temperature drop can be caused by filamentary inclusions composed of titanium atoms in magnesium (insoluble component in magnesium) produced in the process of composite formation and providing the deuterium release from the sample at lower temperature (channels for deuterium diffusion and desorption through the surface barrier). The deuterium desorption data obtained on the example of Mg-Ti composites provide a support for further research into hydrogen storage materials containing not readily soluble chemical elements in the alloy components.

1. Introduction

Magnesium-based alloys are promising in view of the present-day requirements to the metal-hydride hydrogen storage systems. Behavior of hydrogen in the magnesium-based alloys arouses intense scientific and applied interest that is confirmed by many publications [1-8]. However, the use of such alloys is accompanied by some difficulties, one of which is a high hydrogen desorption temperature (550-600 K).

Of particular interest are materials in the nanocrystalline state. An exclusive significance of nanocrystalline materials as hydrogen storage materials is due to their unique structural properties which can provide a high sorption capacitance and potentially high concentrations of hydrogen storage. Therefore, naturally, that the reports about investigations of the hydrogen behavior in the nanocrystalline materials are frequent in the literature [7-9].

One of the methods for obtaining materials in the nanocrystalline state is to introduce nanoformative elements. In our opinion among them are not readily soluble chemical elements or elements which can not interact with components of the composites formed. The literature data confirm that investigations in this direction are promising [5-7].

The purpose of this work is to make Mg-Ti composites by the method of atom-by-atom mixing of components and to study the mechanisms of hydrogen storage and its thermoactivated release from the samples obtained. Note, that Mg and Ti are almost not alloyable, therefore, in the above-mentioned system the immiscibility of liquid and solid phases takes place [10].

2. Materials and methods

To make Mg-Ti composites the plasma evaporation-sputtering method was used providing the atom-by-atom growth of components. The composites with a wide range of ratios insoluble components were obtained [8]. A composite was deposited on the molybdenum foils (0.2 mm thickness, 10 mm width, 250 mm length) placed between the cathode assemblies of the facility. In parallel with the molybdenum foils placed were the copper plates (10×10 mm^2) which served as standards to measure the sample thickness using the gravimetric method and the composite component concentrations was determined by the X-ray fluorescence method.

To introduce deuterium into the samples the ion implantation method was applied. Deuterium desorption temperature ranges and deuterium storage levels were determined using the thermal desorption spectroscopy. In experiments a hydrogen isotope (deuterium) was used to decrease the influence of the background hydrogen contained in the samples and target chambers. The samples were fixed on the foils-heaters made of stainless steel Cr18Ni10Ti ($5\times45\times0.3$ mm^3). The temperature was measured with a chromel-alumel thermocouple fastened to the heater.

3. Results

The most characteristic spectra of ion-implanted deuterium thermal desorption from the Mg-Ti composites are presented in Figure 1. One can see the thermal desorption spectrum (TDS) evolution as a function of the composite composition and the implanted deuterium dose. Also the figure shows the plots of corresponding temperatures of maxima of peaks in the deuterium TDS against the magnesium concentration in the samples under study.

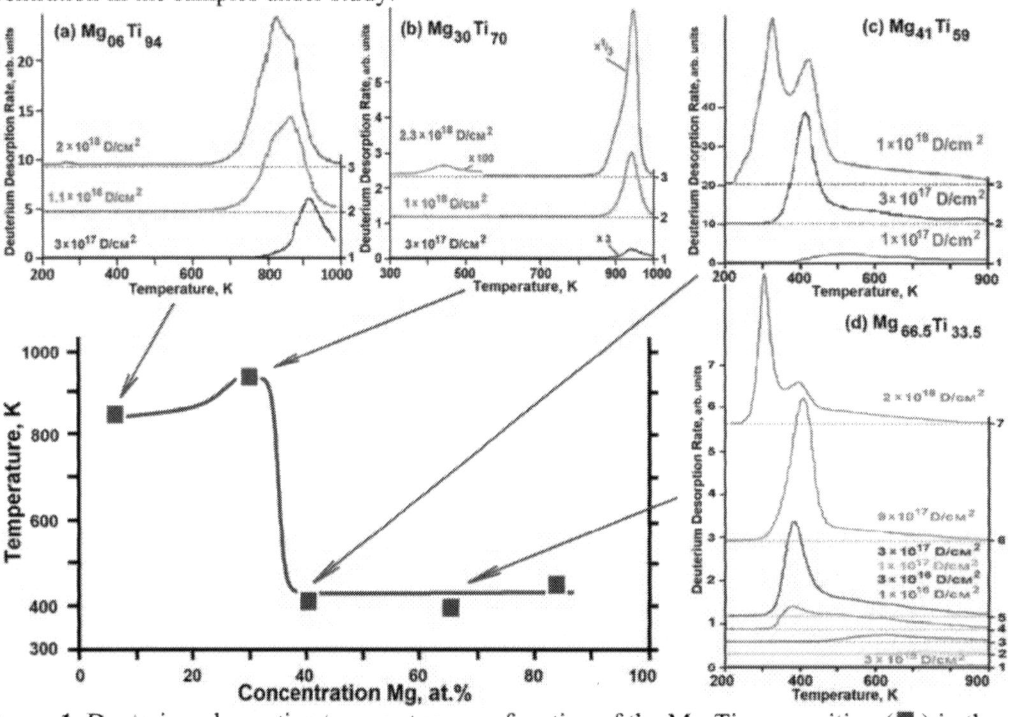

Figure 1. Deuterium desorption temperature as a function of the Mg-Ti composition (■) in the case of a deuterium dose of $\sim1\times10^{18}$ D/cm^2 ($T_{irr.}\sim100$ K) and thermal desorption spectra of deuterium released from composites: (a) $Mg_{06}Ti_{94}$; (b) $Mg_{30}Ti_{70}$; (c) $Mg_{41}Ti_{59}$; (d) $Mg_{66.5}Ti_{33.5}$

A low magnesium concentration and, consequently, a high titanium concentration in the composite are demonstrated in the deuterium TDS as a single peak with a maximum temperature at 800-950 K as a function of the implanted deuterium dose and composite composition (Fig. 1a,b). A single-peak character of the deuterium TDS, observed for magnesium concentration values from 1 to 30 at. %, evidences on the homogeneity of composite structural state in this range. As the magnesium concentration in composites increases to the $Mg_{30}Ti_{70}$ composition, the peak maximum temperature in the deuterium TDS rises to the temperature of 950 K. This is evidence of an increased influence of magnesium on the temperature range of deuterium trapping (see Figure 1b).

The temperature range of deuterium desorption from magnesium-titanium composites, with high titanium content, on the temperature scale is within the region of the deuterium solid solution in α-titanium. Besides, for $Mg_{30-x}Ti_{70+x}$ when the implantation dose increases the peak maximum temperature in the deuterium TDS displaces towards the temperature decrease, that has been earlier observed for the titanium samples fixed at the substrate [11].

For dose of $\sim 2\times10^{18}$ D/cm$_2$ in the deuterium TDS of $Mg_{30-x}Ti_{70+x}$ an additional, not intense, low-temperature slightly-sloped peak with a centre of gravity at ~300-400 K is observed. Thus, in the composite the process of deuterium solid-solution phase formation is finished and the process of hydride formation is started.

As the magnesium concentration in composites increases the deuterium TDS is significantly changing and, as a result, the deuterium desorption temperature for $Mg_{40+x}Ti_{60-x}$ composites sharply decreases (see Fig. 1c,d).

A step-like shape of the maximum temperature curve of thermoactivated deuterium desorption, as a function of the component concentration change, evidences on the existence of two different structural states of the Mg-Ti system depending on the ratio of components.

At the same time, in each of the states only one of components plays a main role – magnesium or titanium. Analysis of curve shapes (Fig. 2), which are changing with concentration of deuterium desorbed from $Mg_{30}Ti_{70}$ (part I) and $Mg_{83.5}Ti_{16.5}$ (part II), shows that they are strongly dependent on the ratio of composite components. In the concentration range, where titanium ($Mg_{30-x}Ti_{70+x}$) is dominant, the amount of trapped deuterium linearly increases with implantation dose increasing. The foregoing shows that in this composite the deuterium diffuses with a high mobility, releases from the implantation region and propagates throughout the sample volume. A similar effect was observed for palladium at a temperature of deuterium implantation of 100 K [12].

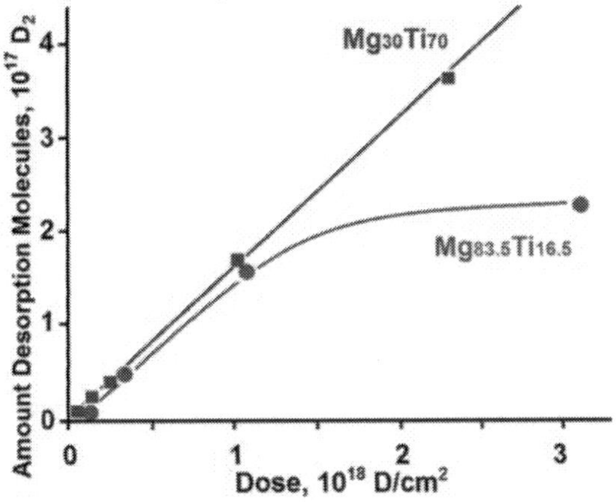

Figure 2. Amount of desorbed deuterium versus the implantation dose for $Mg_{30}Ti_{70}$ (1- ■) and $Mg_{83.5}Ti_{16.5}$ (2- **Error! Objects cannot be created from editing field codes.**)

In the case of $Mg_{40+x}Ti_{60-x}$ the amount of trapped deuterium begins the decrease when the dose is $\sim 1\times10^{18}$ D/cm^2 and reaches the saturation when the implantation dose exceeds 2×10^{18} D/cm^2. It is evident, that in this composite the implanted deuterium has low diffusion mobility and its accumulation occurs, for the most part, in the implantation volume.

The structure of deuterium TDS in the case of $Mg_{40+x}Ti_{60-x}$ strongly depends on the implanted deuterium dose. For low implantation doses (3×10^{15} D/cm^2) the spectrum has only one high-temperature peak with the centre of gravity at 850-900 K. As the implanted deuterium dose increases the deuterium desorption temperature range enlarges towards the temperature decrease (Fig. 1d, curves 2 and 3). Beginning from the dose of $\sim 1\times10^{17}$ D/cm^2 a deuterium desorption region with a centre of gravity at ~400 K arises. The peak intensity is growing up to the dose of 9×10^{17} D/cm^2. As the dose of implanted deuterium increases, the amount of trapped deuterium decreases (see Fig.2, curve 2) and an additional low-temperature region of deuterium desorption with the centre of gravity at 300-330 K takes place (Fig. 1d, curve 7). The low-temperature intensity of the low-temperature peak increases owing to the implanted deuterium and deuterium trapped before at a higher temperature. The changes in the deuterium TDS evidence on the implantation layer structure transformation.

An additional low-temperature peak in TDS, becomes dominant with implanted deuterium dose increasing and reaches the maximum value of ~300 K. This peak arises due to the temperature decreasing in the process of deuterium desorption from the Mg-Ti composites with titanium concentration less than 60 at.%. The deuterium temperature decrease can be caused by filamentary inclusions of insoluble components (here titanium atoms) formed in the process of composite formation that provides the deuterium desorption from the sample at lower temperatures (channels for deuterium diffusion through the surface barrier).

3. Conclusions

The present paper shows that by introducing titanium into magnesium it is possible to decrease significantly the deuterium desorption temperature (namely, to 400-450 K as compared to ~800 K in the case of deuterium desorption from magnesium). A step-like shape of the maximum temperature curve of thermoactivated deuterium desorption, changing as a function of the component concentration, evidences on the existence of two different structural states of the Mg-Ti system depending on the ratio of components.

The deuterium temperature drop can be caused by filamentary inclusions composed of titanium atoms in magnesium (insoluble component in magnesium) produced in the process of composite formation and providing the deuterium release from the sample at lower temperature (channels for deuterium diffusion and desorption through the surface barrier).

The deuterium desorption data obtained using Mg-Ti, Mg-V [7] and Mg-Zr [8] composites provide a support for further research into the hydrogen storage materials containing not readily soluble chemical elements in the alloy components.

Acknowledgement

This study has been done with financial support to the program at the National Academy of Sciences of Ukraine "Hydrogen in the alternative energy and new technologies" (Project #27-13).

References

[1] Huot J, Ravnsbæk D, Zhang J, Cuevas F, Latroche M, Jensen T 2013 Prog Mater Sci 58 30–75
[2] Asano K, Enoki H, Akiba E 2009 J. Alloys Compd. 480 558–63
[3] Shao H, Xin G, Zheng J, Li X, Akiba E 2012 Nano Energy 1 590–601
[4] Anik M, Karanfil F, Kucukdeveci N 2012 Int J Hydrogen Energy 37 299-308
[5] Ershova O, Dobrovolsky V, Solonin Yu, Khyzhun O, Koval A 2008 J. Alloys Compd. 464 212
[6] Tao S, Notten P, Santen R and Jansen 2011 Phys Rev 83B 195403
[7] Neklyudov I, Morozov O, Kulish V, Zhurba V, Galitskiy A , Lomino N, Kuprin O, Ovcharenko V, Reshetnyak E 2012 IOP Conf. Ser.: Mater Sci Eng 38 012061
[8] Neklyudov I, Morozov O, Kulish V, Zhurba V, Lomino N, Ovcharenko V, Kuprin O 2011 IOP Conf. Ser.: Mater. Sci Eng 23 012028
[9] Varin R, Czujko T, Wronski Z 2009 Nanomaterials for Solid State Hydrogen Storage (Springer)
[10] 1999 State diagrams of double metal systems (in Russian) vol. 3-1 ed. N.P. Lyakishev (Moscow: Mashinostroyenie) 334
[11] Neklyudov I, Morozov O and Kulish V 2005 Materialovedenie (in Russian) 11 45-56
[12] Rybalko V, Morozov O, Neklyudov I and Kulish V 2001 Phys Lett 287A 175-182

IOP Conf. Series: Materials Science and Engineering **49** (2013) 012014 doi:10.1088/1757-899X/49/1/012014

Atomic layer deposition of aluminum oxide films on graphene

Raul Rammula, Lauri Aarik, Arne Kasikov, Jekaterina Kozlova, Tauno Kahro, Leonard Matisen, Ahti Niilisk, Harry Alles and Jaan Aarik

Institute of Physics, University of Tartu, Tartu, 51014, Estonia

E-mail: raul.rammula@ut.ee

Abstract. Seed-layer approach was studied to initiate atomic layer deposition (ALD) of Al_2O_3 films on graphene. Low-temperature ALD and electron beam evaporation (EBE) were applied for seed-layer preparation before deposition of the dielectric at 200 °C using trimethyl-aluminum and water or ozone as precursors. To characterize nucleation of the films and possible influence of the ALD processes on the quality of graphene, properties of graphene and Al_2O_3 films were investigated by Raman spectroscopy, X-ray fluorescence and X-ray photoelectron spectroscopy methods. The results suggest that seed layer formation by low-temperature ALD was more efficient in the O_3-based process than in the H_2O-based one while EBE seed layer provided fastest growth of Al_2O_3 together with minimum incubation period.

1. Introduction

Studies of various graphene-containing nanostructures and related thin film materials have been progressively extended since revealing the versatile properties of graphene in 2004 [1]. Field-effect transistors and spintronic devices belong to the most attractive application areas of graphene. These applications require integration of graphene into structures that contain very thin dielectric layers as gate dielectrics and/or tunnel barriers. Atomic layer deposition (ALD) has been considered to be a promising technique for preparation of this kind of dielectrics in the case of electronic devices based on traditional materials. ALD of thin films on graphene is, however, much more complex task because there are no reactive adsorption sites on the perfect graphene surface.

For preparation of monolayer graphene on large areas, chemical vapor deposition (CVD) is widely exploited. Unfortunately, uniformity of graphene obtained by this method is not very high yet [2]. Boundary regions of independently grown graphene flakes in conjunction with step edges, wrinkles and basal plane defects, typical for CVD graphene, are nucleation centers for ALD thin films [3,4]. Furthermore, impurities from the lift-off procedure can contaminate highly reactive graphene edges. These effects together with very low sticking probability of ALD precursors on defect-free areas of graphene make the deposition of ALD films on graphene non-uniform. Thus, direct deposition of Al_2O_3 has resulted in selective growth of the dielectric on graphene edges and defects [3,4]. In order to improve the uniformity of ALD films, a physisorbed seed layer can be added between graphene and dielectric, more effective/reactive precursor chemistry can be used and/or growth temperature (T_G) can be reduced. Typical route to obtain Al_2O_3 film is to apply trimethylaluminum (TMA) and water vapor (H_2O) as precursors [5]. At low T_G (<100 °C), high concentration of hydrogen in the films is a major concern [5] that does not allow obtaining dielectric films with sufficient quality. Furthermore, purge periods that are needed to eliminate CVD like growth should be very long in the low-temperature H_2O-based ALD processes (about 180 seconds at T_G = 33 °C [5]). By replacing water to ozone (O_3),

Al_2O_3 of higher quality can be obtained [6]. However, care must be taken when graphene is treated with ozone, especially at higher temperatures [7,8]. Therefore in order to directly compare nucleation of Al_2O_3 on graphene in different cases, ALD of Al_2O_3 on CVD graphene was studied in this work using 3 seed layer approaches and 2 different oxygen precursors for ALD.

2. Experimental

Monolayer graphene was grown on commercial 25-μm thick polycrystalline copper foils (99.999%, Alfa Aesar) using recipe by van der Zande et al. [9]. Before graphene deposition copper foils were annealed at 960 °C in a CVD reactor for 30 minutes in a flow of Ar/H_2 (99.999%, AGA) gas. For graphene formation copper was exposed to a flow of 10 % CH_4/Ar (99.999%, AGA) gas for about 20 minutes. To avoid possible damages and unintentional functionalization of graphene in the transfer process, Al_2O_3 films were deposited on graphene in its as-grown state on copper foils.

ALD of Al_2O_3 was carried out in a hot-wall flow-type reactor from the sequential pulses of TMA (98%, Aldrich) and H_2O or O_3. BMT Messtechnik equipment was used to produce ozone from O_2 (99.999%, AGA). During the growth experiments O_3 concentration was around 200–220 g/m^3 whereas corresponding partial pressure in the reaction chamber was 34 Pa during O_3/O_2 pulse. Seed layers were deposited by EBE or by applying 1–3 ALD cycles at room temperature, if not stated otherwise. In few experiments, the ALD seed layers were deposited also at temperatures up to 50 °C. The rest of the dielectric layer was grown by applying up to 100 ALD cycles at 200 °C. To compare the film growth on different substrates, Al_2O_3 was simultaneously grown on copper foils and HF etched Si(100) too.

The thicknesses of alumina films were determined from X-ray fluorescence (XRF) data using a reference sample with known thickness and density measured by X-ray reflection (XRR). The Raman spectroscopy method was applied to estimate the quality of graphene before and after ALD processes. The Raman spectra were recorded with a Renishaw inVia micro-Raman spectrometer with an incident laser beam wavelength of 514 nm. A silicon reference sample was used for the calibration of the Raman shift. In addition, X-ray photoelectron spectroscopy (XPS) measurements were carried out using SCIENTA SES 100 electron energy analyser and a non-monochromatic twin anode Thermo XR3E2 X-ray tube with characteristic Al Ka (1486.6 eV) energy to analyze the films.

3. Results and Discussion

3.1. Nucleation and growth rate of Al_2O_3

Figure 1 show that the seed layers deposited on graphene had a significant influence on the dependence of the Al_2O_3 film thickness on the number of ALD cycles applied at 200 °C.

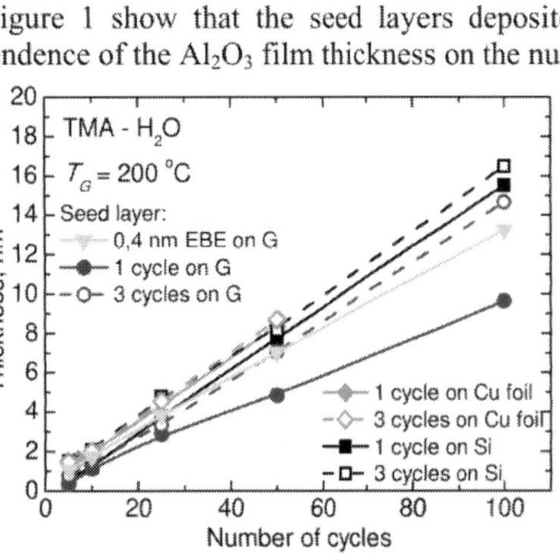

Figure 1. Dependencies of film thickness on number of ALD cycles for Al_2O_3 films deposited on graphene (G), copper foil (Cu) and silicon (Si).

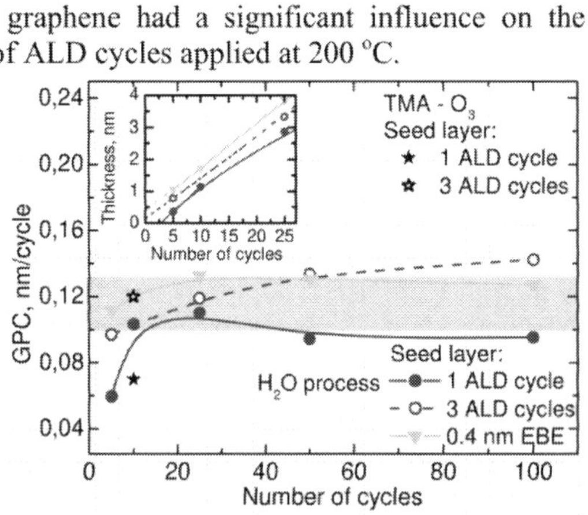

Figure 2. GPC of Al_2O_3 on graphene as a function of the number of ALD cycles.

IOP Conf. Series: Materials Science and Engineering **49** (2013) 012014 doi:10.1088/1757-899X/49/1/012014

Expectedly, nucleation at the initial growth stage (inset of Figure 2) markedly determined the overall film thickness. The results indicate that employment of 1 ALD cycle for deposition of a seed layer was insufficient to initiate uniform growth on graphene in the TMA-H_2O as well as TMA-O_3 process (Figures 1 and 2). Al_2O_3 nucleation on hydrophobic graphene surface was inhibited and the initial growth dead-time (IGDT) was about 3 cycles for the TMA-H_2O process (inset of Figure 2). In contrast, on bare copper foil this effect was not dominant meaning that enough surface sites were present for precursor adsorption already in the beginning of deposition. When 3 ALD cycles were applied for seed layer deposition, IGDT was reduced to almost zero also on graphene and high values of growth per cycle (GPC) were obtained for very thin films in both ALD processes studied. Smallest variation and highest mean value of GPC were observed on EBE seed layer (Figure 2). The GPC values of the films deposited on those sufficiently thick (0.15–0.4 nm) seed layers are very similar to respective values, which have been obtained for the TMA-H_2O and TMA-O_3 ALD processes on Si substrates at 200 ^oC and are highlighted by gray area between 0.10 and 0.13 nm/cycle in Figure 2.

3.2. Structure of graphene

Raman spectra recorded before and after Al_2O_3 deposition (Figure 3) showed characteristic G and 2D bands of graphene on strong background that was caused by the copper substrate. No D band formation at 1340 cm^{-1} could be seen. Thus, the seed layer deposited by applying only 1 ALD cycle protected graphene against marked damages that might have caused by O_3 at 200 ^oC [8]. Nevertheless, Raman spectra of graphene showed blueshift of main bands after deposition of Al_2O_3 by applying 1 TMA-O_3 cycles at 22 ^oC and 10 cycles at 200 ^oC (Figure 3(a)). A possible reason for the shift was partial oxidation of graphene [10]. XPS C1s spectra showing additional C=O, COOH component at higher binding energies (Figure 4) seemed to confirm this conclusion.

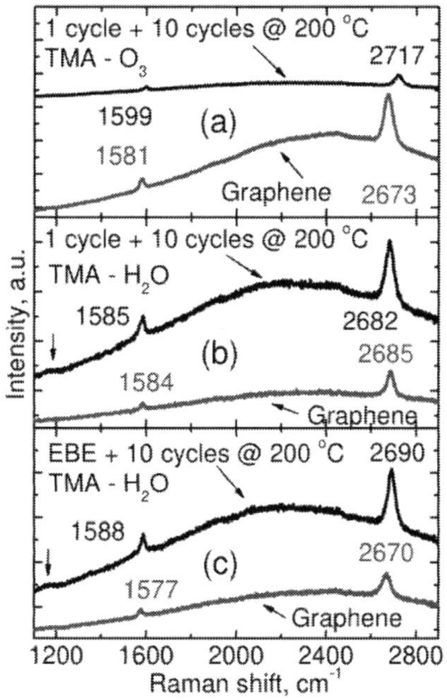

Figure 3. Raman spectra of graphene coated by Al_2O_3 via (a) O_3-based and (b), (c) H_2O-based ALD processes. Pure graphene is indicated as a reference.

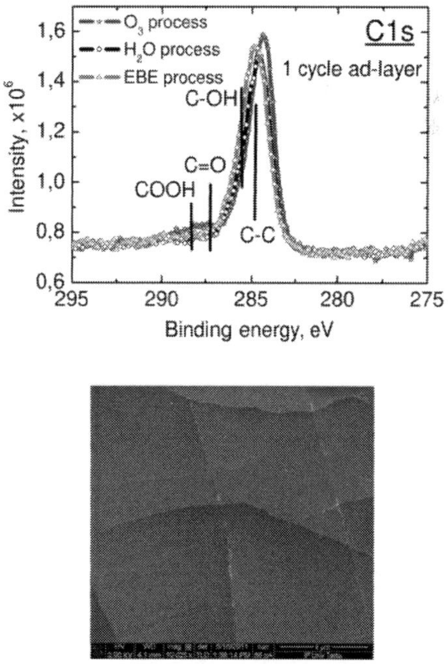

Figure 4. XPS C1s core level spectra of graphene coated by Al_2O_3 (top) and SEM image of graphene (bottom).

Absence of D-band in the Raman spectra (Figure 3) suggests that graphene oxidation and growth defects that were observed by SEM (Figure 4) had relatively weak effect on overall properties of graphene. The increase of seed-layer deposition temperature degraded, however, film uniformity and, according to Raman spectroscopy studies (spectra not shown), caused appearance of honeycomb defects. In Raman spectra of some graphene samples coated by Al_2O_3 in the TMA-H_2O process, a weak band appeared at 1200 cm^{-1} (peaks indicated by arrows in Figures 3 (b) and (c)). The band likely belongs to C-O or C-H vibrations. In the case of samples with EBE seed layers, XPS revealed formation of Cu_2O during Al_2O_3 deposition. Consequently, this seed layer was not dense enough to avoid diffusion of oxygen precursor to the substrate during the following ALD process. As graphene is considered to be a good diffusion barrier, the oxygen precursor most likely diffused into underlying copper through boundary regions in graphene (Figure 4) or areas that were not covered by graphene.

4. Conclusions

This study demonstrated that 0.2–0.4 nm thick Al_2O_3-based seed layers, prepared by ALD or EBE at room temperature, improved markedly Al_2O_3 nucleation on graphene surface during the following ALD of Al_2O_3 at 200 °C. Somewhat better nucleation was observed on EBE seed layers, partially due to greater thickness of those. Compared to the TMA-H_2O ALD process, the TMA-O_3-based one resulted in more uniform growth of the dielectric but caused also some oxidation of graphene. An advantage of the ALD seed layers compared to EBE seed layers is that the former allow deposition of a dielectric on graphene in one process.

Acknowledgements

The authors would like to thank Prof. Väino Sammelselg and Dr. Ilmo Sildos for the access of SEM and Raman measurements. The work was supported by European Science Foundation and Estonian Science Foundation under the EUROCORES Programme EuroGRAPHENE project "Entangled spin pairs in graphene" and by the European Union (FP7) through the program RODIN.

References

[1] K. S. Novoselov, A. K. Geim, S. V. Morozov, D. Jiang, Y. Zhang, S. V. Dubonos, I. V. Grigorieva and A. A. Firsov 2004 *Science* **306** 666-69.

[2] T. Wu, G. Ding, H. Shen, H. Wang, L. Sun, D. Jiang, X. Xie and M. Jiang 2013 *Adv. Funct. Mater.* **23** 198-203.

[3] B. Lee, S.-Y. Park, H.-C. Kim, K. J. Cho, E. M. Vogel, M. J. Kim, R. M. Wallace and J. Kim 2008 *App. Phys. Lett.* **92** 201302-1-3.

[4] H. Alles, J. Aarik, J. Kozlova, A. Niilisk, R. Rammula, V. Sammelselg 2011 *Graphene-Synthesis, Characterization, Properties and Applications INTECH* 99-114.

[5] M. D. Groner, F. H. Fabreguette, J. W. Elam and S. M. George 2004 *Chem. Mater.* **16** 639-45.

[6] Y. Shen, Y. Li, J. Zhang, X. Zhu, Z. Hu, J. Chu 2012 *Optoelect. and adv. Mater.* **6** 618-22.

[7] J. Yuan, L.-P. Ma, S. Pei, J. Du, Y. Su, W. Ren and H.-M. Cheng, *ACS Nano, Accepted Manuscript* DOI: 10.1021/nn400682u.

[8] B. Lee,G. Mordi, M. J. Kim, Y. J. Chabal, E. M. Vogel, R. M. Wallace, K. J. Cho, L. Colombo and J. Kim 2010 *Appl. Phys. Lett.* **97** 043107-1-3.

[9] A.M. van der Zande, R.A. Barton, J.S. Alden, C.S. Ruiz-Vargas, W.S. Whitney, P.H.Q. Pham, J. Park, J.M. Parpia, H.G. Craighead, and P.L. McEuen 2010 *Nano Lett.* **10** 4869-73.

[10] D. L. Duong, G. H. Han, S. M. Lee, F. Gunes, E. S. Kim, S. T. Kim, H. Kim, Q. H. Ta, K. P. So, S. J.Yoon, S. J. Chael, Y. W. Jo, M. H. Park, S. H. Chae, S. C. Lim, J. Y. Choi and Y. H. Lee, 2012, *Nature* **490** 235-40.

Production of mono- and bimetallic nanoparticles of noble metals by pyrolysis of organic extracts on silicon dioxide

V Serga[1], L Kulikova[1], A Cvetkov[1], A Krumina[1], M Kodols[1], S Chornaja[2], K Dubencovs[2] and E Sproge[2]

[1]Institute of Inorganic Chemistry, Riga Technical University, Latvia
[2]Faculty of Material Science and Applied Chemistry, Riga Technical University, Latvia

E-mail: vera_serga@inbox.lv

Abstract. In the present work the influence of the tri-n-octylammonium (Oct_3NH^+) salt anion ($PtCl_6^{2-}$, $PdCl_4^{2-}$, $AuCl_4^-$) nature on the phase composition and mean size of crystallites of the extract pyrolysis products on the SiO_2 nanopowder has been studied. The XRD phase analysis of the composites (metal loading 2.4 wt.%) made under the same conditions, at the pyrolysis of Pt- and Au-containing extracts has shown the formation of nanoparticles of Pt (d_{Pt} = 15 nm) and Au (d_{Au} = 33 nm), respectively. The end-product of the pyrolysis of the Pd-containing extract has an admixture phase of PdO along with the main metal phase (d_{Pd} = 21 nm). At the preparation of bimetallic particles (Pt-Pd, Pt-Au, Pd-Au) on the SiO_2 nanopowder it has been found that the nanoparticles of the PtPd alloy, Pt and Au or Pd and Au nanoparticles are the products of the thermal decomposition of two-component mixtures of extracts. The investigation of catalytic properties of the produced composites in the reaction of glycerol oxidation by molecular oxygen in alkaline aqueous solutions has shown that all bimetallic composites exhibit catalytic activity in contrast to monometallic ones.

1. Introduction

Silicon dioxide is widely used as a carrier at the production of composites, which contain nanoparticles of noble metals, by such methods as sol-gel, ion exchange, impregnation method, deposition-precipitation and others [1]. These composites are used as catalysts in different chemical processes: hydrogenation, CO and alcohol oxidation reactions, manufacture of vinyl acetate, etc.[1, 2].

The promising use of the extractive-pyrolytic method (EPM) for the production of monometallic composites, such as high-dispersed nanoparticles of palladium on α-Al_2O_3 micro granules [3], Al_2O_3, Y_2O_3 nanopowders [4] and platinum on Al_2O_3, γ-AlO(OH),Y_2O_3, CeO_2, SiO_2 nanopowders [5, 6], has been earlier shown. In the reported investigations, solutions of metal-containing tri-n-octylammonium salts ($[Oct_3NH]_2PdCl_4$, $[Oct_3NH]_2PtCl_6$) in toluene were used as precursors, which impregnated the carrier and underwent pyrolysis. Alkylammonium salts are known as typical cationic surfactants, and, as the hydration energy of alkylammonium salt anion increases, its absorption ability from the non-aqueous phase at the hydrophilic interfaces increases as well [7]. The conditions of thermal decomposition of palladium- and platinum-containing tri-n-octylammonium salts, which make it possible to completely remove the organic component of the precursor, have been found in [3, 8]. The influence of the composite production conditions (pyrolysis temperature, annealing time, precursor concentration, the carrier specific surface area) on the mean size of metal crystallites was studied [6]. The composites produced by the EPM have been found [4, 5, 9] to exhibit a high catalytic activity in the reactions of glycerol oxidation by molecular oxygen.

In the course of this research, with the synthesis conditions being equal, individual nanoparticles of gold, platinum and palladium on the SiO_2 nanopowder were produced to study the influence of the Oct_3NH^+ salt anion ($PtCl_6^{2-}$, $PdCl_4^{2-}$, $AuCl_4^-$) nature on the phase composition and mean size of crystallites of the extract pyrolysis products. Additionally, the possibility of EPM application for the production of bimetallic particles by thermal decomposition of two-component mixtures of extracts on the carrier was investigated. The catalytic activity of all produced composites has been examined.

2. Experimental

In order to produce composites, Pt-, Pd- and Au-containing extracts were preliminary produced, in which a solution of tri-n-octylamine in toluene was used as an extractant. The method for the production of metal-containing extracts is described in details in [5, 6]. A weight of a SiO_2 nanopowder (SSA = 190 m^2/g) was impregnated by a solution of the extract or two-component mixtures of extracts. The volume of extract (C_{Me} = 0.4 mol/l) corresponded to metal loading in composite 2.4 wt.%. Then, in order to remove the solvent, the sample was dried for 5-10 min at t = 80-105 ^{0}C. Thermal treatment involved the heating of the sample in the air from room temperature to 300 ^{0}C at the 10^0/min rate and the annealing within 5 min.

The phase composition of the extract pyrolysis products on the carrier was defined by an X-ray diffraction method using a diffractometer D8 Advance (Bruker Corporation) with CuK_α radiation (λ = 1.5418Å). The mean size of metal crystallites (d_{Me}) was defined from the (111) peak width by the Scherrer method. SEM measurements were made using the MIRA\TESCAN operating at 15 kV. The specific surface area (SSA) of the composites was measured using the HROM-3 chromatograph by the BET method at the temperature of liquid nitrogen.

Glycerol was oxidized by molecular oxygen in the presence of the produced composites in alkaline aqueous solutions in a thermo-stated slurry bubble column reactor operated in a batch mode. In order to determine the concentration of the reaction products, liquid samples were collected periodically from the reaction mixture. The filtered samples were analyzed by a high performance liquid chromatograph (WATERS 2487).

3. Results

The XRD phase analysis of the produced monometallic composites has shown (Fig. 1) that the pyrolysis of $[Oct_3NH]_2PtCl_6$ and $[Oct_3NH]AuCl_4$ salts results only in the nanoparticles of platinum (PDF ICDD 00-004-0802) with d_{Pt} = 15 nm (Fig. 1, curve 1) and of gold (PDF ICDD 00-004-0784) with d_{Au} = 33 nm (Fig. 1, curve 3) on the carrier. Along with the main metal phase (d_{Pd} = 21 nm, PDF ICDD 01-088-2335), the end-product of the thermal decomposition of $[Oct_3NH]_2PdCl_4$ salt contains also an admixture phase of palladium oxide (PDF ICDD 00-043-1024, fig. 1, curve 2). The PdO formation seemed to be attributed to the thermal treatment performed in the air, not in an inert atmosphere. Hence, d_{Au} greatly exceeds d_{Pt} and d_{Pd} in the produced composites.

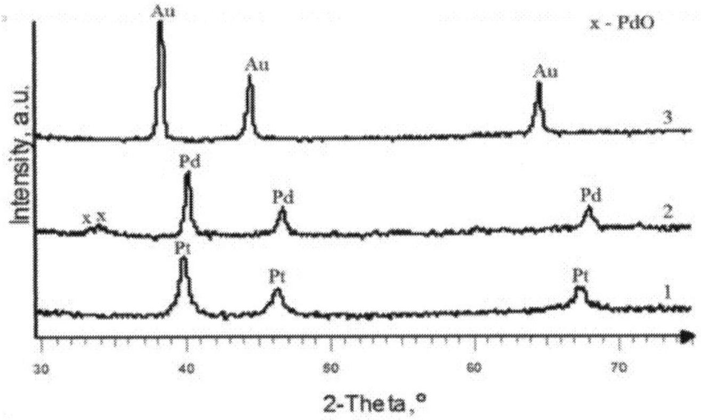

Figure 1. XRD patterns of the composites:
1 - Pt/SiO_2; 2 – Pd/ SiO_2; 3 -Au/ SiO_2.

It is known [7] that the hydration energy of $AuCl_4^-$ anion is less than that of $PtCl_6^{2-}$ and $PdCl_4^{2-}$ anions, therefore, the hydrophilic surface of the carrier is worse wetted by the Au-containing extract at the stage of impregnation than by the Pt- and Pd-containing extracts that probably contributes to the formation of larger metal crystallites. The SSA value of the Au-containing composite (165 m^2/g) is smaller that those of the Pt- (178 m^2/g) and Pd-containing (185 m^2/g) composites.

The results obtained by SEM have revealed (Fig. 2) that the gold nanoparticles form larger agglomerates on the carrier surface, which are distributed less uniformly (Fig. 2a) that the agglomerates of palladium (Fig. 2b) and platinum (Fig. 2c).

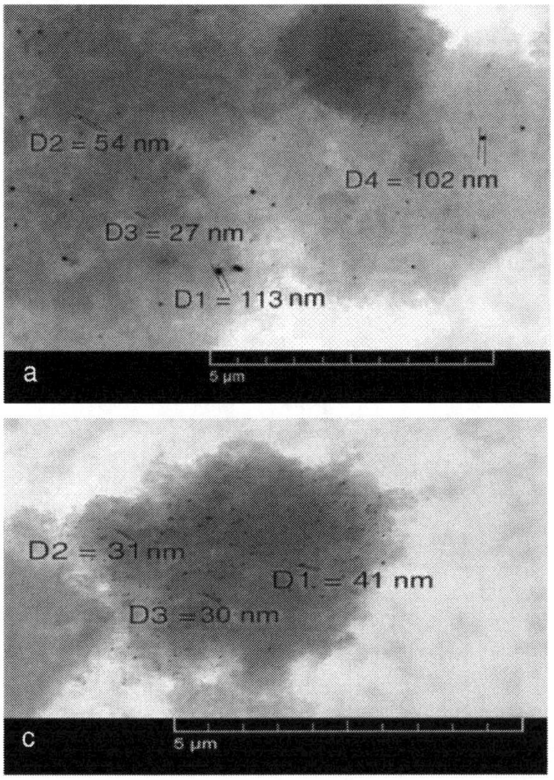

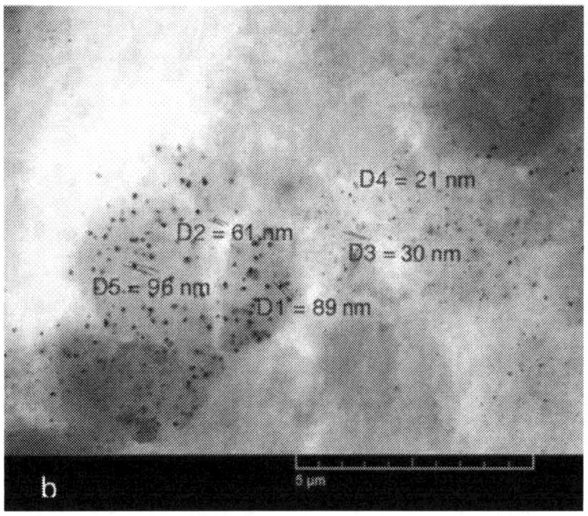

Figure 2. SEM images of monometallic composites:
a - Au/ SiO_2; b - Pd/ SiO_2; c - Pt/SiO_2.

The phase composition of the pyrolysis products the of two-component mixtures of extracts on the carrier was studied at the production of bimetallic composites (Fig. 3). It has been found that the nanoparticles of the PtPd alloy (PDF ICDD 01-072-2838, d_{PtPd} = 20 nm) and of PdO as the admixture phase are the end-products of the thermal decomposition of the mixture of Pt- and Pd-containing extracts (Fig. 3, curve 1); those of the Pd- and Au-containing mixture are the nanoparticles of palladium, gold (d_{Me} = 20 nm) and PdO (Fig. 3, curve 2); those of the Pt-and Au-containing mixture give the nanoparticles of platinum (d_{Pt} = 12 nm) and gold (d_{Au} = 33 nm) (Fig. 3, curve 3).

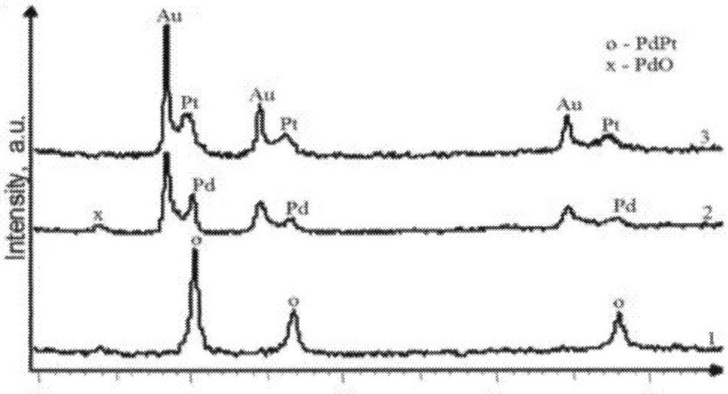

Figure 3. XRD patterns of the composites produced from two-component mixtures of extracts: 1 – Pt-Pd/ SiO_2; 2 - Pd-Au/ SiO_2; 3 - Pt-Au/ SiO_2.

The catalytic activity of all produced composites was tested in the reaction of glycerol oxidation by molecular oxygen in alkaline aqueous solutions (Table 1). The data in the table evidence that the largest catalytic activity is typical for the platinum-containing composites. Moreover, the bimetallic composites are much more active – the conversion of glycerol is by 26-30% higher if compared with the conversion in the presence of a Pt/SiO_2 monometallic composite. The main product of oxidation is glyceric acid, the yield selectivity of which is practically similar in the presence of all platinum-containing composites and varies within 52-57%. Monometallic Pd/SiO_2 and Au/SiO_2 are practically non-active. The disadvantage of the composites under investigation is the partial dissolution of the carrier during the oxidation of glycerol in alkaline aqueous solutions. Therefore, those composites as catalysts should be preferably used in neutral aqueous solutions.

Table 1. Catalytic activity of mono- and bimetallic catalysts at glycerol oxidation.

Composite	Glycerol Conversion	Selectivity,%					
		Glyceric acid	Tartronic acid	Lactic acid	Glycolic acid	Oxalic acid	Formic acid
Pd/SiO_2	1	55	0	0	45	0	0
Pt/SiO_2	62	56	4	29	8	1	2
Au/SiO_2	3	10	0	32	51	0	7
*Pd-Au/SiO_2	18	90	2	2	4	1	1
Pt-Au/SiO_2	92	52	8	32	8	1	2
Pt-Pd/SiO_2	88	57	12	24	5	2	1

Reaction conditions: c_0(glycerol) = 0.3 M, c_0(NaOH) = 1.5 M, n(glycerol)/n(Pt) = 300 mol/mol, 60 °C, pO_2 = 1 atm, reaction time 4 h.*Oxidation time 7 h.

4. Conclusions

The investigations on the nature of Oct_3NH^+ salt anion ($PtCl_6^{2-}$, $PdCl_4^{2-}$, $AuCl_4^-$) effect on the phase composition and mean size of the crystallites of the pyrolysis extracts' products on the SiO_2 have shown that the exchange of $PtCl_6^{2-}$ anion by $PdCl_4^{2-}$ one results in the increase of the mean size of metal crystallites from 15 nm to 21 nm and in appearance of the admixture phase of metal oxide (PdO), and the exchange by $AuCl_4^-$ results in the increase of the mean size of metal crystallites to 33 nm. It has been found that the use of two-component mixtures of extracts makes it possible to produce bimetallic composites.

The study of the catalytic properties of the produced composites has revealed that the monometallic platinum-containing and all bimetallic composites exhibit catalytic activity in the reaction of glycerol oxidation by molecular oxygen in alkaline aqueous solutions.

Acknowledgment. This work was supported by the European Regional Development Fund – project No. 2010/0304/2DP/2.1.1.1.0/10/APIA/VIAA/087.

References

[1] Ertl G, Knözinger H, Schüth F and Weitkamp J 2008 *Handbook of heteregeneous catalysis* (Wiley-VCH Verlag GmbH& Co. KGaA, Germany)
[2] Heiz U, Landman U 2007, 2008 *Nanocatalysis* (Springer-Verlag Berlin Heidelberg) pp 377-387
[3] Serga V, Kulikova L, Krumina A, Chornaja S, Dubencov K and Maiorov M 2013 *Journal of Materials Science and Engineering A* **3** (2) 104-108
[4] Cornaja S, Kulikova L, Serga V, Kampars V, Dubencovs K, Zizkuna S and Muravjova O EU Pat.: 10164874.9 (15.06.2011)
[5] Palcevskis E, Kulikova L, Serga V, Cvetkov A, Čornaja S, Sproge E and Dubencovs K 2012 *Journal of the Serbian Chemical Society* **77** (12) pp 1799-1806 JSC121116147P
[6] Serga V, Kulikova L, Cvetkov A and Krumina A 2012 *IOP Conf. Series: Materials Science and Engineering* **38** 012062
[7] Kazarinov VE 1987 *The Interface Structure and Electrochemical Processes at the Boundary Between Two Immiscible Liquids* (Springer: Berlin and Heidelberg) pp 179-205
[8] Serga V, Kulikova L, Grekhov V, Kalnacs J, Murashov A and Vilken A 2011 *Abstracts of International conference "Functional materials and nanotechnologies"* 245
[9] Chornaja S, Dubencov K, Kampars V, Stepanova O, Zhizhkun S, Serga V and Kulikova L 2013 *Reaction Kinetics, Mechanisms and Catalysis* **108** 341-357

Examination of dielectric dispersion of complex oxides on the basis of bismuth-containing titanates

M.S. Shashkov[1], O.V. Malyshkina[1], E.V. Barabanova[1], M.S. Korolyova[2], I.V. Piyr[2]

[1]Tver State University, Tver, Russia
[2] Komi Institute of Chemistry of Ural branch of Russian Academy of Sciences, Syktyvkar, Russia

e-mail: Olga.Malyshkina@mail.ru

Abstract. In the present work the samples of complex oxide compounds on the basis of bismuth titanates with chalcolamprite structure type and layered perovskite doped with Cr, Fe and Co were studied at room temperature by the method of dielectric spectroscopy in the frequency range of 30 to 10^6 Hz.

1. Introduction

One trend of modern materials science is the replacement of lead-containing materials with lead-free equivalents. In the present work we study the samples of bismuth-containing compounds which are of interest with respect to utilizing their dielectric properties. Complex bismuth titanates with layered perovskite structure are ferroelectric with high Curie temperature. They are used for the production of radio capacitors, piezoelectric transmitters, filters, sonar devices, etc.

Complex oxides of bismuth titanate with a structure of layered perovskite and pyrochlore doped by Cr, Fe and Cu in accordance with the stoichiometric formulas $Bi_{1.6}Ti_2Cr_{0.16}O_{6.6}$, $Bi_{1.55}Ti_{1,97}Cu_{0.78}O_7$ and $Bi_4Ti_2Fe_1O_{11.5}$ were studied. The dielectric properties were measured in the frequency range from 30 Hz to 10^6 Hz. The low value of the voltage applied to the samples allowed to exclude the nonlinear polarization processes.

2. Sample preparation

The polycrystalline samples of iron containing bismuth titanate with designated composition were prepared by the procedure of solid phase synthesis in air using Bi_2O_3 (99.99%), TiO_2 (99.999%) and Fe_2O_3 (99.9%) Cr_2O_3(99.9%), CuO (99.9%) in stoichiometric ratios. Before sintering, the starting oxides were mixed in an agate mortar for 30 minutes, pressed into pellets and put into a corundum crucible. The pellets were first sintered for 6 h at 650°C (below 825°C which is the melting point of Bi_2O_3 and is within the temperature range of its phase transition). Then several sintering processes with intermediate grinding and pelleting were carried out at 950 °C (5 h), 1050 °C(5 h) and 1100 °C (5 h) for Cr-doped pyrochlore. The samples were cooled down to room temperature with the furnace. The evaporation of Bi_2O_3 was not detected neither by the thermogravimetric measurements nor by the elemental analysis of the samples.

3. Theory

Analysis and presentation of the experimental data were performed using the fractal-power universal law for the dielectric response. According to this law the response is presented by a power function, whereby ε'(f), ε"(f), as well as the σ'(f) and σ"(f) are described as a power function of the applied field frequency [1]. In this case the responses due to either dipole or mobile charge carriers are separated. It is to be noted that there are two independent processes of the dielectric response. The dipole relaxation is characterized by a maximum in the ε"(f) dependence, while there is no maximum in the response due to mobile charge carriers.

For the case of dipole relaxation, the determination of relaxation parameters was carried out by an analysis of the experimental data, which consists of the following:

1) determination of the frequency shift corresponding to the maximum of ε"(f),
2) analysis of the ε"(T) dependence in Arrhenius coordinates.

In the case of conductivity relaxation such an analysis is not possible, so an empirical calculation method [2] was applied. The idea behind this method is to introduce a value $\beta^* = 1/\varepsilon^*$, where the expressions for β' and β'' have the form

$$\beta' = \frac{\varepsilon'}{(\varepsilon')^2 + (\varepsilon'')^2}$$

$$\beta'' = \frac{\varepsilon''}{(\varepsilon')^2 + (\varepsilon'')^2}.$$

Whereas there is no maximum in the curves $\varepsilon''(f)$, it may be observed in the $\beta''(f)$ curves. In view of the relaxational type of the conductivity the analysis of its dispersion was performed making use of the relation $\varphi(f)$ = Im σ/Re σ, which is the inverse of the dielectric loss tangent ($tg\delta = \varphi^{-1}(f)$).

4. Results and Discussion

The frequency dependence of the complex dielectric characteristics for $Bi_{1.6}Ti_2Cr_{0.16}O_{6.6}$, $Bi_{1.55}Ti_{1.97}Cu_{0.78}O_7$ and $Bi_4Ti_2Fe_1O_{11.5}$ ceramics was obtained. The results for the dielectric permittivity, conductivity, dielectric module and dielectric loss tangent are shown in Figures 1 and 2, respectively.

There is no difference in the form of dielectric spectra of investigated materials (Fig.1, a). The only disparity is in the magnitudes of the real parts of permittivity $\varepsilon'(f)$. The absence of a maximum in the frequency dependence of the imaginary parts of permittivity $\varepsilon''(f)$, which characterizes the dielectric loss, allows to assume that the nature of the response is due to the presence of mobile charge carriers. In this case the use of the empirical calculation method of finding the relaxation times becomes validated.

Frequency behaviour of the dielectric module is illustrated in Figure 1(b). The presence of the maxima on the graphs of β"(ω) allows to determine the relaxation times as the reciprocal of the corresponding cyclic frequencies for all the samples under study (Table 1).

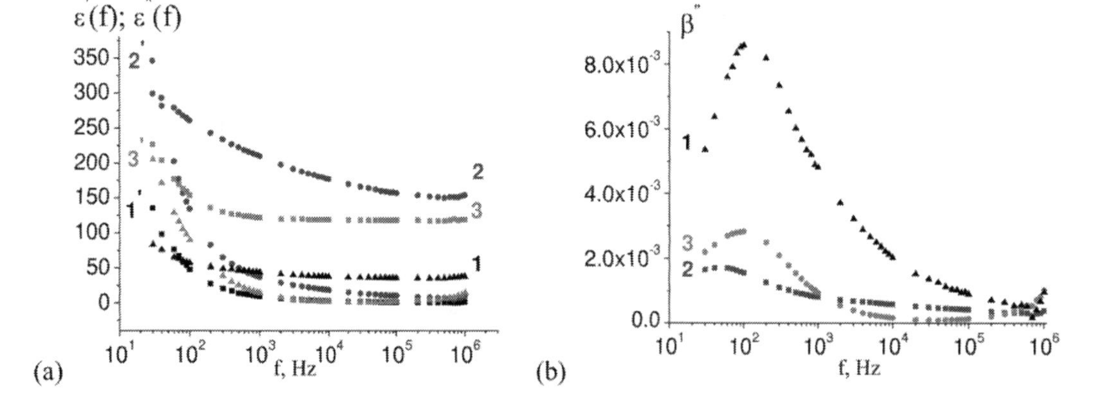

Figure 1. (a) Frequency dependence of the real (1, 2, 3) and imaginary (1′, 2′, 3′) parts of the dielectric permittivity; (b) imaginary part of the dielectric module for $Bi_{1.6}Ti_2Cr_{0.16}O_{6.6}$ (1, 1′), $Bi_{1.55}Ti_{1.97}Cu_{0.78}O_7$ (2, 2′) and $Bi_4Ti_2Fe_1O_{11.5}$ (3, 3′) compounds

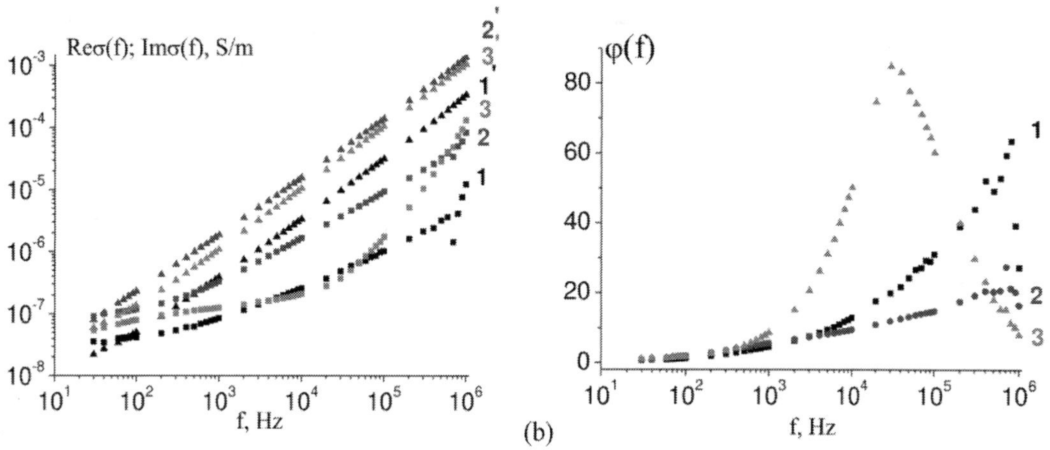

(a) (b)

Figure 2. (a) Frequency dependence of the real and imaginary parts of the complex conductivity; (b) $\varphi(f) = \mathrm{Im}\,\sigma / \mathrm{Re}\,\sigma$ for $Bi_{1.6}Ti_2Cr_{0.16}O_{6.6}$ (1, 1′), $Bi_{1.55}Ti_{1.97}Cu_{0.78}O_7$ (2, 2′) and $Bi_4Ti_2Fe_1O_{11.5}$ (3, 3′) compounds

Table 1. The relaxation times for $Bi_{1.6}Ti_2Cr_{0.16}O_{6.6}$, $Bi_{1.55}Ti_{1.97}Cu_{0.78}O_7$ and $Bi_4Ti_2Fe_1O_{11.5}$ compounds

Compound	$Bi_{1.6}Ti_2Cr_{0.16}O_{6.6}$	$Bi_{1.55}Ti_{1.97}Cu_{0.78}O_7$	$Bi_4Ti_2Fe_1O_{11.5}$
Relaxation time (τ), s	$1.59 \cdot 10^{-3}$	$3.98 \cdot 10^{-3}$	$1.59 \cdot 10^{-3}$

The frequency dependence of the conductivity is of the power function type (Figure 2 (a)), $\sigma \sim \omega^s$. Most of the authors interpreted this dependence in terms of the hopping conductivity model in the two-site approximation, independently of the investigated frequency range. It is obvious that with decreasing frequency, when the characteristic the hopping length is comparable to the average distance between the centers of localization, a transition to the hopping of the charge carriers by multi-site clusters (the multiplet conductivity) should be observed. In the theoretical works [2] it was shown that this transition does not lead to the anomalous changes in the behavior of $\sigma(f)$. The only difference in the multiplet conductivity is the increase of the index s with ω.

There are no anomalies in the behavior of Re $\sigma(f)$ and Im $\sigma(f)$ graphs (Fig.2(a)). However, the relations $\varphi(f)$ (Fig.2 (b)) have maxima for all compositions, so the relation $\varphi(f) = \mathrm{Im}\,\sigma / \mathrm{Re}\,\sigma$ seems to be most appropriate for the analysis of the conductivity. The presence of the maximum on the graph of $\varphi(f)$ indicates that the dielectric losses are due to the relaxation nature of hopping transport.

5. Conclusion

The influence of the conductivity on the low frequency dielectric spectra for the compositions $Bi_{1.6}Ti_2Cr_{0.16}O_{6.6}$, $Bi_{1.55}Ti_{1.97}Cu_{0.78}O_7$ and $Bi_4Ti_2Fe_1O_{11.5}$ was observed experimentally. A suggestion on the hopping type mechanisms of conductivity was deduced relying on the frequency dependence analysis. In the frequency range with weak dependence of $\varphi(f)$, hopping carrier occur in clusters of intermediate size (with a small, though greater than two number of sites). In the range where $\varphi(f)$ increases with frequency, the hopping conductivity is realized in the clusters having large though finite size. For the region where $\varphi(f)$ decreases with increasing the frequency, the conductivity is realized by a two-site hopping.

References

[1] Jonscher AK 1996 *Universal relaxation law* (London: Chelsea Dielectrics Press Ltd)
[2] Bryksin VV, Ljakonov MN, Muzdba VM, Hanin SD 1981 *Phys. of Solid State* **23** 1516

IOP Conf. Series: Materials Science and Engineering **49** (2013) 012017 doi:10.1088/1757-899X/49/1/012017

Ethylene vinylacetate copolymer and nanographite composite as chemical vapour sensor

Santa Stepina, Gita Sakale, Maris Knite

Azenes street 14/24 Riga, LV-1007, Latvia

Santa.Stepina@rtu.lv

Abstract. Polymer-nanostructured carbon composite as chemical vapour sensor is described, made by the dissolution method of a non-conductive polymer, ethylene vinylacetate copolymer, mixed with conductive nanographite particles (carbon black). Sensor exhibits relative electrical resistance change in chemical vapours, like ethanol and toluene. Since the sensor is relatively cheap, easy to fabricate, it can be used in air quality monitoring and at industries to control hazardous substance concentration in the air, for example, to protect workers from exposure to chemical spills.

1. Introduction

In nowadays scientific development has increased amount and the variety of used chemicals in manufacturing [1], therefore there is need to investigate and create sensor which can detect different kind of chemicals and warn factory workers about potential leakage of chemicals.

Ethanol and toluene are most popular chemicals. Ethanol is polar solvent and is used in factories to produce large amount of goods, like alcohol, perfume, in disinfectants. Toluene is non-polar solvent used to produce paint, paint thinners, adhesives and nail polishers.

Still these two chemicals are not the only ones. There are a wide range of hazardous chemicals in manufacturing and consequently in the human environment. Although there are several methods described (photo/flame-ionization detectors, mobile gas chromatographs, semiconductor based gas sensors) to detect different chemicals vapours, however there is still no ideal device invented. That is why in the present work we developed a chemical vapour sensor, whose active layer is a thin film of a non-conductive polymer – ethylene vinylacetate copolymer – mixed with conductive nanographite particles, in order to develop new sensor device. With the term "nanographite" can be marked following fillers: extra-conductive highly structured carbon black (EHSCB), carbon nanotubes (CNT), thermally exfoliated graphite (TEG) as well as recently discovered graphenes because all of them have a sp^2-hybridized crystal structure like graphite, but at least one dimension of nanographite is smaller then 100 nm. In this research we used EHSCB as composite filler. Their primary nanoparticles are like polyhedron of mean size 30 nm whose faces are made of graphite platelets. Therefore this kind of carbon black has very high conductivity and is named as "extra-conductive".

2. Materials and Methods

2.1. Materials

To produce composite ethylene vinylacetate copolymer (Sigma Aldrich; figure 1) is used as matrix. It is known that amount of vinylacetate impacts the copolymer crystallization degree and flexibility. In this case we used ethylene vinylacetate copolymer with 40 % vinylacetate content. Copolymer consists of ethylene and vinylacetate repeating units, where ethylene unit is non-polar and vinylacetate – polar. Copolymer complex structure indicates that the sensor could detect both polar and non-polar organic solvent vapours.

Nanographite particles (EHSCB: PRINTEX XE-2) with average particle size 30 nm are used as conductive filler. Particles specific surface: 950 m^2/g and DBP (dibutyl phthalate) adsorption: 380 ml/100 g. PRINTEX XE-2 as mentioned before has high electrical and it has been used in many electrical rubber creation. This type of carbon black has no additional surface treatment as mentioned in [2].

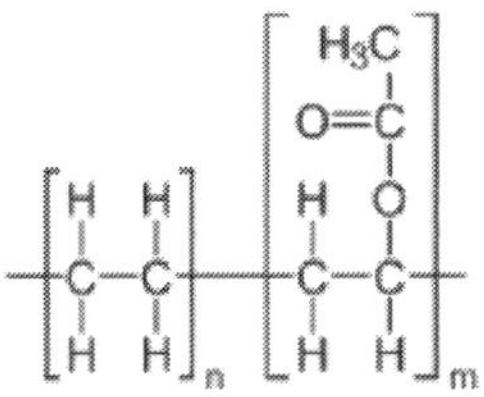

Figure 1. Structure of ethylene vinylacetate copolymer

2.2. Composite production

Sensor production is based on dissolution method, where ethylene vinylacetate copolymer was dissolved in chloroform solution. Then different mass parts of EHSCB were homogenized in chloroform using ultrasonic homogenizer. The obtained carbon black suspension was added to polymer solution and stirred on magnetic stirrer for 2 hours. After that obtained homogenous composite mixture spotted on epoxy laminate plate with cooper electrodes by dip-coating method.

2.2.1. Sensing mechanism

Ethylene vinylacetate-nanostructured carbon composites (EVA-NCC) chemical sensoreffect can be explained on the basis of conductive filler percolation theory in the matrix. In the vicinity of the percolation threshold, the electrical conductivity σ of the composite changes according to:

$$\sigma = \sigma_0 / \Phi - \Phi_C /^t, \qquad (1)$$

where σ_0 is scaling factor, Φ_C is critical concentration or percolation threshold and t is the critical index [3]. In our previous studies [4] we have shown the existence of tunnelling conductivity between two adjacent conductive nanoparticle aggregates. We also showed the best sensitivity of such kind composites in the vicinity of the percolation threshold [4,5]. Absorption of the solvent vapour molecules by polymer induces polymer matrix swelling and the distance between carbon nano-particle aggregates increases. Consequently, the tunnelling currents between carbon aggregates decreases and the electrical resistance of the composite increases.

3. Results and discussions

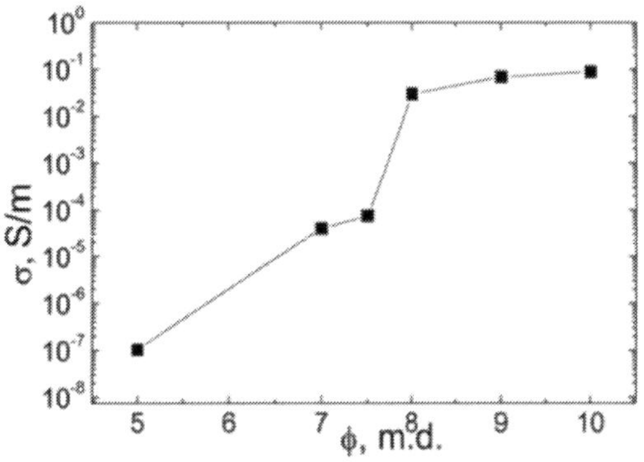

Figure 2. Conductivity of EVA-NCC as function of CB concentration in the composite

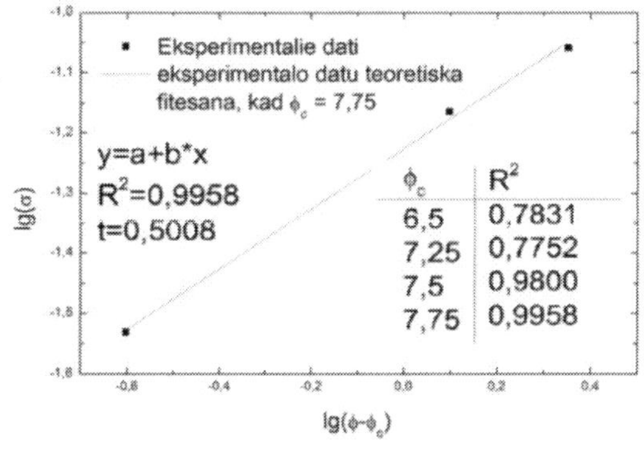

Figure 3. Approximation of EVA-NCC experimental data with percolation scaling law

In order to start testing sensoreffect, there is necessary to determine the percolation threshold of the EVA-NCC, because the percolation threshold transition predicts the most sensitivity of chemicals sensor. Experimentally measured percolation curve is shown on Figure 2. Theoretical fit of experimental data by equation (1) is shown on Figure 3. From approximation (Figure 3) of EVA-NCC experimental data with percolation scaling law one can decide, that critical concentration is 7,75 mass parts of carbon black. This fitting result correlates with ethanol vapour sensing experiments (Figure 4) where sensoreffect of each obtained composite sample was measured.

Figure 5 show how the polarity of VOC affects EVA-NCC VOC sensoreffect. For example, the EVA-NCC with 8 mass parts of EHSCB presents remarkably better sensitivity to nonpolar VOC.

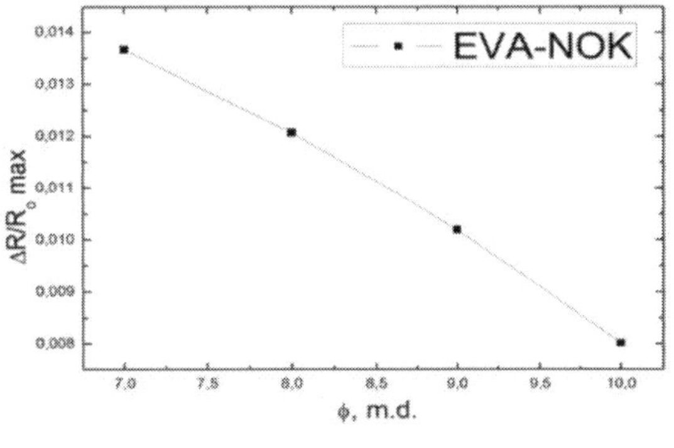

Figure 4. Maximal relative electrical resistance change of EVA-NCC versus carbon black concentrtion in the composite. Samples were exposed to ethanol vapours (9069 ppm) for ... seconds. Thickness of the samples 70 μm.

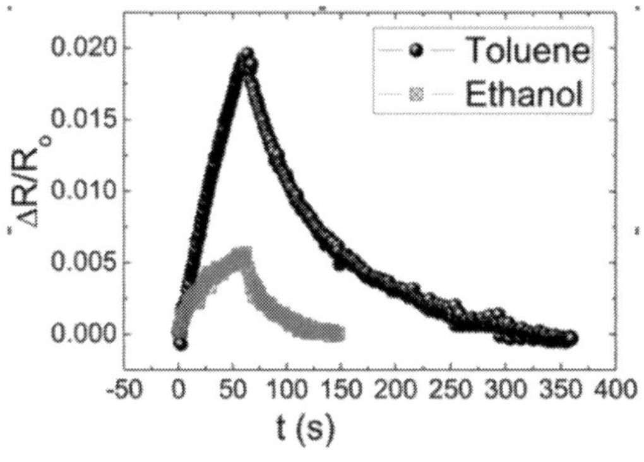

Figure 5. EVA-NCC electrical resistance change versus time in toluene and ethanol vapours. EVA-NCC contains 8 mass parts of carbon black. Vapour concentration was 500 ppm.

Experimental data (Figure 6) shows that EVA-NCC is able to detect VOCs at elevated relative humidity. It was found that at the water-ethanol mixture ratio 70:30 changes the composite response mechanism to vapour: the contribution of proton conductivity of the water layer on the surface of EVA-NCC sample overcomes the contribution of tunnelling currents inside of composite.

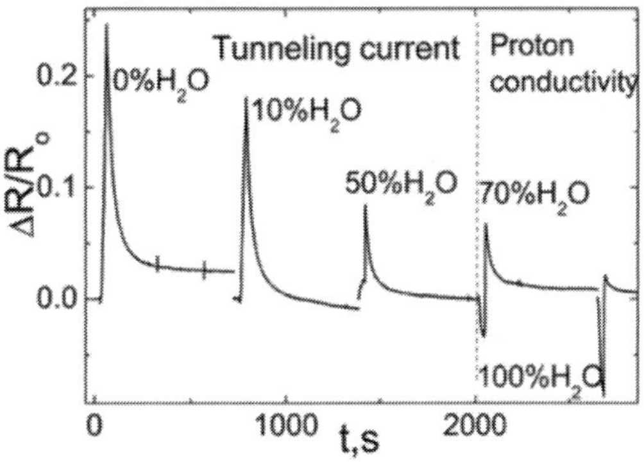

Figure 6. EVA-NCC (8 mass parts) relative electrical resistance change in time, when ethanol ratio in water-ethanol mixture decreases.. Sample thickness 70 µm and samples exposure time in ethanol vapour 90 seconds

4. Conclusions

There are investigated EVA-NCC percolation threshold, volatile organic compounds (VOC) sensoreffect and its influencing factors.

EVA-NCC ethanol vapour sensoreffect has been determined. The highest relative electrical resistance change in vapour is observed at 7 mass parts.

Experimental data shows that EVA-NCC is able to detect VOCs in high relative humidity. It was found that at the water-ethanol mixture of 70:30 ratios the composites electrical response mechanism varies from tunelling conductivity to the proton conductivity.

5. Acknowledgements

This work has been supported by the Latvian National Research Programme IMIS.

References

[1] Manassaram D M, Orr M F, Kaye W E 2003 Hazardous substances events associated with the manufacturing of chemicals and allied products *J. Hazard. Mater.* **104** 123–135

[2] Wichmann H G 2009 *Electrically conductive polymer nanocomposite matrix systems with load and health monitoring capabilities* (Hamburg: Technical University of Hamburg) p 19-22

[3] Stauffer D and Aharony A, Introduction to Percolation theory, 4th ed. (Taylor & Francis, London, 1985).

[4] Sakale G, Knite M, Teteris V, Tupureina V, Stepina S, Liepa E 2010 The investigation of sensing mechanism of ethanol vapour in polymer-nanostructured carbon composite. *Cent. Eur. J. Phys.* **9** 307-312.

[5] Knite M, Zavickis J, Teteris V, Linarts A 2011 Polyisoprene-multi wall carbon nanotube composite structure for flexible pressure sensor application. *J. Nanosci. Nanotechnol.* **11** 1-5

Replacing Chlorine with Hydrogen Chloride as a Possible Reactant for Synthesis of Titanium Carbide Derived Carbon Powders for High-Technology Devices

Indrek Tallo, Thomas Thomberg, Alar Jänes and Enn Lust

Institute of Chemistry, University of Tartu, 14 a Ravila Str., 50411, Tartu, Estonia

E-mail: indrek.tallo@ut.ee

Abstract. Micro- and mesoporous carbide-derived carbons were synthesized from titanium carbide (TiC) powder via gas phase reaction by using different reactants (Cl_2 and HCl) within the temperature range from 700 to 1100 °C. Analysis of XRD results show that TiC-derived carbons (TiC-CDC) consist mainly of graphitic crystallites. The first-order Raman spectra showed the graphite-like absorption peaks at ~1577 cm^{-1} and the disorder-induced peaks at ~1338 cm^{-1}. The energy-related properties of supercapacitors based on 1 M $(C_2H_5)_3CH_3NBF_4$ in acetonitrile and carbide–derived carbons (TiC-CDC (Cl_2) and TiC-CDC (HCl)) as electrode materials were also investigated using cyclic voltammetry, electrochemical impedance spectroscopy, galvanostatic charge/discharge and constant power methods. The Ragone plots for carbide-derived carbons prepared by using different reactants (Cl_2, HCl) are quite similar and at high power loads TiC-CDC (Cl_2) material synthesized at 900 °C, i.e. materials with optimal porous structure, deliver higher power at constant energy.

1. Introduction

Selective etching of carbides is an attractive technique for controlled synthesis of various carbon nanostructures at the atomic level [1-7]. Development of carbide-derived carbon (CDC) with unique micro- and mesoporous structure, narrow pore size distribution and the possibility to fine-tune pore sizes, confirmed previously [3,8], has noticeably forced the development of applications requiring the CDC materials [9-12]. Previously it was found that the micropore formation is strongly influenced by initial carbide density while the formation of mesopores is influenced mainly by the carbide structure and chemical properties, but not raw carbide density [1]. In the present study, extraction of Ti atoms from TiC was performed by high temperature treatment with Cl_2 or HCl gases. The structure of CDC was characterized in detail using X-ray diffraction (XRD), Raman spectroscopy and high-resolution transmission electron microscopy (HRTEM) methods. The energy-related properties of supercapacitors were also investigated using constant power methods.

2. Experimental part and discussion

2.1. Preparation of TiC-CDC materials, XRD, Raman spectroscopy and HRTEM analysis

Titanium carbide TiC (99,5% purity, -325 mesh powder, Sigma-Aldrich), was placed onto a quartz stationary bed reactor and reacted with Cl_2 or HCl (99.99%) at chosen fixed reaction temperatures from 700 to 1100 °C. Details of the etching technique have been reported previously [6].

XRD analysis [6] of the powder samples was carried out to investigate the structural changes in TiC-CDC that occurred with different etching agents at different reaction temperatures. The XRD measurements were performed using CuKα radiation (45 kV, 35 mA, λ= 0.154056 nm) with a step size 0.02° of glancing angle θ and with the holding time of 5 s at fixed θ on Bruker D8 Advance diffractometer. The diffraction spectra were recorded at 25 °C and treated by the AXES 3.0 software. The XRD patterns (Fig. 1) of synthesized CDCs showed reflections corresponding to the graphite 002 and 100/101 planes at $2\theta\sim$ 26° and ~43°, respectively. The 002 diffraction peak at 2θ ~26° corresponding to parallel graphene layers can be observed in Fig. 1. The 100/101 diffraction peaks at 2θ~43° characterize the 2D in-plane symmetry along the graphene layers. As the reaction temperature of TiC increased from 700 to 1100 °C, the 002 peak becomes narrower, indicating the ordering processes in the mainly amorphous carbon. Comparison of the XRD patterns for TiC-CDC (Cl_2) and TiC-CDC (HCl) materials shows that TiC-CDC (HCl) carbons are more ordered characterized by 002 and 100/101 more intense peaks. However, TiC-CDC (HCl) materials synthesized at lower temperatures (T < 900 °C) contain some traces of initial titanium carbide.

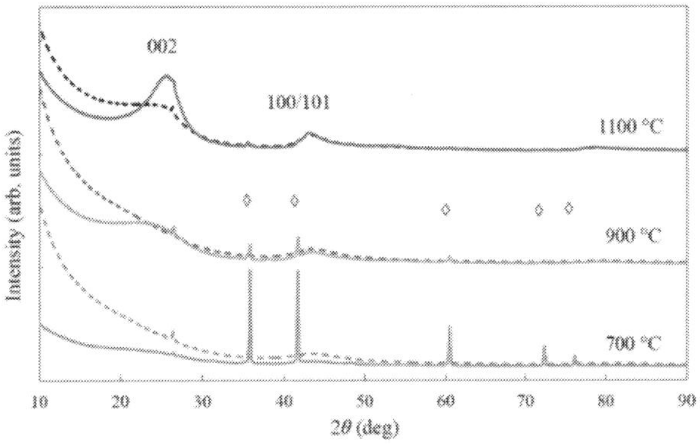

Fig. 1. Characteristic XRD patterns for porous TiC-CDC prepared by using different reactants (dashed line – Cl_2, solid line – HCl) at different reaction temperatures. Symbol ◊ denotes crystallographic orientations of residual TiC.

The Raman spectra were recorded using a Renishaw inVia micro-Raman spectrometer using Ar laser excitation (λ_L= 514 nm). The first-order Raman spectra of perfectly ordered monocrystalline graphite, shows one peak at ~1577 cm^{-1}, whereas disordered amorphous carbon generally demonstrates two peaks [6]: the so-called graphite (G) peak at ~1577 cm^{-1} and the disorder-induced (D) peak at ~1338 cm^{-1} (Fig. 2). The G-peak corresponds to graphite in-plane bond-stretching motion of pairs of C atoms in sp^2 configuration with E_{2g} symmetry. Thus, this mode does not require the presence of sixfold C rings, and it occurs at all sp^2 sites, not only for those atoms located in hexagonal structure. The D-peak is a breathing mode with A_{1g} symmetry which is forbidden in perfect graphite and only becomes active in the presence of disorder in graphite structure. The spectrum also shows the second-order peak of D-band (2D) at ~2670 cm^{-1}. The increase of 2D peak is related with the crystallographic ordering of the graphitic structure [1]. According to the data in Figs. 2 very small narrowing of the G-band takes place indicating the beginning of graphitization of porous CDCs.

For the morphological studies, the TiC-CDC were examined using HRTEM on a Tecnai 12 instrument operated at the 120 kV accelerating voltage. Reaction with Cl_2 or HCl at 700 °C results in a formation of completely amorphous carbon (Fig. 2). Ordering of carbon and formation of graphitic structures starts at ~900 °C. However, only at 900 °C and above, both Raman spectroscopy and XRD methods detected noticeably increased ordering of mainly amorphous carbon. A HRTEM micrograph of a sample processed at 1100 °C shows a network of well-ordered graphitic sheets within amorphous carbon.

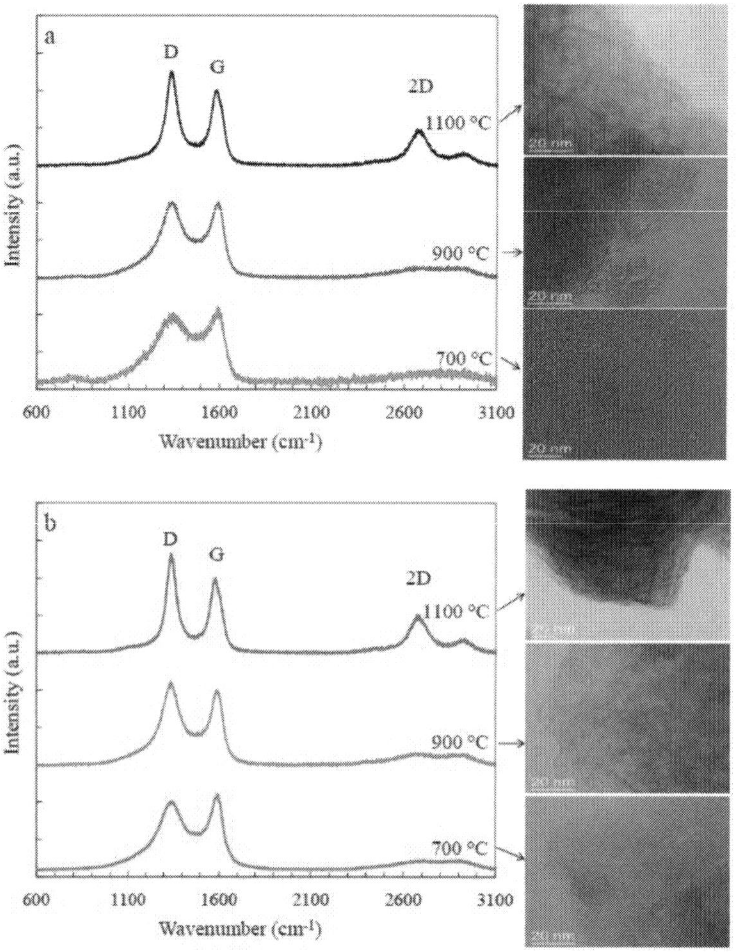

Fig. 2. Raman analysis and HRTEM images for porous TiC-CDC prepared by using different reactants (a – Cl_2 and b – HCl) at different reaction temperatures.

2.2. Specific energy and power, Ragone plots

The maximum specific energy (E_{max} (W h kg^{-1}) and power P_{max} (kW kg^{-1}) for supercapacitors studied can be estimated using Eq. (1):

$$E_{max} = \frac{C\Delta U^2}{2m} \qquad P_{max} = \frac{\Delta U^2}{4\text{ESR}\, m} \tag{1}$$

where C is the capacitance of the cell in F cm^{-2}, ESR is equivalent series resistance in Ω cm^2, obtained from Nyquist plots at $f \rightarrow \infty$, and m is the total active material weight of two electrodes (g cm^{-2}) [6]. The specific energy, calculated at cell potential U = 3.0 V, are maximal for TiC-CDC (Cl_2 800 °C) and TiC-CDC (HCl 900 °C), 43.1 W h kg^{-1} and 31.1 W h kg^{-1}, respectively. The specific power, calculated at cell potential U = 3.0 V, are maximal for TiC-CDC (Cl_2 1000 °C) and TiC-CDC (HCl 1000 °C), 805.2 kW kg^{-1} and 847.5 kW kg^{-1}, respectively.

The Ragone plots (calculated to the total active material weight of two electrodes) of the supercapacitors based on TiC-CDC (Cl_2) and TiC-CDC (HCl) electrodes have been obtained from constant power tests within the cell potential range from 3.0 V to 1.5 V and are shown in Fig. 3. The Ragone plots for carbide-derived carbons prepared by using different reactants (Cl_2, HCl) are quite similar and at high power loads TiC-CDC (Cl_2) material synthesized at 900 °C, i.e. materials with optimal mico- and mesopore fraction, deliver higher power at constant energy [6].

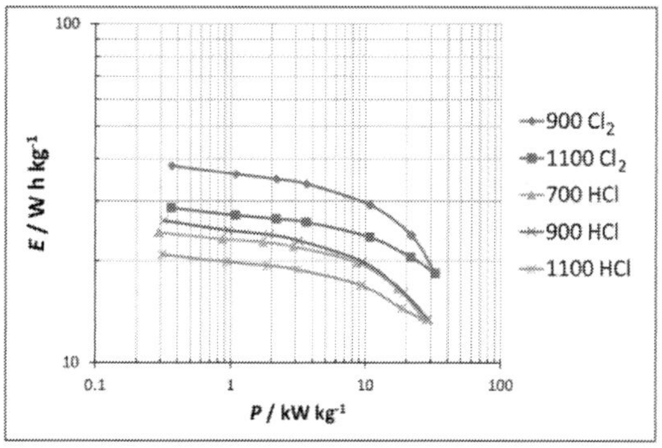

Fig. 3. Ragone plots for the supercapacitors based on 1 M $(C_2H_5)_3CH_3NBF_4$ in acetonitrile and completed using TiC–CDC electrodes, prepared at different synthesis temperatures by using different reactants.

3. Conclusions

Micro- and mesoporous carbide-derived carbons were synthesized from titanium carbide (TiC) powder via gas phase reaction by using different reactants within the temperature range from 700 to 1100 °C. Analysis of XRD results show that TiC-derived carbons (TiC-CDC) consist mainly of graphitic crystallites. The first-order Raman spectra showed the graphite-like absorption peaks at ~1577 cm^{-1} and the disorder-induced peaks at ~1338 cm^{-1}. The Ragone plots (calculated to the total active material weight of two electrodes), of the supercapacitors based on TiC-CDC (Cl_2) and TiC-CDC (HCl) electrodes have been obtained from constant power tests within the potential range from 3.0 V to 1.5 V are quite similar. However, at higher power loads TiC-CDC (Cl_2) material synthesized at 900 °C, i.e. materials with optimal pore structure, deliver higher power at constant energy.

Acknowledgements

This work was supported in part by the Estonian Ministry of Education and Research (project SF0180002s08) by graduate school „Functional materials and technologies" receiving funding from the European Social Fund under project 1.2.0401.09-0079 in Estonia and by the Estonian Science Foundation under Project No. 8172.

References

[1] Yushin G, Nikitin A and Gogotsi Y (2006) *Nanomaterials Handbook* (CRC Press, Taylor & Francis Group, Boca Raton) pp 239-282
[2] Jänes A, Permann L, Arulepp M and Lust E 2004 *Electrochem. Commun.* **6** 313
[3] Jänes A, Kurig H and Lust E 2007 *Carbon* **45** 1226
[4] Thomberg T, Kurig H, Jänes A and Lust E 2011 *Micropor. Mesopor. Mater.* **141** 88
[5] Tallo I, Thomberg T, Kontturi K, Jänes A and Lust E 2011 *Carbon* **49** 4427
[6] Tallo I, Thomberg T, Kurig H, Kontturi K, Jänes A and Lust E 2013 *J. Solid State Electrochem.* **17** 19
[7] Tallo I, Thomberg T, Jänes A and Lust E 2012 *J. Electrochem. Soc.* **159** A208
[8] Jänes A, Thomberg T, Kurig H and Lust E 2009 Carbon **47** 23
[9] Thomberg T, Jänes A and Lust E *Electrochim. Acta* 2010 **55** 3138
[10] Jänes A, Permann L, Nigu P and Lust E 2004 *Surf.Sci.* **560** 145
[11] Jänes A, Permann L, Arulepp M and Lust E 2004 *J. Electroanal. Chem.* **569** 257
[12] Torop J, Sugino T, Asaka K, Jänes A, Lust E and Aabloo A 2012 *Sensors and Actuators B: Chem.* **161** 629

IOP Conf. Series: Materials Science and Engineering **49** (2013) 012019 doi:10.1088/1757-899X/49/1/012019

Atomic Layer Deposition of Zirconium Oxide on Carbon Nanoparticles

A Tamm[1], A-L Peikolainen[2], J Kozlova[1], L Aarik[1], K Roosalu[1], I Kärkkänen[3], Mändar[1], J Aarik[1], K Kukli[1,4]
[1]University of Tartu, Institute of Physics, Department of Materials Science, Riia 142, EE-51014 Tartu, Estonia,
[2]IMS Lab, Institute of Technology, University of Tartu, Nooruse 1, EE-50411 Tartu, Estonia,
[3] Peter Grünberg Institut, Forschungszentrum Jülich, Leo-Brandt Strasse, 52425 Jülich, Germany,
[4]University of Helsinki, Department of Chemistry, P.O.Box 55, FI-00014, Univ. Helsinki, Finland

E-mail: aile.tamm@ut.ee

Abstract. In this report we describe preparation of structures containing carbon nanoparticles for potential applications in nonvolatile memories. The carbon nanoparticles were synthesized from 5-methylresorcinol and formaldehyde via base catalysed polycondensation reaction, and were distributed over substrates by dip-coating the substrates into an organic solution. Before deposition of nanoparticles the substrates were covered with 2 nm thick Al_2O_3 layer grown by atomic layer deposition (ALD) from $Al(CH_3)_3$ and O_3. After deposition of nanoparticles the samples were coated with ZrO_2 films grown from $C_5H_5Zr[N(CH_3)_2]_3$ and H_2O. Both dielectrics were grown in two-temperature ALD processes starting deposition of Al_2O_3 at 25 °C and ZrO_2 at 200 °C, thereafter completing both processes at a substrate temperature of 300 °C. Deposition of ZrO_2 changed the structure of C-nanoparticles, which still remained in a $Si/Al_2O_3/C/ZrO_2$ structure as a separate layer. Electrical characterization of nanostructures containing Al_2O_3 as tunnel oxide, C-nanoparticles as charge traps and ZrO_2 as control oxide showed hysteretic flat-band voltage shift of about 1V.

1.Introduction

Nonvolatile semiconductor memory based on storing electric charge in non-contacted floating gates is the dominant memory concept for portable devices [1] that assumes design of memories containing conductive nanodots in (multilayer) gate dielectrics. Applications discussed include the use of ZrO_2 with nanocrystal-embedded non-volatile flash memory devices [2].

In this work we describe the structures with carbon nanoparticles as possible charge traps between ZrO_2 as the control oxide and Al_2O_3 as the tunnel oxide layer grown by atomic layer deposition (ALD). The C-nanoparticles were embedded in $Si/Al_2O_3/C/ZrO_2$ structures as a separate two-dimensional layer. The idea about the inclusion of the C-nanoparticles into nanostructure was based on the knowledge that carbon is an easily accessible and inexpensive material with high work function

(4.6 eV) as well as on the expectation that C-nanoparticles may act like electron-trapping nanoparticles possessing metallic conductivity.

2. Experimental

The films were deposited in an in-house built atomic layer deposition (ALD) reactor [3] on planar SiO_2/Si and TiN_x substrates. Before deposition of C-nanoparticles all substrates were covered with 2 nm thick Al_2O_3 layer grown from $Al(CH_3)_3$ (98% Strem Chemicals) and O_3 in two-temperature ALD process starting deposition at 25°C (5 cycles) and completing at 300°C (20 cycles). The C-nanoparticles were covered with ZrO_2 films grown from $C_5H_5Zr[N(CH_3)_2]_3$ (Air Liquide, Inc.) and H_2O in two temperature ALD process applying first two cycles at 200°C and then 60 cycles at 300°C. Cycle times for $C_5H_5Zr[N(CH_3)_2]_3$ and H_2O process were 2-2-1-5 s. For $Al(CH_3)_3$ and O_3 process cycle times of 2-2-2-5 s were applied. The film thicknesses were around 2 nm for Al_2O_3 and 7 nm for ZrO_2. The carbon nanoparticles, deposited by dip-coating in organic solution, were synthesized from 5-methylresorcinol (99.9%, Carboshale Ltd., Estonia) and formaldehyde (37% solution in water, stabilized with 10-15% of methanol, Sigma-Aldrich) via base (Na_2CO_3, 99.8%, Sigma-Aldrich) catalysed synthesis in aqueous media [4] polycondensation reaction. The size of C-nanoparticles measured by small angle X-ray scattering (SAXS) was approximately 50 nm in the solution [5]. After dip-coating and following heating in the reactor up to 300°C the size of C-nanoparticles was 25-30 nm [5]. The film thicknesses were measured using a spectroscopic ellipsometer (SE) GES5-E (SEMILAB, Sopra). The phase composition of the $Si/Al_2O_3/C/ZrO_2$ structure was evaluated by grazing incidence (GIXRD) X-ray diffraction analysis. The phase composition of C-nanoparticles before and after deposition of ZrO_2 was studied by Raman spectroscopy using a Renishaw inVia micro-Raman spectrometer. The (surface) morphology was investigated by scanning electron microscopy (SEM) using a Dual Beam® equipment Helios NanoLab 600 (FEI Company). The focused ion beam (FIB) of the equipment was used for the preparation of samples for cross-section studies of the $Si/Al_2O_3/C/ZrO_2$ structures. Electrical characteristics were measured on $TiN_x/Al_2O_3/C/ZrO_2/Ti/Al$ capacitors at room temperature using Hewlett Packard 4282A Precision LCR meter with a frequency range of 100 kHz to 1 MHz, whereby the capacitance-voltage (C-V) scan was made using voltage step of 0.1V and maximum bias voltage of 2 or 3V. The Al/Ti top electrodes of the capacitor structures with an area of 0.204 mm^2 each were prepared by e-beam evaporation of the metals through a shadow mask.

3. Results and discussion

The SEM analysis was used to explore the morphology of $Si/Al_2O_3/C$ substrates covered by C-nanoparticles before (not shown) and after the ALD of ZrO_2 (Figure 1, left panel). The adhesion and coverage of C-nanoparticles was quite uniform all over the substrate. The mean diameter of the particles was ~45-60 nm. In order to evaluate the conformality, cross-section images of the $Si/Al_2O_3/C/ZrO_2$ structure were taken (Figure 1, right panel). C-nanoparticles were uniformly covered with ZrO_2 films and remained in the structure as a separate layer between Al_2O_3 and ZrO_2.

Figure 2 represents the grazing incidence (GIXRD) X-ray diffraction patterns of $Si/Al_2O_3/C/ZrO_2$ and $Si/Al_2O_3/ZrO_2$ nanostructures, i.e. the structures with and without C-nanoparticles, respectively. The ZrO_2 films grown to thicknesses of 7 nm otherwise consisted essentially of tetragonal (or cubic) phase. The film deposited on a sample without the C-nanoparticles between Al_2O_3 and ZrO_2 layers showed preferential (100) orientation while that deposited on C-nanoparticles did not (Figure 2). This difference is not surprising because complex surface shape of the sample with nanoparticles, differently from that of a planar surface of the $Si/Al_2O_3/ZrO_2$ structure, makes available all possible orientations of the thin film deposited on this kind of surface.

IOP Conf. Series: Materials Science and Engineering **49** (2013) 012019 doi:10.1088/1757-899X/49/1/012019

Figure 1. Representative bird-eye view (left panel) and SEM –FIB cross-section image (right panel) of the $Si/Al_2O_3/C/ZrO_2$ structure. C-nanoparticles were layered on 2 nm thick Al_2O_3 pre-deposited on Si substrate, and then covered with 7 nm thick ZrO_2.

Raman spectra taken from C-nanoparticles before and after deposition of zirconium oxide in $Si/Al_2O_3/C/ZrO_2$ nanostructures (Figure 3) demonstrated shifts in the peak positions after deposition of ZrO_2. Before deposition of ZrO_2 the characteristic peak apparent at 1590 cm^{-1} indicated the formation of extremely small crystals of carbon-based materials which may be referred to as a monocrystalline graphite first described by Tunistra and Koenig [6] in 1970. In the $Si/Al_2O_3/C/ZrO_2$ structure the peak was much weaker and broader, and additional peaks at 1540 and 1640 cm^{-1} appeared. A reason for these additional peaks is the activation of otherwise forbidden vibration modes by structural disorder [7,8]. Although the structure of C-nanoparticles changed during deposition of ZrO_2 the particles still formed of a separate layer in the structure similarly to that observed in our previous work [5].

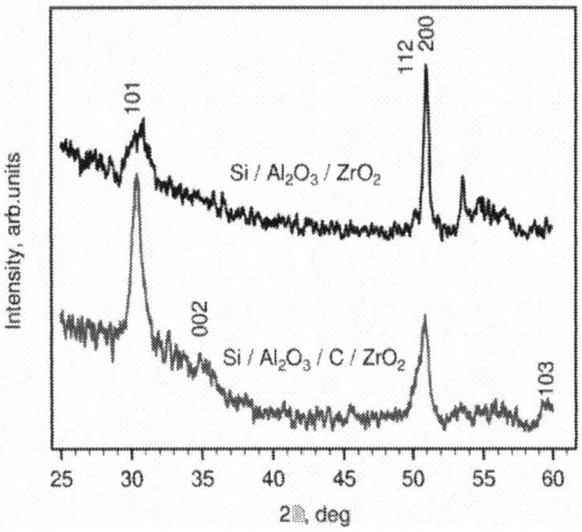

Figure 2. GIXRD patterns of zirconium oxides in $Si/Al_2O_3/C/ZrO_2$ and $Si/Al_2O_3/ZrO_2$ nanostructures. Miller indices of tetragonal phase are shown at the positions of corresponding reflections.

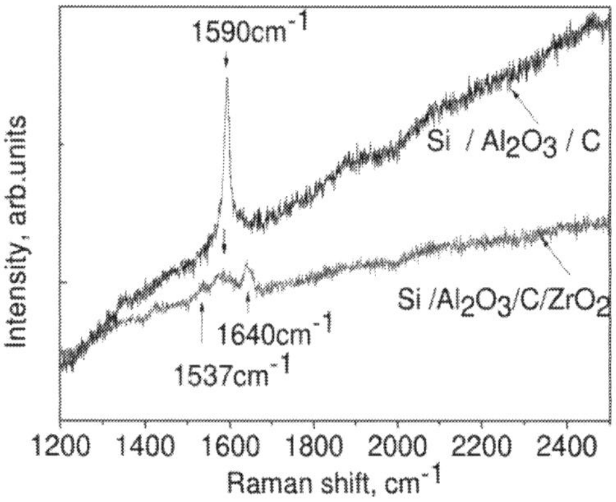

Figure 3. Raman spectra obtained from C-nanoparticles in $Si/Al_2O_3/C$ and $Si/Al_2O_3/C/ZrO_2$ nanostructures. Only the region implying characteristic wavenumbers of carbon are shown.

Figure 4 shows the C–V curves of the $TiN/Al_2O_3/C/ZrO_2$ capacitors with the measurement frequency as the parameter. Capacitance density at 500 kHz of the structure at accumulation was 1.72 nF/mm^2 and hysteresis at 500 kHz was approximately 1V (Figure 4, inset). For comparison, capacitor structures with a single ZrO_2 film grown from $C_5H_5Zr[N(CH_3)_2]_3$ and O_3 did not show that wide hysteresis at the same frequencies [8]. The flat-band shift in nanostructures with C-nanoparticles between tunnel and control oxides refers to certain memory effect.

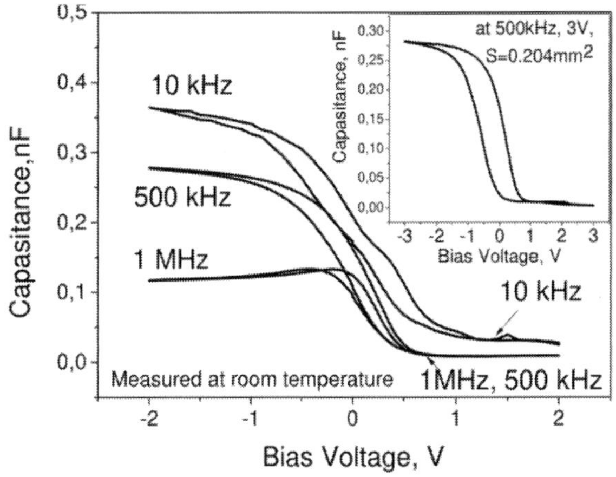

Figure 4. Capacitance-voltage characteristics of the $TiN_x/Al_2O_3/C/ZrO_2$ capacitors, the measurement frequencies are indicated by labels. Flat-band shift measured at 500 kHz are depicted by inset.

4.Summary

$Si/Al_2O_3/C/ZrO_2$ nanostructures consisting of Al_2O_3 adhesion layer, C-nanoparticles and ZrO_2 top layer were prepared. Upon deposition of ZrO_2 from $C_5H_5Zr[N(CH_3)_2]_3$ and H_2O at 300°C, the structure of C-nanoparticles was changed, but the carbon particles remained in a $Si/Al_2O_3/C/ZrO_2$ structure as a separate layer. C-nanoparticles were covered by ZrO_2 with good conformity and uniformity. The preferred growth phase of the ZrO_2 film was tetragonal (or cubic) in $Si/Al_2O_3/C/ZrO_2$ nanostructures. Electrical characterization of $TiNx/Al_2O_3/C/ZrO_2$ nanostructures was conducted, revealing flat-band voltage shift, i.e. hysteresis in C-V curves reaching 1V.

Acknowledgments

The authors are thankful to Prof. V. Sammelselg and Dr. I. Sildos for the access to SEM equipment and Raman spectrometer, respectively, and to Dr. J.Niinistö from the University of Helsinki, and Air Liquide, Inc. for providing the $C_5H_5Zr[N(CH_3)_2]_3$ precursor. The study was partially supported by Estonian Research Council Research targeted projects PUT170 and Estonian Centre of Excellence in Research Project TK117.

References

[1] Kuesters K H, Beug M F, Schroeder U, Nagel N, Bewersdorff U, Dallmann G, Jakschik S, Knoefler R, Kudelka S, Ludwig C, Manger D, Mueller W and Tilke A 2009 *Adv. Eng. Mater.* **11** 241
[2] Panda D, Tseng T-Y 2013 *Thin Solid Films* **531** 1-20
[3] Aarik J, Aidla A, Uustare T, Sammelselg V 1995 *J. Cryst. Growth* **148** 268
[4] Pérez-Caballero F, Peikolainen A-L, Uibu M, Kuusik R, Volobujeva O, Koel M 2008 *Micropor. Mesopor. Mater.* **108** 230-236
[5] Tamm A, Peikolainen A-L, Kozlova J, Mändar H, Aidla A, Rammula R, Aarik L, Roosalu K, Lu J, Hultman L, Koel M, Kukli K, Aarik J 2013 *Thin Solid Films,* **538** 16-20
[6] Tunistra F and Koenig J L 1970 *J. Chem. Phys.* **53** 1126-1130
[7] Wilson J I B,Walton J S and Beamson G 2001 *J. of Elect. Spectros. and Relat. Phenomena* **121** 183–201
[8] Kukli K, Niinistö J, Tamm A, Lu J, Ritala M, Leskelä M 2009 *J. Vac. Sci. Technol. B* **27** 226-229.

IOP Conf. Series: Materials Science and Engineering **49** (2013) 012020 doi:10.1088/1757-899X/49/1/012020

Carbon materials for supercapacitor application by hydrothermal carbonization of D-glucose

T. Tooming, T. Thomberg, T. Romann, R. Palm, A. Jänes and E. Lust

Institute of Chemistry, University of Tartu, 14 a Ravila Str., 50411, Tartu, Estonia

E-mail: thomas.thomberg@ut.ee

Abstract. Hydrothermal carbonization (HTC) of 1 M D-glucose solution was carried out in a stainless steel autoclave, which was introduced into a tube furnace and maintained at the temperature 260 °C for 24 hours for HTC reaction. Obtained carbonaceous materials were pyrolyzed and activated using carbon dioxide to fine-tune the porosity of HTC carbons. Raman spectroscopy and high resolution transmission electron microscopy data revealed that synthesised carbon materials were mainly amorphous. Scanning electron microscopy studies demonstrated that carbon materials were consisted of micrometer scale mainly spherically shaped particles maintaining their shape and structure throughout pyrolysis and activation processes. Based on the low-temperature N_2 sorption experiments the specific surface areas up to 1540 $m^2\ g^{-1}$ were measured for differently activated HTC. Various electrochemical methods were used to study influence of carbon materials physical characteristics on the supercapacitor electrochemical performance based on HTC carbon electrodes in 1 M $(C_2H_5)_3CH_3NBF_4$ solution in acetonitrile.

1. Introduction

Decrease of fossil fuels resources and impact of combustion of fuels to the environment has been forced materials chemists to develop new technologies and materials for novel energy storage and conversion devices. Recently, some attention has been paid how to produce functional carbonaceous materials by hydrothermal carbonization (HTC) method and various carbonaceous materials with different particle size, shape and functionality have been synthesised [1-3]. However, these carbonaceous materials obtained, possess only a small number of micropores and therefore have limited micro/mesoporosity and micropore surface area. The simplest possibility to increase the surface area and tune the pore size of carbonaceous materials is pyrolysis, where organic molecules are removed and porosity is generated. The pore size and the surface area of the carbonized raw powder can also be increased by using additional activation methods based on the application of air, carbon dioxide, water steam, KOH, NaOH, H_3PO_4 etc. as an activation agents [4].

In the present work, activated carbons (ACs) were synthesised from D-glucose by using HTC method followed by pyrolysis and activation with carbon dioxide at 900 °C with different activation step lengths to prepare carbon materials with variable specific surface area, pore size distribution and optimal ratio of micro- and mesopores for supercapacitor application [5-16]. Synthesised carbon materials were noted as GDAC (glucose derived activated carbon). As a result of different activation step lengths the burn-off values of 0%, 18%, 24%, 32%, 41% and 51% have been calculated and materials were noted as GDAC-0, GDAC-1, GDAC-2, GDAC-3, GDAC-4, GDAC-5, respectively.

The electrodes were composed of an aluminium current collector and of the GDAC material layer. The GDAC material layer used was prepared from HTC carbon (94 wt%) and from the mixture of binder (6 wt%) (PTFE, 60% dispersion in H_2O). The two–electrode standard Al test cell with two identical electrodes (~2.0 cm^2) was completed inside the glove box (O_2 and H_2O concentrations lower than 0.1 ppm) and all electrochemical experiments were carried out at temperature T = 20 °C. Carefully vacuum dried 25 μm thick TF4425 (Nippon Kodoshi) separator sheet was used for mechanical separation of the working electrodes from each other. Various electrochemical methods (CV, EIS, constant power discharge) were used to study the influence of carbon materials physical characteristics on the supercapacitor electrochemical performance, i.e. on the region of ideal polarizability, high and low frequency series resistance and capacitance, characteristic relaxation time constant, etc.. 1 M $(C_2H_5)_3CH_3NBF_4$ solution in acetonitrile (AN) has been selected as an electrolyte for GDAC based supercapacitors.

2. Experimental part and discussion

2.1. Hydrothermal carbonization, pyrolysis and activation

The hydrothermal carbonization of 50 ml 1 M D-(+)-glucose solution in H_2O was carried out in a 60 ml stainless steel autoclave. Autoclave was placed into a tube furnace, which was heated up at the rate of 10 °C min^{-1} and was maintained at the temperature 260 °C for 24 hours for hydrothermal reaction. Thereafter, autoclave was cooled down to room temperature and the carbonaceous material was collected and washed for several times with ultra-pure (Milli-Q$^+$) water. Finally, resulting carbonaceous material was dried in a vacuum oven (80 °C, 100 mbar) overnight.

The dried raw HTC carbonaceous material was pyrolysed under Ar flow in a tube furnace. After the carbonization step additional activation of the samples were carried out with carbon dioxide at 900 °C with different activation step lengths. During heating up and cooling down steps the reactor was flushed with argon.

2.2. Physical characterization

The Raman spectra were recorded using a Renishaw inVia micro-Raman spectrometer using Ar laser excitation (λ = 514 nm) and the fitting of the D- and G-bands were performed by the fitting of the two Lorentzian curves to the G- and D-band, respectively.

The porous structure of GDAC was characterized by using the low-temperature N_2 adsorption/desorption experiments at the boiling temperature of liquid nitrogen (−195.8 °C) using the ASAP 2020 system. The specific surface area, micropore surface area (S_{micro}), micropore volume (V_{micro}) and total pore volume (V_{tot}) and other parameters for porous carbon materials were calculated according to the Brunauer-Emmett-Teller and *t-plot* method. The pore size distribution was determined using non local density functional theory assuming a slit-shaped pores model [7-9,13,16].

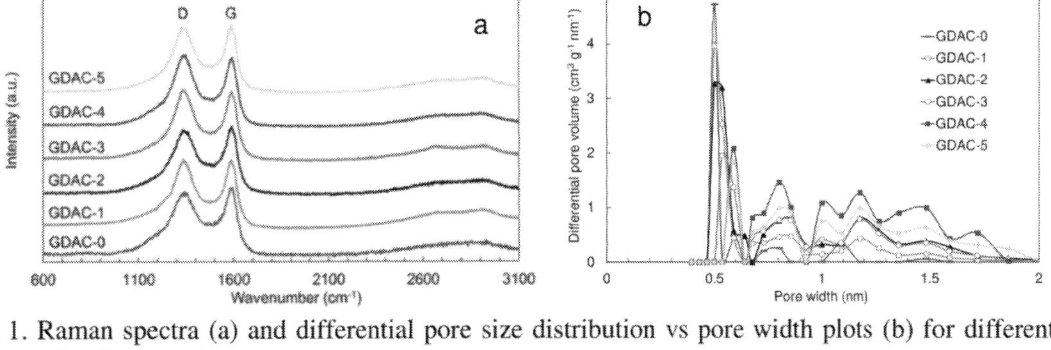

Fig. 1. Raman spectra (a) and differential pore size distribution vs pore width plots (b) for different porous GDAC materials.

2.3. SEM and HRTEM data for GDAC materials

The surface structure and morphology of HTC carbons was examined by scanning electron microscopy (SEM) method using a FIB–SEM Helios™ Nanolab 600 system and high resolution transmission electron microscopy (HRTEM) using a Tecnai 12 instrument operated at a 120 kV accelerating voltage, respectively.

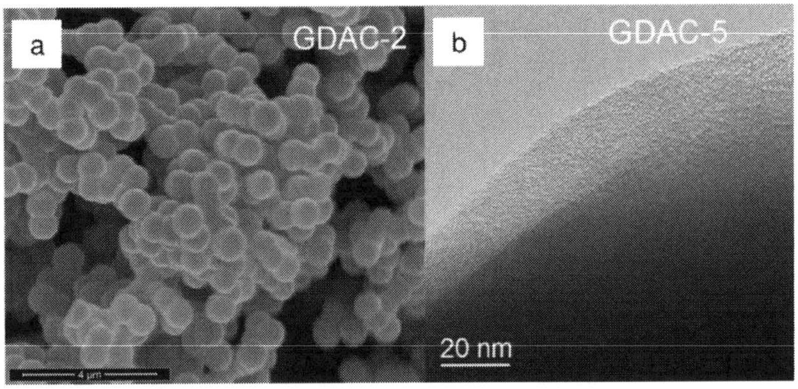

Fig. 2. SEM (a) and HRTEM (b) images for porous GDAC materials.

As shown in Fig. 2a, the carbon produced from 1 M D-(+)-glucose solution exhibited interconnected spheres with a narrow size of about 1.0 μm and relatively smooth surface structure. After activation, the spheres size reduced only slightly as a result of burn-off of carbon during activation step. Fig. 2b, exhibit the HRTEM image for porous GDAC-5 prepared during activation, show small segments of curved graphene monolayers with no evidence of interlayer connection corresponding to mainly amorphous carbon. From the HRTEM image, it is clear, that there is no obvious change in GDAC structure and activation has practically no influence on the crystal structure, which are in a good agreement with powder Raman spectroscopy results (Fig. 1a).

2.4. Specific energy and power, Ragone plots

The specific energy (*E*) and power (*P*) relationship, i.e. Ragone plots (calculated to the total active material weight of two electrodes), for the supercapacitors based on GDAC electrodes have been obtained from constant power tests within the cell potential range from 3.0 V to 1.5 V and are shown in Fig. 3.

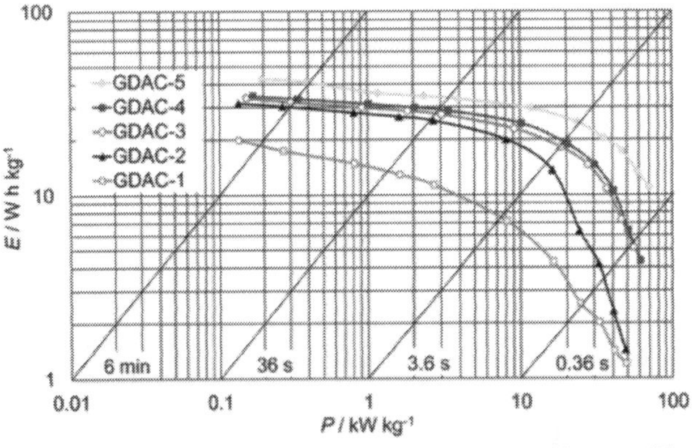

Fig. 3. Ragone plots for the supercapacitors completed using different GDAC electrodes in 1 M $(C_2H_5)_3CH_3NBF_4$ + AN solution, obtained from constant power tests within the cell potential range from 3.0 V to 1.5 V.

The Ragone plots for GDAC materials, having different porosity, micropore and total pore volume, pore size distribution, etc., show noticeable influence of activation condition on the stored energy and delivered power values. Stored energy values for supercapacitors were increased with the increase of the average pore width, because all pores are accessible for partially desolvated ion adsorption, and can be applied for effective energy storage. GDAC materials with higher porosity and larger average pore width (Fig. 1b) ensure fast ions transport in pores and deliver higher power at constant energy. These energy and power values are even higher compared with values established for supercapacitors based on WC–CDC (tungsten carbide derived carbon) electrodes in the same electrolyte system [16], thus, showing excellent characteristics for electrode materials applicable for ultra-high rate supercapacitor devices.

3. Conclusions

Comparison of the data for the cells completed using different GDAC electrodes indicated a strong influence of the electrode material activation conditions (i.e. porosity characteristics) on the relaxation time constant τ_R values decrease with increasing of the average pore size and decrease of the pore resistance of electrolyte in micropores. The Ragone plots for GDAC material with different micropore and total pore volume, pore size distribution etc. showed noticeable dependence of the stored energy and delivered power values on the conditions of the activation step. Stored energy and delivered power values for supercapacitors based GDAC materials were increased with increase of average pore width, because all pores are accessible for partially desolvated ion adsorption, and can be applied for effective energy storage. GDAC materials with higher porosity and larger average pore width ensure fast ions transport in pores and deliver higher power at constant energy. Thus, the energy and power values established for supercapacitors based on GDAC electrodes in 1 M $(C_2H_5)_3CH_3NBF_4$ in acetonitrile solution show excellent characteristics applicable for ultra-high power devices.

Acknowledgements

This work was supported by the Estonian Science Foundation under Projects Nos. 8172 and 9184, Estonian Ministry of Education and Research project SF0180002s08, European Regional Development Fund Project SLOKT10209T and Estonian Centre of Excellence in Research Project TK117 3.2.0101-0030 "High-technology Materials for Sustainable Development".

References

[1] Titirici M M, White R J, Falco C and Sevilla M 2012 *Energy Environ. Sci.* **5** 6796
[2] Sevilla M and Fuertes A B 2009 *Chem. Eur. J.* **15** 4195
[3] Li M, Li W and Liu S 2011 *Carbohydate Research* **346** 999
[4] Marsh H and Rodriguez-Reinoso F 2006 *Activated Carbon* (Elsevier, Amsterdam)
[5] Thomberg T, Jänes A and Lust E 2009 *J. Electroanal. Chem.* **630** 55
[6] Thomberg T, Jänes A and Lust E 2010 *Electrochim. Acta* **55** 3138
[7] Thomberg T, Kurig H, Jänes A and Lust E 2011 *Micropor. Mesopor. Mater.* **141** 88
[8] Tallo I, Thomberg T, Kontturi K, Jänes A and Lust E 2011 *Carbon* **49** 4427
[9] Tallo I, Thomberg T, Kurig H, Jänes A, Kontturi K and Lust E 2013 *J. Solid State Electrochem.* **17** 19
[10] Eskusson J, Jänes A, Kikas A, Matisen L and Lust E 2011 *J Power Sources* **196** 4109
[11] Laheäär A, Jänes A and Lust E 2011 *Electrochim. Acta* **56** 9048
[12] Laheäär A, Jänes A and Lust E 2012 *Electrochim. Acta* **82** 309
[13] Jänes A, Thomberg T, Kurig H and Lust E 2009 *Carbon* **47** 23
[14] Jänes A, Permann L, Arulepp M and Lust E 2004 *J. Electroanal. Chem.* **569** 257
[15] Tõnurist K, Jänes A, Thomberg T, Kurig H and Lust E 2009 *J. Electrochem. Soc.* **156** A334
[16] Tallo I, Thomberg T, Jänes A, Kontturi K and Lust E 2012 *J. Electrochem. Soc.* **159** A208

IOP Conf. Series: Materials Science and Engineering **49** (2013) 012021 doi:10.1088/1757-899X/49/1/012021

Neutron scattering study of structural and magnetic size effects in NiO

A M Balagurov[1], I A Bobrikov[1], J Grabis[2], D Jakovlevs[3], A Kuzmin[4], M Maiorov[5], N Mironova-Ulmane[4]

[1] Frank Laboratory of Neutron Physics, Joint Institute for Nuclear Research, RU-141980 Dubna, Russian Federation

[2] Institute of Inorganic Chemistry, Riga Technical University, Miera street 34, LV-2169 Salaspils, Latvia

[3] Riga Technical University, Kalku street 1, LV-1048 Riga, Latvia

[4] Institute of Solid State Physics, University of Latvia, Kengaraga street 8, LV-1063 Riga, Latvia

[5] Institute of Physics, University of Latvia, Miera street 32, LV-2169 Salaspils, Latvia

E-mail: nina.ulmane@cfi.lu.lv

Abstract. Nickel oxide powders with the grain size of 13-1500 nm have been studied by neutron scattering, scanning electron microscopy and vibrating sample magnetometry. We have found that the atomic structure and the antiferromagnetic ordering are nearly independent of the average size of grains. The existence of the uncompensated spins in nanoparticles with the grain size below 100 nm has been detected.

1. Introduction

Effect of crystallite size on the atomic and magnetic structures of antiferromagnetic (AFM) simple oxides such as MnO [1], CoO [2] and NiO [3] remains an interesting topic of research. In particular, the existence of more than two magnetic sublattices as in bulk antiferromagnetic (AFM) nickel oxide has been proposed in NiO nanoparticles [3, 4] and still remains the topic of discussions. Besides antiferromagnetically ordered core, NiO nanoparticles are supposed to have magnetically disordered uncompensated surface spins [3, 5], which can yield a detectable net magnetic moment.

The existence of AFM ordering at room temperature in NiO nanopowders with the crystallite sizes of 13-1500 nm has been recently confirmed in [6] by several experimental techniques, including neutron diffraction. It was also found that the crystal lattice of nanoparticles in AFM phase experiences weak rhombohedral distortion as in the bulk NiO [6].

In this study we have employed the neutron scattering method to probe a correlation between the size of the coherently scattering structural and magnetic regions in NiO nanopowders.

2. Experimental

Four NiO samples (S1, S2, S3, S4) were prepared according to the procedure described in [6]. The grain average sizes determined from the BET specific surface area measurements were 13 nm (S1), 100 nm (S2), 138 nm (S3) and 1500 nm (S4). The morphology of all samples

IOP Conf. Series: Materials Science and Engineering **49** (2013) 012021 doi:10.1088/1757-899X/49/1/012021

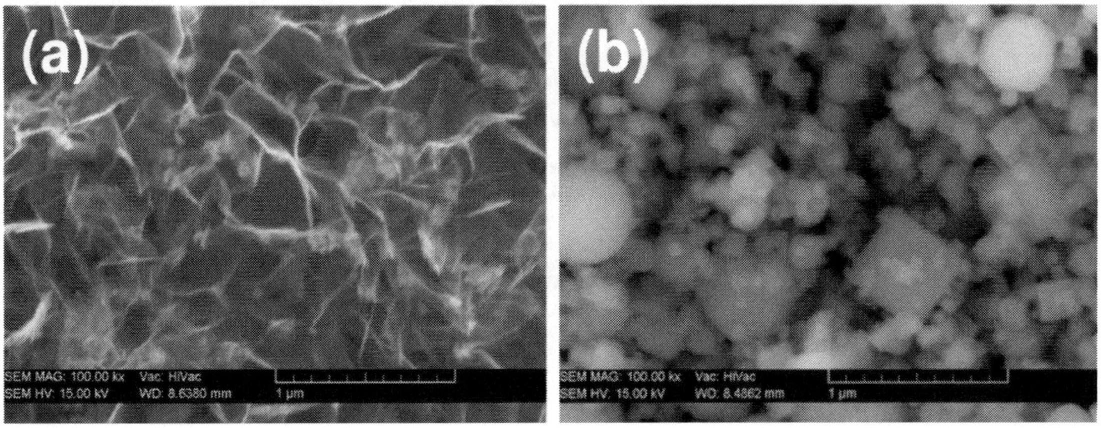

Figure 1. SEM images of NiO samples S1 (left) and S3 (right).

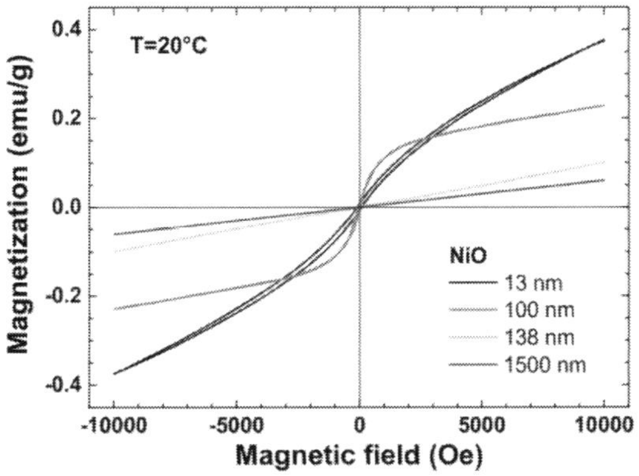

Figure 2. The magnetization curves for NiO samples.

Table 1. NiO grain size L_{BET} (in nm) determined from the BET specific surface area measurements. The sizes (in nm) of coherent neutron scattering regions L_{nuc} and L_{mag} calculated from the width of the nuclear and magnetic reflections, respectively.

Sample	S1	S2	S3	S4
L_{BET}	13	100	138	1500
L_{nuc}	9.3-10.6	94-102	89-123	>380
L_{mag}	<10	37-43	101-133	>280

was studied by scanning electron microscopy (SEM) using TESCAN MIRA microscope with secondary electron detector (Fig. 1). Static magnetization (Fig. 2) was measured by vibrating sample magnetometry (VSM Lake Shore Cryotronics, Inc., model 7404 VSM) using magnetic field in the range from −10 kOe to +10 kOe. The diffraction patterns were measured at room temperature using high resolution Fourier diffractometer (HRFD) at the IBR-2 pulsed reactor in Dubna [7]. Their analysis was performed using the Rietveld method by the MRIA code [8]. The widths of the diffraction peaks were analysed using the UPEAK code [9].

3. Results and discussion

SEM images for two selected samples S1 and S3 are shown in Fig. 1. A homogeneous agglomeration of nanoparticles is observed for the sample S1, however their average size of grains (L_{BET} = 13 nm) is too small to be resolved. In the case of the sample S3, the nanoparticles are well visible showing a broad distribution (30-500 nm) of sizes and shapes.

Magnetic field dependence of magnetization, measured at 20°C, is shown in Fig. 2. A hysteretic behavior is observed for samples S1 and S2, having smallest grain size (Table 1). It is consistent with previous findings and can be explained by the presence of the uncompensated

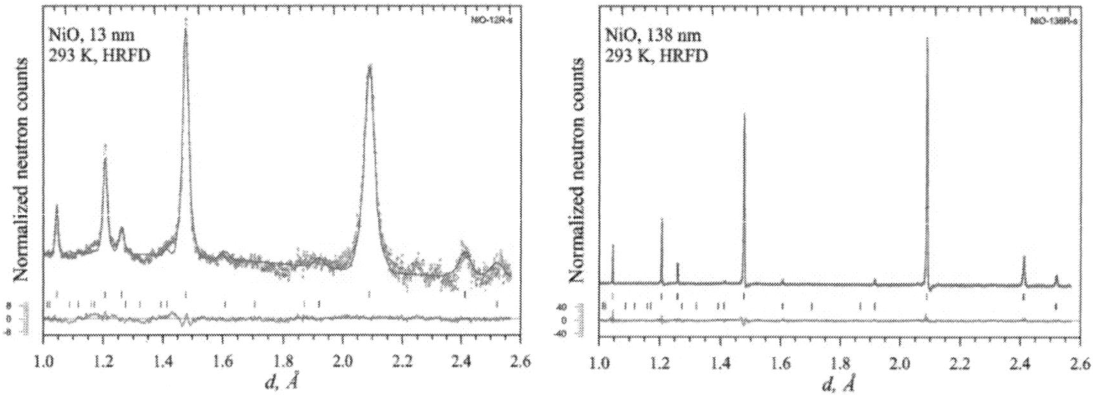

Figure 3. Rietveld refinement of the neutron diffraction patterns measured at room temperature for NiO samples S1 and S3 having the grain size of 13 nm (left) and 138 nm (right). The solid circles correspond to the experimental data, and the solid line – to the calculated pattern. The residual curve is shown at the bottom. The vertical ticks indicate the calculated nuclear (upper row) and magnetic (lower row) reflection positions.

spins in the nanoparticle cores and at the nanoparticle surface [10, 5].

Above the Nèel temperature $T_N = 523$ K bulk NiO is in the paramagnetic phase and has a rock-salt cubic structure (space group $Fm\bar{3}m$, No. 255) with the lattice parameter $a_c \approx 4.177$ Å. Below T_N magnetic moments of the Ni^{2+} ions order antiferromagnetically leading to rhombohedral distortion with $a_R \approx a_c/\sqrt{2} \approx 2.954$ Å, $\alpha \approx 60.07°$ (space group $R\bar{3}m$, No.166) [11, 12]. The rhombohedral magnetic cell is twice as large as the crystallographic one $a_{mag} = 2a_R = 5.907$ Å.

Since the lattice distortion in NiO is very small, it can be detected only using high-resolution diffractometer. In standard operation (Fourier chopper maximum rotation speed $V_{max} = 4000$ rpm), the resolution of HRFD is $\Delta d/d \approx 0.0015$ for $d = 2$ Å, and it improves with increasing d_{hkl}. At this resolution the reliably determined microstrain in crystallites is $\Delta a/a \approx 0.0008$ and more, and the average size of coherently scattering domains is $L_{coh} \approx 250$ nm or less.

The HRFD diffraction patterns for two selected NiO samples S1 and S3 are shown in Fig. 3. Both structural and magnetic diffraction peaks are well resolved due to the high symmetry of NiO and high-resolution measurement mode used. No impurity phases have been detected.

Processing of the spectra by the Rietveld method using the MRIA code [8] allowed us to use the experimentally measured line shape. The processing was carried out in a two-phase version (crystal and magnetic structure) in a limited range d_{hkl} with the ability to set different widths of the peaks in phases. The convergence of the minimization process for all spectra was good, except for the magnetic phase in the sample S1.

Upon a decrease of the crystallite size, the unit cell parameter a_R remains practically constant, but the width of the diffraction peaks increases. However, a small (0.16%) increase in the cell volume was observed for the sample S1 with $L_{BET} \approx 13$ nm. This result agrees with an increase of the lattice parameter in nanosized NiO, which has been found previously by x-ray diffraction [10, 13] and x-ray absorption spectroscopy [14]. Note that while the widths of the structural and magnetic peaks are close in the sample S4, the broadening of the magnetic peaks occurs upon crystallites size reduction (Fig. 4). The sizes of the coherently scattering blocks, estimated from the peak widths of the nuclear and magnetic reflections, are given in Table 1.

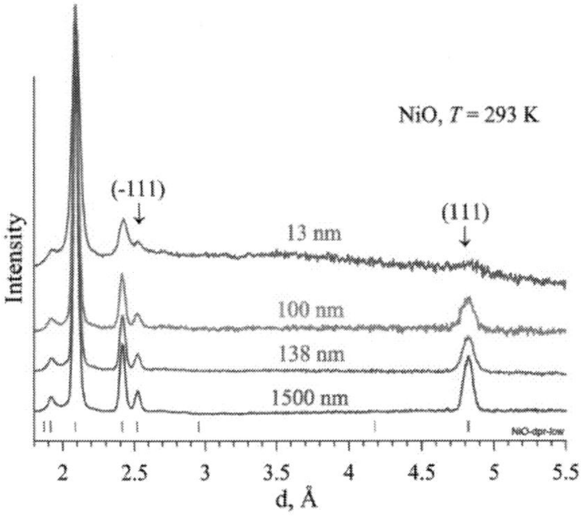

Figure 4. Low resolution diffraction patterns for large d_{hkl}. The positions of the magnetic peaks (111) and (-111) are indicated.

4. Conclusions

Nickel oxide nanoparticles with the average size of grains in the range of 13-1500 nm have been studied by neutron scattering, scanning electron microscopy and magnetometry.

We found that the atomic structure and the type of magnetic ordering are nearly independent of the average size of grains. There is only a small (0.16%) increase in the cell volume when the grain size decreases down to 13 nm. In the samples with the grain sizes of 138 and 1500 nm, the average size of the coherently scattering structural and magnetic regions coincides and is large, but smaller than the grain size. When the grain size is reduced down to 100 nm, it becomes comparable to that of the crystallites, whereas the magnetic coherence length is about 3 times smaller. Finally, the long-range magnetic order becomes largely destroyed in the sample with the smallest grain size of 13 nm.

Acknowledgments

This work was supported by Latvian Science Council Grant No. 187/2012 and ERDF project Nr. 2010/0204/2DP/2.1.1.2.0/10/APIA/VIAA/010.

References

[1] Gołosovsky I V, Mirebeau I, Sakhnenko V P, Kurdyukov D A and Kumzerov Y A 2005 *Phys. Rev.* B **72** 144409
[2] Dobrynin A N, Ievlev D N, Hendrich C, Temst K, Lievens P, Hörmann U, Verbeeck J, Van Tendeloo G and Vantomme A 2006 *Phys. Rev.* B **73** 245416
[3] Kodama R H, Makhlouf S A and Berkowitz A E 1997 *Phys. Rev. Lett.* **79** 1393
[4] Kodama R H and Berkowitz A E 1999 *Phys. Rev. B* **59** 6321
[5] Winkler E, Zysler R D, Mansilla M V, Fiorani D, Rinaldi D, Vasilakaki M, Trohidou K N 2008 *Nanotechnology* **19** 185702
[6] Mironova-Ulmane N, Kuzmin A, Grabis J, Sildos I, Voronin V I, Berger I F and Kazantsev V A 2011 *Solid State Phenom.* **168-169** 341
[7] Balagurov A M 2005 *Neutron News* **16** 8
[8] Zlokazov V B and V. V. Chernyshev V V 1992 *J. Appl. Cryst.* **25** 447
[9] Zlokazov V B 1978 *Comp. Phys. Commun.* **13** 389
[10] Li L, Chen L, Qihe R and Li G 2006 *Appl. Phys. Lett.* **89** 134102
[11] Roth W L 1958 *Phys. Rev.* **110** 1333
[12] Massarotti V, Capsoni D, Berbenni V, Riccardi R, Marini A and Antolini E 1991 *Z. Naturforsch.* **46a** 503
[13] Makhlouf S A, Kassem M A and Abdel-Rahim M A 2009 *J. Mater. Sci.* **44** 3438
[14] Anspoks A, Kalinko A, Kalendarev R and Kuzmin A 2012 *Phys. Rev.* B **86** 174114

IOP Conf. Series: Materials Science and Engineering **49** (2013) 012022 doi:10.1088/1757-899X/49/1/012022

Improvement of UV Protection Properties of the Textile from Natural Fibres by the Sol-gel Method

S Vihodceva[1], **S Kukle**[2]

[1,2] Riga Technical University Institute of Textile Materials and Design, 14/24 Azenes street, LV-1048, Riga, Latvia

E-mail: Svetlana.Vihodceva@rtu.lv

Abstract. In this research pure cotton textile was successfully modified by zinc oxide nanosol prepared by the sol-gel method. The cotton fabric was dipped in the nanosol solution for 10 minutes, dried at 90 °C for 10 minutes with further thermal post-treatment at 120 °C, 140 °C or 160 °C for 2 minutes. Comparison of coating of samples prepared using different thermal post-treatments was made. Before and after laundering tests ultraviolet protective properties of the textile samples were determined according to the standards, results show that textiles after treatment with nanosol have excellent ultraviolet protection properties, as well treated samples after laundering tests (50 washing-drying cycles) still provide excellent ultraviolet protection. Analyses based on the scanning electron microscopy and spectrophotometer measurements show that obtained textile coatings are distributed evenly, not only on the surface of yarns but in the depth of textile material as well, and are resistant to exploitation process that indicates about very good adhesion between the coating and the fabric surface.

1. Introduction

Natural fibres cannot be replaced by synthetic fibers due to their intrinsic biodegradable and wear comfort properties [1]. Nowadays, an advance in the knowledge of the fine structure and chemical composition of natural fibers has stimulated research and development into methods of improving their chemical and physical properties [1]. Special attention has been focused recently on the ultraviolet (UV) transmission of textile because of the growing demand in the marketplace for light-weight apparel that offers to human's protection from UV radiation (UVR). Zinc oxide is wide direct band gap (3.2 - 3.3 eV) semiconductor, which are transparent in the visible and absorb in the UV range [2-4]. Zinc oxide absorbs UVR via process of electron excitation called band-gap absorption. In comparison with the organic absorbers conventionally used in the textile industry, inorganic materials show no degradation under UV exposure and are therefore extremely stable and the oxides are classified as non-toxic materials [2-4]. For the abovementioned reasons, zinc oxide seems to be ideal for the preparation of highly UV-blocking, nanosol based coatings [2]. The present paper describes the deposition of zinc oxide thin-coatings by the sol-gel method on cotton textile substrates.

2. Materials and Methods

In the present study sol-gel method was evaluated in terms of the deposition of zinc oxide thin films on the commercial plain weave 100% cotton textile substrate.

2.1. Nanosol Preparation and Fabric Treatment

Tetraethoxysilane (TEOS, $C_8H_{20}O_4Si$), ethanol (C_2H_5OH), hydrofluoric acid (HF), deionized water and $Zn(CH_3COO)2{\bullet}2H_2O$ have been used for nanosols preparation. Nanosols were prepared by a controlled hydrolysis, by adding ethyl alcohol slowly into TEOS with continuous stirring, after adding deionized water and hydrofluoric acid, stirred for 30 minutes at 50 °C temperature, after mixed with the zinc acetate with continuous stirring 10 minutes. Samples were prepared zinc acetate concentration 5%. The fabric samples were dipped into the prepared nanosol, immersed for 10 minutes at room temperature. Subsequently, the samples were dried at 90 °C for 10 minutes with further thermal post-treatment at 120 °C, 140 °C or 160 °C for 2 minutes.

3. Results and Discussion

The nanosol process has to be adapted for the treatment of textiles because of textile low heat resistance; a thermal post-treatment at higher temperatures should be avoided to reduce degradation and deformation of the textile materials, and has to be adapted with regard to the type of textile coated and to the applied nanosol. Curing is an important step in the sol-gel coating process and curing temperatures of between 100 - 170 °C and different time from 1 to 30 minutes reported in the literature [5-8].

3.1. UV Protection Measurements

The UV protection of prepared samples was measured using the Varian Cary50 Solascreen spectrophotometer (Australia) according to the Australian/New Zealand AS/NZS 4399:1996 [9] and European EN 13758-1:2001+A1:2006 standards[10]. From each sample group 4 samples were scanned for UV transmission, 4 scans (one scan totally consist of 6 measurements, 3 measurements from each of 2 different points of the sample surface) per group were acquired.

Graphs in figure 1 and figure 2, and data table 1 indicate that thermal post-treatment time has influence on the calculated UV protection factor (UPF), UPF is the ration of the average effective UVR irradiance calculated for unprotected skin to the average effective UVR irradiance calculated for skin protected by the test fabric [9]. Data indicates that the best results of UPF of samples after treatment with nanosol were obtained with the thermal post-treatment at 120 °C for 2 minutes, achieved excellent UPF results as well as after 50 washing-drying cycles. According to the experiment data can be concluded that moderate thermal post-treatment at 120 °C for 2 minutes is appropriate for cotton samples coated by nanosol and it is inexpedient to increase the thermal post-treatment time.

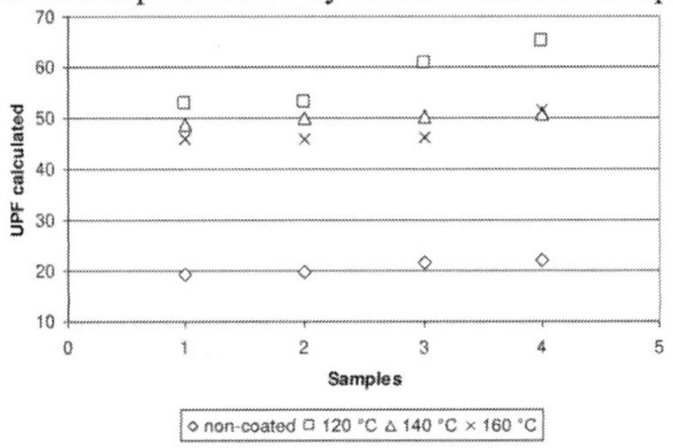

Figure 1. Comparison of calculated UPF of samples with different thermal post-treatments.

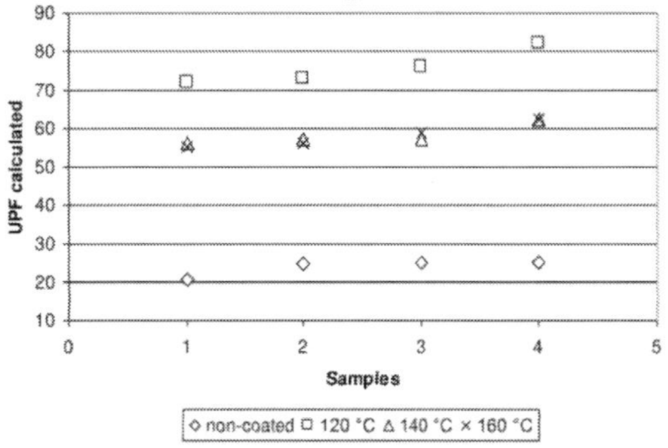

Figure 2. Comparison of calculated UPF of samples with different thermal post-treatments after 50 washing-drying cycles.

The UPF of cotton textile coated with zinc oxide nanosol with thermal post-treatment at 120 °C for 2 minutes increased approximately by 2.9 times in comparison with UPF values of uncoated textile (table 1) and the UPF range corresponds to the standards [9, 10] rating "50+ UPF" that mean excellent protection (table 1). According to the requirements of standard [10] for all materials used for clothing assembly, the lowest UPF value shall be not less than 40 UPF.

Table 1. UV Protection Measurements

Samples	Calculated UPF	UVA transmission, %	UVB transmission, %	UPF rating
	After treatment with zinc oxide nanosol			
Non-coated	19.929	3.874	4.058	15/20
Post-thermal treatment 120 °C	57.901	1.405	1.238	50+
Post-thermal treatment 140 °C	50.030	1.834	1.734	45/50+
Post-thermal treatment 160 °C	47.398	1.703	1.290	45/50+
	After 50 drying-washing cycles			
Non-coated	22.961	3.407	3.438	20/25
Post-thermal treatment 120 °C	75.680	1.698	1.081	50+
Post-thermal treatment 140 °C	58.107	1.531	1.412	50+
Post-thermal treatment 160 °C	58.304	1.704	1.373	50+

The durability and stability of the treatment to multiple home laundering it indicates that is achieved very good adhesion between the coating and the fabric surface, as well treatment imparted excellent UVR protection to the cotton textile in both regions of UVR – UVA and UVB (figure 3).

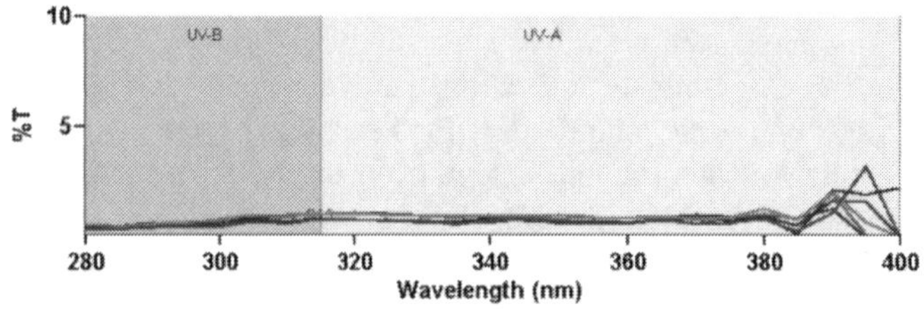

Figure 3. UVA and UVB transmission after sample treatment

3.2. Scanning Electron Microscopy (SEM) Micrograph Analysis

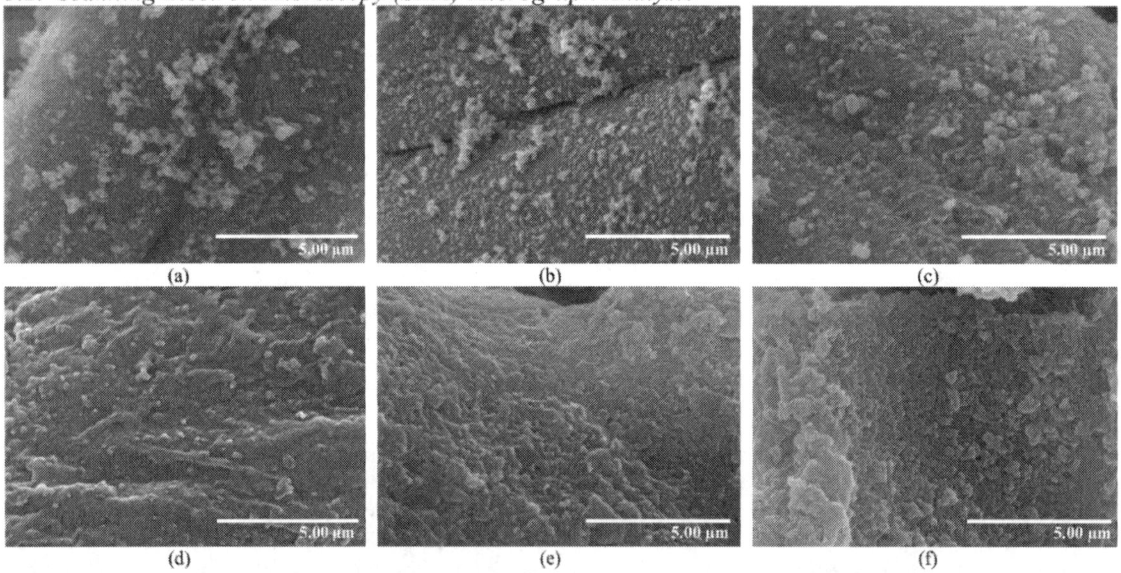

Figure 4. SEM micrographs of samples with different post-thermal treatment: coated - a) 120 °C, b) 140 °C, c) 160 °C; after 50 washing-drying cycles - d) 120 °C, e) 140 °C, f) 160 °C.

Morphological changes as a result of sol-gel treatment of cotton fabric were investigated using scanning electron microscopy (SEM) Hitachi S-3400N (Japan). Figure 4 (a, b, c) shows the morphological changes of the cotton textile surface after treatment by nanosol with different thermal post-treatment time. At high magnification, the formation of a thin layer is observed on the fiber surface (figure 4 a,b,c); surface of untreated cotton fabric has grooves and cracks in contrast, such characteristics completely disappeared on the surface of the treated cotton fabric. Figure 4 (a,b,c) indicates that samples treated by nanosol have sufficiently evenly distributed coating, on fibres can notice the presence of silica particles that was proofed by energy dispersive x-ray (EDX) analysis of treated samples in previous experiment [11].

SEM micrographs figure 4 (d, e, f) reveals that received zinc oxide thin-coatings are resistant to exploitation impact (50 washing-drying cycles) the coatings on the cotton samples still are distributed evenly on the fibres surface without significant defects, coating became thicker, as well increased the roughness of coating, it can explain the increasing of UPF after washing-drying cycles.

4. Conclusions

Spectrophotometer and SEM analyses show that a drying at 90 °C for 10 minutes and a moderate thermal post-treatment at 120 °C for 2 minutes after the nanosol with zinc acetate concentration 5% applied to the cotton textile is appropriate for the cotton textile sample coating by nanosol, the deposited coating on textile is distributed evenly and is resistant to exploitation process (laundry tests – 50 washing-drying cycles), show UPF rating 50+ after coating as well as after multiple washing processes. If compared to the standards, it is an excellent fabric protection towards UV radiation.

Results show that the UPF increase after washing cycles the possible reasons of increasing are: that the process of washing led to compaction due to shrinkage presumably decreasing porosity, decreasing of silica element in coating composition during washing-cycles, because during the washing cycles from textile fibers rewashed silica particles that were formed during the coating formation, as well thermal treatment below 200°C form xerogel and xerogels are amorphous networks and received coating is amorphous it is not so stable like crystalline coating, because of that during washing was occurred coating changes - coating became thicker and increased the roughness of coating.

5. Acknowledgements

This work has been supported by the European Social Fund within the project «Support for the implementation of doctoral studies at Riga Technical University».

ESF EIROPAS SOCIĀLAIS FONDS INVESTING IN YOUR FUTURE EIROPAS SAVIENĪBA

References

[1] Rippon J A and Evans D J 2012 Improving the properties of natural fibres by chemical treatments *Handbook of natural fibres* ed Kozlowski M (London: Woodhead Publishing Limited) pp 63–140
[2] Textor T 2009 Modification of textile surface using sol-gel technique *Surface modification of textiles* ed Wei Q (London: CRC Press) pp 185–214
[3] Karmali R S, Bartakke A, Borker V P and Rane K S 2011 *J. Bioint. Rsrch.in Appl. Chem.* **1** 57
[4] Nikolaeva N S, Ivanov V V Shubin A A 2010 *J. Of Siberian Federal Univ. Chem.* **2** 153
[5] Mahlting B and Textor T 2008 *Nanosols and textiles* (Singapore: World Scientific Publishing)
[6] Abidi N, Cabrales L, Hequet E 2009 *J. Appl. Mat.and Interfaces* **1** 2141
[7] Farouk A, Textir T, Schollmeyer E, Tarbuk A, Grancacic A M 2010 *AUTEX Rsrch. J.* **10** 58
[8] Brzezinski S, Kowalczyk D, Borak B, Jasiorski M 2011 *Fib.& Text. In East.Eur.* **19** 83
[9] Australian/New Zealand Standard *Sun Protective clothing-Evaluation classification* AS/NZS 4399:1996
[10] European standard *Solar UV protective properties – Part 1: Method of test for apparel fabrics* EN 13758-1:2001+A1:2006:E
[11] Vihodceva S, Kukle S 2012 *J. of Riga Tech. Un. Mat. Sci. and Clothing Tech.* **7** 19

IOP Conf. Series: Materials Science and Engineering **49** (2013) 012023 doi:10.1088/1757-899X/49/1/012023

Magnetic moment of single vortices in YBCO nano-superconducting particle: Eilenberger approach

I. Zakharchuk[1,2], A. Sharafeev[1,2], P. Belova[1,3], M. Safonchik[1,4], K. B. Traito[1], E. Lähderanta[1]

[1] Lappeenranta University of Technology, P.O.Box 20, FI-53851, Lappeenranta, Finland
[2] Saint-Petersburg Electrotechnical University, Popov str. 5, RU-197376, St.Petersburg, Russia
[3] Petrozavodsk State University, Lenin str. 33, RU-185640, Petrozavodsk, Russia
[4] A. F. Ioffe Physico-Technical Institute, St. Petersburg, RU194021, Russia

E-mail: ivan.zakharchuk@lut.fi

Abstract. Temperature dependence of single vortex magnetic moment in nanosize superconducting particles is investigated in the framework of quasiclassical Eilenberger approach. Such nanoparticles can be used for preparation of high-quality superconducting thin films with high critical current density. In contrast to bulk materials where the vortex magnetic moment is totally determined by flux quantum, in nano-sized specimens (with characteristic size, D, much less than effective penetration depth, λ_{eff}) the quantization rule is violated and magnetic moment is proportional to $D^2/\lambda^2_{eff}(T)$. Due to strong repulsion between vortices in nanoparticles only a single vortex can be trapped in them. Because of small size of particles the screening current of the vortex is located near the vortex core where the current is quite high and comparable to depairing currents. Therefore, the superconducting electron density, n_s, depends on the current value and the distance from the vortex core. This effect is especially important for superconductors having gap nodes, such as YBCO.

The current dependence of n_s in nanoparticles is analogous to the Volovik effect in flux-line lattice in bulk samples. The magnitude of the effect can be obtained by comparing the temperature dependence of magnetic moment in the vortex and in the Meissner states. In the last case the value of screening current is small and superconducting response to the external field is determined by London penetration depth. Because of importance of nonlinear and nonlocal effects, the quantum mechanical Eilenberger approach is applied for description of the vortex in nanoparticles. The flattening of $1/\lambda^2_{eff}(T)$ dependence has been found. A comparison of the theoretical results with experimental magnetization data in Meissner and mixed states of YBCO nanopowders has been done. The presence of nonlinear and nonlocal effects in vortex current distribution is clearly visible. The obtained results are important for the description of pining in nanostructured high-T_c thin films.

1. Introduction

Quasiclassical calculation of the density of states of a single vortex in an anisotropic superconductor was performed in Ref. [1], and it was shown that if the gap itself is not highly anisotropic, the Fermi surface anisotropy dominates, preventing direct observation of superconducting gap features. This serves as a cautionary message for the analysis of scanning tunneling spectroscopy data on the vortex state on Fe-based superconductors, in particular LiFeAs, which was treated explicitly. It was found that Doppler shift (DS) on quasiparticle excitations on extended states (so-called "Volovik effect" [2]) immediately induce zero energy excitations proportional to H from a smaller gap band in $s^{\pm}$ symmetry [3]. The structure of a single vortex in a FeAs superconductor was studied in Ref. [4] in the framework of two formulations of superconductivity for the recently proposed sign-reversed s-wave ($s^{\pm}$) scenario: (i) a continuum model taking into account the existence of an electron and a hole band, with a repulsive local interaction between the two; (ii) a lattice tight-binding model with two orbitals per unit cell and a next-nearest-neighbour attractive interaction. An impurity located outside the vortex core has little effect on the local density of states peak, but an impurity close to the vortex core can almost suppress it and modify its position. In Ref. [5] a linear regime at higher fields and a limiting square root H behavior at very low fields were found. The crossover from a Volovik-like $\sqrt{H}$ to a linear field dependence can be understood from a multiband calculation in the quasiclassical approximation assuming gaps with different momentum dependence on the hole- and electron-like Fermi surface sheets.

In a direct scanning tunneling spectroscopy experiment the problem of the quantum vortex phases in strongly confined superconductors was addressed [6]. The strong confinement regime is achieved in grown ultrathin single nanocrystals of Pb by tuning their lateral size (diameter ≈ 140 nm and height ≈ 2.8 nm) to a few coherence lengths. Cren *et. al.* [6] used the phenomenological Ginzburg-Landau theory for explanation of obtained results. However, the aim of our paper is to apply microscopical Eilenberger approach to high-T_c nanoparticles. The asymptotical behaviour of the amplitude and the phase of the pairing potential is obtained. The Eilenberger equations can be obtained from a full quantum mechanical approach (the Bogoliubov-de Gennes equations) using an expansion in terms of α^{-1}, where $\alpha = v_F/v_\Delta$ is the Dirac cone anisotropy, v_F is the Fermi velocity, and v_Δ is the quasiparticle velocity tangential to the Fermi surface at the node. This expansion is quite reasonable for the description of high-T_c superconductors, where $\alpha = 14$ for YBaCuO and 20 for BiSrCaCuO [7]. The Eilenberger equations have been solved previously [8] in the vortex core region. Here, we find the behaviour of the amplitude and the phase of the pairing potential $\Delta(r)$ at long distances r from the vortex core. It is found that the DS method works reasonably well at distances $r \geq \xi_0$ at low temperatures. The nonlocal effects are important inside the core and for the description of effects of the fourfold vortex symmetry outside the core. It is also shown that at higher temperatures the DS method should be modified by including a pairing potential $\Delta(r)$ calculated self-consistently.

2. Quasiclassical approach

We consider an isolated two-dimensional vortex in a d-wave superconductor. The center of the vortex is taken as the origin. The Fermi surface is assumed to be isotropic and cylindrical. To obtain the quasiclassical Green functions we solve the quasiclassical Eilenberger equations for the pairing potential $\Delta(\theta, r) = \bar{\Delta}(r)\cos(2\theta)\exp(i\phi)$ [8, 9], where θ is the angle between the $\mathbf{k}$ vector and the a axis (or x axis) and $\exp(i\phi) = (x + iy)/r$. It should be noted here that the spatial variation of the supercurrent and the d-wave order parameter induce small subdominant s and d_{xy} components in the pairing order parameter [10]. We are not considering these effects because they can be included in a straightforward way in our calculations. Throughout this paper, the energies and the lengths are measured in units of the uniform gap Δ_0 at $T = 0$ and the coherence length $\xi_0 = v_F/\Delta_0$, respectively.

For calculation it is convenient to parametrize the quasiclassical Green function via [9]

$$\bar{f} = \frac{2\bar{a}}{1+\bar{a}\bar{b}}, \quad \bar{f}^{\dagger} = \frac{2\bar{b}}{1+\bar{a}\bar{b}}, \quad g = \frac{1-\bar{a}\bar{b}}{1+\bar{a}\bar{b}}, \tag{1}$$

where the anomalous Green functions $\bar{f}$ and $\bar{f}^{\dagger}$ are related to the usual notations as $f = \bar{f}\exp(i\phi)$ and $f^{\dagger} = \bar{f}^{\dagger}\exp(-i\phi)$. The functions $\bar{a}$ and $\bar{b}$ satisfy the independent nonlinear Ricatti equations

$$\partial_{\parallel}\bar{a}(\omega_n, \theta, \mathbf{r}) = \bar{\Delta}(\theta, \mathbf{r}) - \{2\omega_n + i\partial_{\parallel}\phi + \bar{\Delta}^*(\theta, \mathbf{r})\bar{a}(\omega_n, \theta, \mathbf{r})\}\bar{a}(\omega_n, \theta, \mathbf{r}), \tag{2}$$

$$\partial_{\parallel}\bar{b}(\omega_n,\theta,\mathbf{r}) = -\bar{\Delta}(\theta,\mathbf{r}) + \{2\omega_n + i\partial_{\parallel}\phi + \bar{\Delta}(\theta,\mathbf{r})\bar{b}(\omega_n,\theta,\mathbf{r})\}\bar{b}(\omega_n,\theta,\mathbf{r}), \quad (3)$$

where $\omega_n = (2n+1)\pi T$ is the fermionic Matsubara frequency, $\partial_{\parallel} = d/dr_{\parallel}$ and $\partial_{\parallel}\phi = -r_{\perp}/r^2$. Here, we use the coordinate system $\hat{\mathbf{u}} = \cos\theta\hat{\mathbf{x}} + \sin\theta\hat{\mathbf{y}}$, $\hat{\mathbf{v}} = -\sin\theta\hat{\mathbf{x}} + \cos\theta\hat{\mathbf{y}}$. Thus, a point $\mathbf{r} = x\hat{\mathbf{x}} + y\hat{\mathbf{y}}$ is denoted as $\mathbf{r} = r_{\parallel}\hat{\mathbf{u}} + r_{\perp}\hat{\mathbf{v}}$. Eqs. (2) and (3) include both the nonlocal effects ($\partial_{\parallel}\bar{a}$ and $\partial_{\parallel}\bar{b}$ terms) and the nonlinear effects ($\bar{a}$ and $\bar{b}$ are nonlinear functions of $\partial_{\parallel}\phi$). Since we consider an isolated vortex in extreme type-II superconductors, the vector potential in Eqs. (2) and (3) can be neglected.

As has been shown [11] the solution of Eqs. (2) and (3) is quite stable: after integration over a length of a few ξ_0 it becomes almost independent of the initial values. This solution corresponds to a simple exponential relaxation of the functions $\bar{a}$ and $\bar{b}$ to their local "steady-state" values defined by the local values of the order parameter. Therefore, to find $\bar{\Delta}$ at a given point, one does not need the values of $\bar{\Delta}$ at distances larger than several ξ_0 along the trajectory. This peculiarity is used for integration of Eqs. (2) and (3) at long distances. First, we find some approximative solution at the distance of several ξ_0 from a given point and consider it as the boundary condition. Next, we make the integration up to the given point by the Runge-Kutta method with a variable step. To find this approximative boundary condition a linear expansion $\bar{a} = a_0 + a_1 r_{\parallel}$, $\bar{b} = b_0 + b_1 r_{\parallel}$, and $\bar{\Delta} = \Delta_0 + \Delta_1 r_{\parallel}$ is used near the given point. Substituting this expansion to Eqs. (2) and (3) and equating the coefficients under the same power of $r_{\parallel}$ we obtain the set of equations for a_0, a_1, b_0 and b_1.

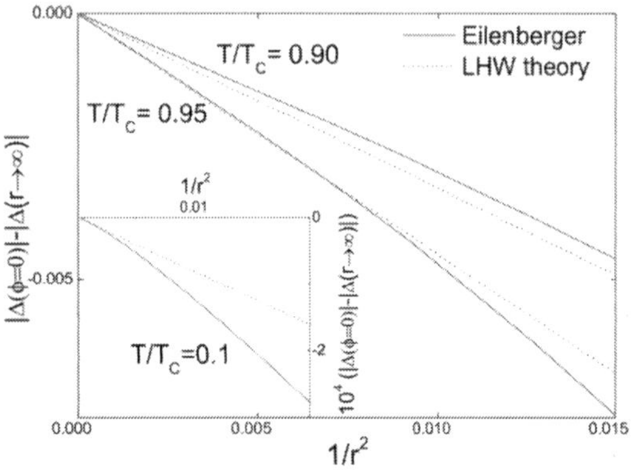

Figure 1. Asymptotical behaviour of the amplitude of the pairing potential $|\Delta|$ at long distances for $\phi = 0$ and $T/T_c = 0.9, 0.95$ and at $T/T_c = 0.1$ (solid lines). The dotted lines in the main panel and the inset show the values $|\Delta(T,r)| - |\Delta(T)|$ obtained from Eq. (4) at $T/T_c = 0.9$ and 0.95 and from Eq. (5) at $T/T_c = 0.1$, respectively.

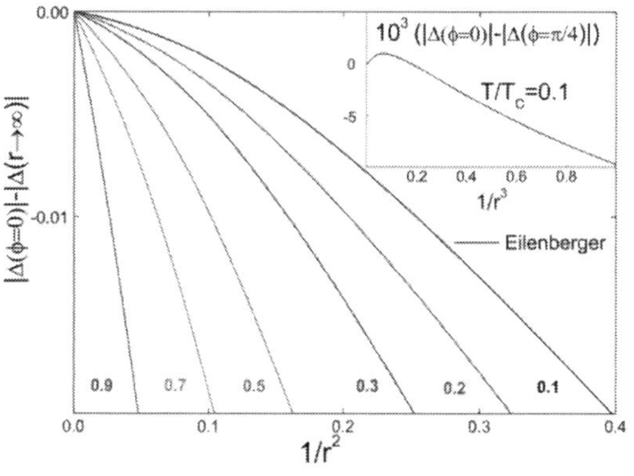

Figure 2. Behaviour of the amplitude of the pairing potential at intermediate distances for $\phi = 0$ and six values of T/T_c between 0.1–0.9. The inset depicts the asymptotical behaviour of the difference of the amplitude along 0 and $\pi/4$ directions, $|\Delta(\phi = 0)| - |\Delta(\phi = \pi/4)|$ at $T/T_c = 0.1$.

Figure 1 shows the asymptotical behaviour of the amplitude of the pairing potential at long distances for $\phi = 0$, $T/T_c = 0.9, 0.95$ in the main panel and $T/T_c = 0.1$ in the inset (solid lines), obtained using Eilenberger approach. As can be seen from this figure $|\Delta|$ relaxes to its bulk value ($r \to \infty$) as $1/r^2$ with the power of r being independent of the temperature. The same law of the relaxation has been obtained in the numerical solution of the Bogoliubov-de Gennes equations [12] at $T = 0$ K. This relaxation law is different from the law [8], $|\Delta(r)| \propto \tanh(r)$.

The analytical expansion of the BCS solution near the infinity point has been obtained perturbatively by Li, Hirschfeld and Wölfle (LHW theory) [10]. In Ginsburg-Landau regime near T_c they found

$$|\Delta(T,r)| - |\Delta(T)| = -\frac{1}{3/4 + \epsilon_F m v_s^2(\mathbf{r})/\Delta^2(T)} \frac{\epsilon_F m v_s^2(\mathbf{r})}{\Delta^2(T)}, \quad (4)$$

where ϵ_F is the Fermi energy and $\mathbf{v_s}(\mathbf{r})$ is the superconducting electron velocity. At low temperatures and long distances $mv_s v_F \ll T \ll \Delta(T)$ the solution [10] is

$$|\Delta(T,r)| - |\Delta(T)| = -\frac{2\ln 2}{1 - [9\zeta(3)T^3 - 4\ln 2\epsilon_F T m v_s^2(\mathbf{r}]/\Delta^3(T)} \times \frac{\epsilon_F m v_s^2(\mathbf{r})}{\Delta^2(T)} \frac{T}{\Delta(T)}, \quad (5)$$

where $\zeta(3)$ is the Riemann function. The dotted lines in the main panel and the inset of Fig. 1 shows the values $|\Delta(T,r)| - |\Delta(T)|$ obtained from Eq. (4) at $T/T_c = 0.9$ and 0.95 and from Eq. (5) at $T/T_c = 0.1$, respectively. As can be seen from this figure the slopes of asymptotics in LHW theory agrees well with those obtained numerically by us (solid lines).

Figure 2 shows the behaviour of the amplitude of the pairing potential at intermediate distances for $\phi = 0$ and six values of T/T_c between 0.1 – 0.9. The behaviour of $|\Delta(\phi = 0)| - |\Delta(\phi = \pi/4)|$ at $T/T_c = 0.1$ and the change of the sign of this quantity is clearly visible in the inset to Fig. 2.

3. Conclusions

The quasiclassical Eilenberger equations are solved numerically for an isolated two-dimensional vortex for high-T_c nanoparticles. In the core area our results reproduce those obtained previously [8]. New asymptotical behaviour of the amplitude is obtained. Taking into account the suppression of the pairing potential as in the self-consistent Doppler-shift method, good agreement with the exact calculation is observed over the whole range of the radius. The temperature dependencies of the magnetic moment for superconducting YBaCuO nanoparticles was investigated (i) in the Meissner state and (ii) when the magnetic flux is trapped in the particles. It was found that the ratio $\lambda_{ab}^2(0)/\lambda_{ab}^2(T)$ obeys the power law behaviour with the exponent $n = 2.0 \pm 0.2$ and $n = 2.5 \pm 0.05$ in these two cases, respectively [13]. We connect this behaviour with Volovik effect in nanoparticles.

References

[1] Wang Y, Hirschfeld P J and Vekhter I 2012 *Phys. Rev. B* **85** 020506(R)
[2] Volovik G E 1993 *JETP Lett.* **58** 469
[3] Bang Y 2010 *Phys. Rev. Lett.* **104** 217001
[4] Araújo M A N, Cardoso M and Sacramento P D 2009 *New J. Phys* **11** 113008
[5] Wang Y, Kim J S, Stewart G R, Hirschfeld P J, Graser S, Kasahara S, Terashima T, Matsuda Y, Shibauchi T and Vekhter I 2011 *Phys. Rev. B* **84** 184524
[6] Cren T, Serrier-Garcia L, Debontridder F and Roditchev D 2011 *Phys. Rev. Lett.* **107** 097202
[7] Chiao M, Hill R, Lupien C, Taillefer L, Lamber P, Gagnon R and Fournier P 2000 *Phys. Rev. B* **62** 3554
[8] Ichioka M, Hayashi N, Enomoto N and Machida K 1996 *Phys. Rev. B* **53** 15316
[9] Schopohl N and Maki K 1995 *Phys. Rev. B* **52** 490
[10] Li M, Hirschfeld P J and Wölfle P 2001 *Phys. Rev. B* **63** 054504
[11] Amin M H S, Rashkeev S N, Coury M, Omelyanchouk A N and Zagoskin A M 2002 *Phys. Rev. B* **66** 174515
[12] Franz M and Tešanović Z 1998 *Phys. Rev. Lett.* **80** 4763
[13] Blinov E V, Fleisher V G, Laiho R, Lähderanta E, Stepanov Yu P and Traito K B 1997 *Physica C* **281** 17

IOP Conf. Series: Materials Science and Engineering **49** (2013) 012024 doi:10.1088/1757-899X/49/1/012024

Surface relief grating recording in azo polymer films

J. Aleksejeva, J. Teteris
Institute of Solid State Physics, University of Latvia, Kengaraga str. 8, Riga, Latvia, LV-1063

E-mail: aleksejeva.jelena@gmail.com

Abstract. In this work holographic recording of surface relief grating in organic Poly(Disperse Red 1 – methacrylate) thin films was studied. In this compound azo dye is chemically bonded to monomers making mass transport very efficient in the presence of optical field gradient. Sample preparation was performed using different solvents and coating methods. In the obtained samples holographic recording was performed by solid-state diode pump laser with 532 nm wavelength. Polarization grating and surface relief grating formation was studied by diode laser with 660 nm wavelength. Diffraction efficiency's dependence on read-out laser beam polarization state was investigated. The depth of surface relief grating was measured by AFM and studied in dependence on recording beam intensity, exposure and thickness of the sample.

1. Introduction

In this work surface relief grating formation in azo polymer films was studied. Azo compound moieties can be aligned in the presence of electric field of light radiation. Due to that process photoinduced birefringence and dichroism appears. If we use interference pattern of two +45/-45 orthogonally linearly polarized beams resulting electric field vector has polarization modulation on the surface of the thin film, and surface relief grating (SRG) can be recorded. Holographic recording of SRG is particularly interesting due to its possible applications in optoelectronics, data storage and telecommunications [1, 2, 3].

We studied holographic recording in (Disperse Red 1 - methacrylate) thin films where well known PMMA polymer is used only methyl group is substituted by Disperse Red 1 azo-dye. SRG formation in these films is fast and doesn't require great exposures. The depth of obtained SRG exceeds the thickness of film, although SRG formation process strongly depends on intensity and thickness of the film.

2. Experimental

2.1. Sample preparation

Research was made using commercially available material Poly(Disperse Red 1 – methacrylate) – (P(DR1-MA)) purchased at Sigma-Aldrich. The chemical structure of used material is shown in figure 1. Maximal absorption of P(DR1-MA) polymer is at λ_{max} = 467 nm. For sample preparation P(DR1-MA) was dissolved in toluene or trichloroethylene. In order to obtain films with different thickness solvents with different concentration (1.0±0.2%, 2.0±0.4%, 5.0±1%, 7.0±1.4%, 10.0±2.0%) were prepared. Solvents were purified using PTFE filters with pore size of 0.45 μm. Material was applied on substrate by spin-coating method or using applicator with thickness 120 μm. Before applying glass substrate was cleaned in dichromate - sulphuric acid solution. Films were dried in ambient conditions.

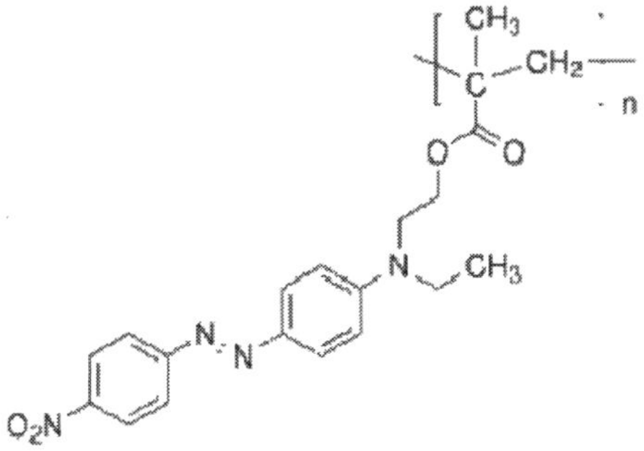

Figure 1. Chemical structure of Poly(Disperse Red 1 - methacrylate)

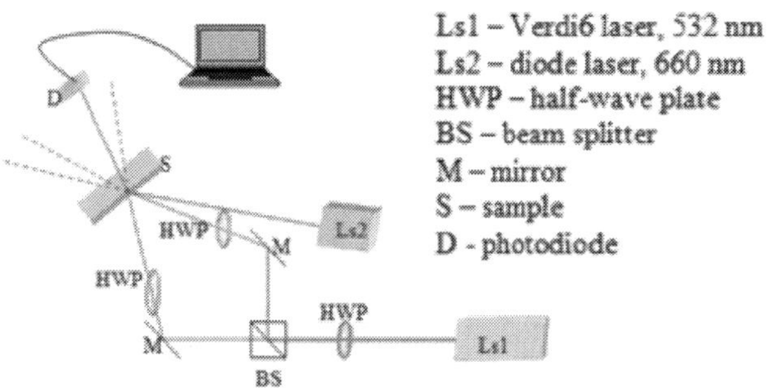

Figure 2. Experimental set-up

2.2. Experimental set-up

Holographic recording was performed on experimental set-up shown in Fig. 2. For recording solid-state diode-pumped laser Verdi-6 (Ls1) with wavelength λ_1 = 532 nm was used. It was split into two equal intensities beams by beam splitter (BS). Using mirrors (M) beams were deflected toward a sample (S). Half-wave plates (HWP) were used in order to set polarization state. The angle between recording beams was θ = 30°, thus grating period was Λ = 1 μm according to Bragg's law. Read-out was made using diode laser with wavelength λ_2 = 660 nm. The laser was placed at Bragg's angle and intensity of diffracted beam of the first order was measured by photodiode (D) which was connected to PC.

Polarization state of recording beams was set to orthogonally linear - *+45/-45*. In this case on the surface of the film there is polarization modulation, but no intensity modulation [4].

3. Results and discussion

A kinetics of SRG formation and diffraction efficiency (DE) of recording with intensity I = 0.95 W/cm^2 in film with thickness d = 490 nm is shown in Fig. 3. As long as the depth of SRG does not exceed the thickness of the film, DE and the depth of SRG increase linearly. The speed of SRG formation reaches its maximal value in the first seconds of recording and strongly depends on intensity of recording beams. If in the first t=*15s* at intensity I = 0.95 W/cm^2 the average speed is u=7.8 nm/s, then at intensity I = 1.86 W/cm^2 it is – u = 12.3 nm/s. Average speed was evaluated using formula: $u = (h_t - h_{0s})/t$, where h_t – the depth of SRG after t = 15s recording. Further SRG formation is slowed down by surface tension of the film and SRG forms slower. When maximal value is reached, low decrement of the depth of SRG is observed due to high surface tension and then saturation is reached. DE curve corresponds to SRG depth changes with a slight shift (figure 3). The shift between h and DE curves is caused by the phase shift between volume and surface relief gratings due to read-out was made with s-polarized probe beam [5]. Thus, the main contribution in the value of DE gives SRG, while volume grating is implicit.

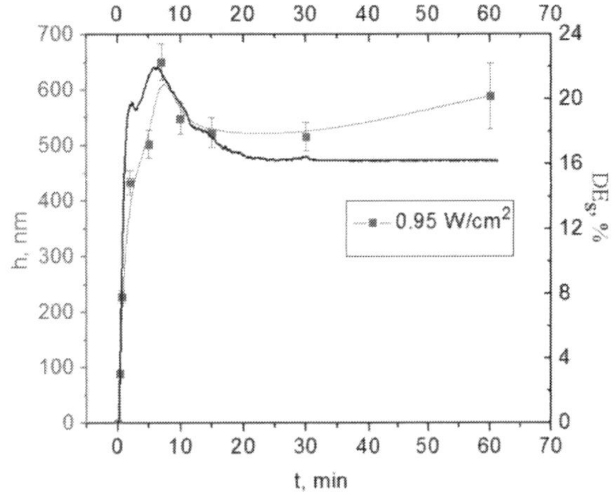

Figure 3. Diffraction efficiency and kinetics of SRG formation in film with a thickness of d = 490 nm

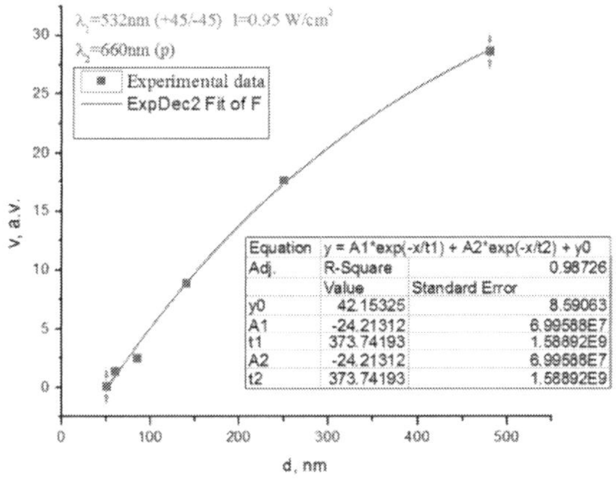

Figure 4. Recording velocity's dependence on thickness of the sample.

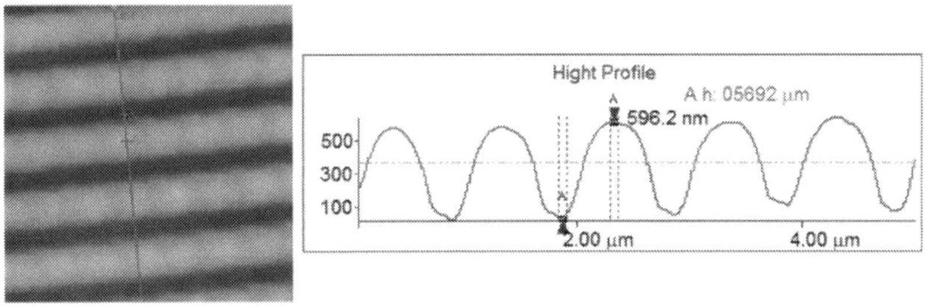

Figure 5. Profile of SRG recorded by intensity of I = 0.95 W/cm^2; t = 3 min; d = 500 nm

The depth of SRG exceed the thickness of the sample 1.3 times, but ratio h/Λ = 0.65. Obtained result shows a very efficient and fast SRG formation during holographic recording. Chemically attached azo-dye to the main polymer chain provides effective mass transport, since azo-dye moiety is moving along with polymeric chain. Glass transition temperature of P(DR1-MA) is low – T_g = 82° C, thus there is low viscosity and deep SRG can be recorded. Gratings stay permanent after holographic recording is done.

The formation of SRG strongly depends on thickness of the P(DR1-MA) film (figure 4). If we want to discuss the velocity of grating formation, we have to measure a tangent of linear part of diffraction efficiency curve: $tg\alpha = v$ [4]. The velocity increases according to exponential decay function shown in figure 4. The saturation of the curve is expected at d = 1.5 μm, due to high absorption. The radiation can not penetrate the bottom layer of the film and velocity of SRG formation does not increase for thicker films.

The profile of obtained SRG by intensity I = 0.95 W/cm^2 recorded for t = 3 min in the film with the thickness of d = 500 nm is shown in figure 5. The depth h of SRG exceed the thickness of the film 1.2 times, but ratio $h{:}d$ strongly depends on the thickness d. In this case, ratio $h{:}d$ is limited by period of grating. Using thinner films (d = 120 nm), $h{:}d$ = 2.

4. Conclusions

In this work experimental study of holographic recording in Poly(Disperse Red 1 - methacrylate) thin films is present. We observed efficient surface relief formation in this material. We showed that formation process strongly depends on the thickness of the film itself as well on the intensity of recording beams. The study of SRG formation kinetics showed that the fastest SRG formation takes place in the first seconds of recording, when the average speed reaches 12.3 nm/s. The depth of SRG exceeds the thickness of the sample 1.3 times, but the value strongly depends on the thickness of the sample.

In Poly(Disperse Red 1 - methacrylate) thin film all the material applied on the glass substrate forms SRG making material usage maximal. The depth of SRG is limited by grating period and surface tension.

Acknowledgement

This work has been supported by the European Social Fund within the project «Support for Doctoral Studies at University of Latvia» and ERAF project No 2010/0204/2DP/2.1.1.2.0/10/APIAA/VIAA/010

References

[1] A. Raicu Luca, L. Rocha, A.-M. Resmerita, A. Macovei, M. Hamel, A.-M. Macsim, N.Nichita, N. Hurduc, 2011, eXPRESS Polymer Letters Vol.5, No.11 959–969

[2] O. Kulikovska, K. Gharagozloo-Hubmann, J. Stumpe, B. D Huey and V. N Bliznyuk, 2012, Nanotechnology 23,485309 (14pp)

[3] J. Aleksejeva, A. Gerbreders, M. Reinfelde, J. Teteris, JOAM, 2011, Vol. 13, No. 11-12, p. 1577 – 1580

[4] J. Teteris, J. Aleksejeva, U. Gertners, 2011 IOP Conf. Series: Materials Science and Engineering23 012002

[5] F. Lagugne Labarthet, T. Buffeteau, and C. Sourisseau, 1999, J. Phys. Chem. B,103,6690-6699

IOP Conf. Series: Materials Science and Engineering **49** (2013) 012025 doi:10.1088/1757-899X/49/1/012025

Characterization of functional groups of airborne particulate matter

M Baitimirova[1], **J Katkevics**[1,2], **L Baumane**[3], **E Bakis**[1], **A Viksna**[1,2]
[1]Department of Chemistry, University of Latvia, 48 Kr. Valdemara Street, Riga, LV-1013, Latvia
[2]Institute of Chemical Physics, University of Latvia, 4 Kronvalda boulevard, Riga, LV-1586, Latvia
[3] Latvian Institute of Organic Synthesis, 21 Aizkraukles Street, Riga, LV-1006, Latvia

E-mail: arturs.viksna@lu.lv

Abstract. Particulate matter of organic combustibles burning consists of various hydrocarbons and radicals, which may cause harmful impact to human health. In this study solid particulate matter were collected on the filters from burning of various combustibles in a burning chamber and from atmosphere of city of Riga by dichotomous impactor. FTIR spectra were obtained before and after samples' treatment. Absorptions associated with aliphatic and aromatic hydrocarbons and alcohol functional groups were observed in the FTIR spectra. Free radicals of particulate matter were detected by electron paramagnetic resonance (EPR).

1. Introduction

When organic combustibles such as candle, kerosene, gasoline or diesel are burning, combustible breaks up and forms other substances by thermal decomposition process. The precursor of formation of black carbon is the formation of first aromatic ring, which molecular weight growth with the formation of polycyclic aromatic hydrocarbons. When the fine particles are formed, they grow up by surface reactions and by agglomeration [1]. Particles of black carbon are not solid and compact, but look like a spongy, which is formed from very fine particles. Airborne particulate matter from atmosphere of city and particles from candles burning are agglomerates of Aitken (1nm - 0.1 μm) and accumulation (0.1-2 μm) mode particles [2]. Particulate matter of candles burning consists of about 27% weight carbon and 73% weight of oxygen [3]. Ambient particulate matter may contain sulfate, ammonium, silicate, inorganic nitrate and organic compounds, which functional groups are hydroxyl, aliphatic carbon and carbonyl [4; 5]. It has been suggested that the nervous system may be more susceptible to attack by organic particles, which appear to have a greater tendency to cross into the olfactory nerve and may pass into the olfactory bulb than inorganic species [6]. Burning of organic combustibles also produces active products of termo destruction – radicals ($\dot{C}H$; $\dot{C}H_2$; $\dot{C}H_3$; $\dot{C}HO$; $\dot{C}_2$; $\dot{C}H\dot{C}H$ and other) [7], which increase harmful impact to the human health. The chemical characterization of particulate matter to which humans are exposed to provides information important to the understanding of our chemical environment and associated health risks. The aim of current studies is to characterize functional groups of airborne particulate matter by Fourier-transform infrared (FTIR) spectroscopy and electron paramagnetic resonance (EPR).

2. Experimental

Airborne particulate matter was collected on the glass fiber filters from burning of various combustibles in a burning chamber and from atmosphere of city of Riga by dichotomous impactor. The dichotomous impactor was placed in Riga, on Kr. Valdemara 48 (~8 m above ground level and ~1 m from building wall). Dichotomous impactor fractionated aerosol particles by size: coarse (2.5-10 μm, which designate

PM_{10}) and fine (≤2.5 μm, which designate $PM_{2.5}$). Particulate matter from atmosphere of city was sampled in February and March, 2013. Candles were used for the burning to obtain fine particulate matter. Kerosene was used in the burning experiment for the larger particles formation. The closed system for different candles and kerosene combustion was laboratory built (Figure 1). Candle or kerosene *(1)* was burning in the closed burning chamber *(2)*. A flow of purified air *(3)* was controlled by reductor and rotameter *(4)*. The pressure in the chamber was controlled by manometer *(5)*. Particulate matter from candles and kerosene combustion was collected on filters, using the particle sampler which consists of filter holder *(6)*, pump *(7)*, rotameter *(8)* and flowmeter *(9)*.

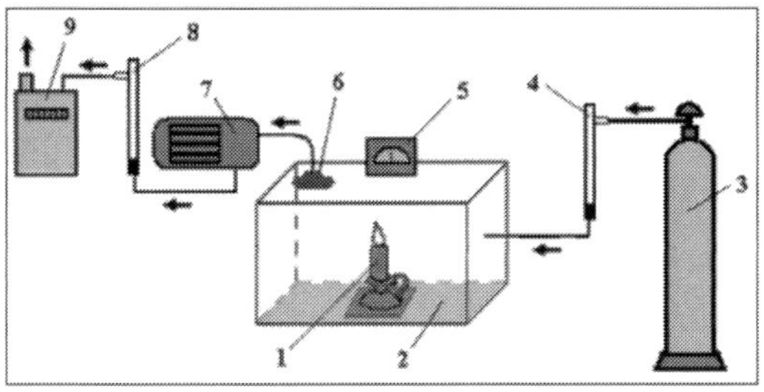

Figure 1. The burning system for the obtaining particulate matter from combustion process (1 – candle, 2 – burning chamber, 3 - bottled air, 4 and 8 – rotameter, 5 – manometer, 6 – filter holder, 7 – pump, 9 – flowmeter)

The functional groups of particulate matter were determined by Fourier-transform infrared spectroscopy (FTIR). The top layer of collected particulate matter was scraped from glass fibber filters. A thin tablet was formed from particulate matter and KBr powder. FTIR spectra were measured using Spectrum Two IR spectrometer (PerkinElmer, USA) with computer program PerkinElmer Spectrum v.10.03.07. The spectral signal was measured from 4000 to 400 cm^{-1} wave lengths. Each spectrum was taken by averaging 6 scans at a resolution of 4 cm^{-1}.

All samples for electron paramagnetic resonance (EPR) analysis were placed in flat dismountable cell WG 806-B-Q. EPR spectra were recorded using an EMX-plus EPR spectrometer (Bruker, Germany). Reference marker ER 4119HS-2100 (g-factor 1.9800, radical concentration $1.15 \cdot 10^{-3}$ %) was used for quantitative EPR. The EPR instrumental settings for field scan were as follows: field sweep, 200G; microwave frequency, 9.78 GHz; microwave power, 0.2 mW; modulation amplitude, 5 G; conversation time 163 ms; time constant, 327 ms; sweep time 167 s; receiver gain, $1 \cdot 10^3$; resolution, 1024 points, 5 scan. The concentration of free radicals in the samples was calculated using the double integration method of the first derivative signal and comparison with the marker (using Bruker's WINEPR program). All treatments of blank filter (without particulate matter) were done parallel with treatments of samples of particulate matter. The spectrum of blank filter was subtracted from obtained EPR spectrum of sample. So, in this paper showed EPR spectra pertain only to particulate matter.

The FTIR spectra of airborne particulate matter were obtained before and after treatment. Samples were dried over phosphorus pentoxide 2 weeks. Tar products were removed by acetone and benzene extraction. Samples of particulate matter were carbonized in nitrogen (N_2) flow at 450° C to remove volatile compounds and compounds, which evaporate at this temperature. Presence of carbonyl group was checked by reduction by sodium borohydride.

3. Results and discussion

Figure 2 show FTIR spectra of airborne particulate matter, collected from burning of various combustibles. Functional groups associated with infrared absorption were identified by previously published FTIR analyses of ambient particulate matter [4; 5; 8] and by guidelines of functional groups identification [9]. Absorptions associated with hydroxyl (3430 cm^{-1}), aromatic C-H (3055 cm^{-1}), ammonium ion (3242 cm-1), aliphatic C-H (2925 and 2857 cm^{-1}), C=C (1631 cm^{-1}), C-H (1435 cm^{-1}) and

C-O (1055 cm^{-1}) functional groups are indicated in figure 2. The sharp peak of aliphatic carbon C-H group at 2925 cm^{-1} was chosen as internal standard, because it has presence in all obtained spectra and it has the same intensity of absorption in all spectra. A baseline was drawn from 4000 cm^{-1} to 500 cm^{-1}. The intensity of peaks was measured from baseline. Ratio of peak intensity to intensity of internal standard was calculated.

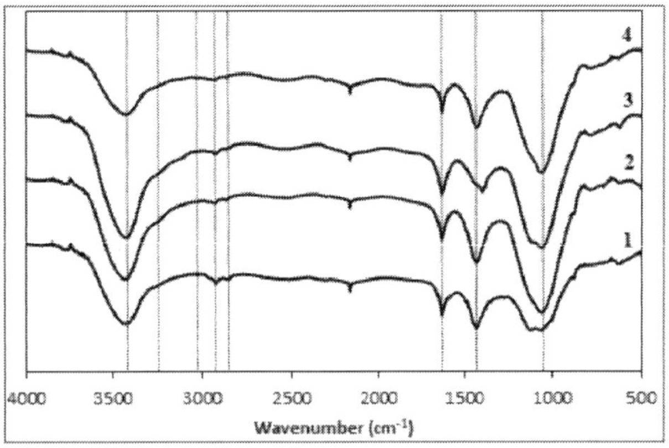

Figure 2. FTIR spectra of particulate matter from burning of various combustibles (1-kerosene, 2 – from atmosphere of city, 3 – paraffin candle, 4 - stearin candle). The dotted lines represent (from left to right) C-OH (3430 cm^{-1}), NH_4^+ (3242 cm^{-1}), aromatic C-H (3055 cm^{-1}), aliphatic C-H (2925 and 2857 cm^{-1}), aromatic C=CH (1631 cm^{-1}), C-H (1435 cm^{-1}) and C-O (1055 cm^{-1})

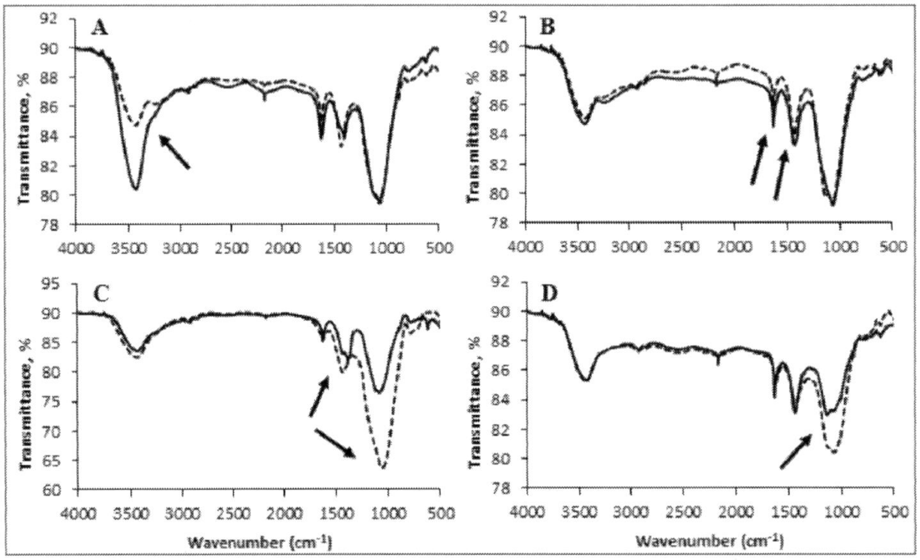

Figure 3. FTIR spectra of particulate matter from kerosene combustion before (solid line) and after (dotted line) treatment (A – drying over P_2O_5, B – extraction, C – reduction, D – carbonization)

Figure 3 show FTIR spectra of particulate matter from kerosene combustion depend of treatment. After particulate matter drying over phosphorus pentoxide, peak intensity of C-OH group (3430 cm^{-1}) was decreased, but not got lost (Figure 3, A). Water, which was produced in the burning process, was removed after drying over P_2O_5. Remained signal at this region indicated about -OH group, which is chemically connected. The peak of ammonium vibrations at 3242 cm^{-1} was uncovered after drying. After extraction by acetone intensity of the peaks of C-H bending (1435 cm^{-1}) and aromatic C=C (1631 cm^{-1}) were decreased (Figure 3, B). That is products of tar were successfully removed. Intensity of C-H bending

(1435 cm^{-1}) and C-O stretching (1055 cm^{-1}) was increased after treatment by sodium borohydride, but intensity at region from 1640 to 1850 cm^{-1} was not changed (Figure 3, C). That means, analyzed particulate matter did not contain carbonyl group, in despite of literature data [4; 5; 8]. After carbonization (Figure 3, D) only intensity of peak of C-O group (1055 cm^{-1}) was changed (it was increased). For the most part in the obtained FTIR spectra intensity of C-O stretching (1055 cm^{-1}) is higher than an intensity of O-H stretching (3430 cm^{-1}). It was assumed that the residuary intensity of C-O stretching may be from radicals, which are formed in the burning process. The same differences of functional groups absorptions were obtained in FTIR spectra of particulate matter from candles burning and from atmosphere of city.

Obtained EPR spectra (Figure 4) show, that all samples of particulate matter from various sources contain organic radicals. In Figure 4 can see a tendency - the smaller particles the higher is concentration of radicals. Particulate matter with diameter ≤2.5 µm ($PM_{2.5}$) contained more radicals than particulate matter with diameter 2.5-10 µm (PM_{10}). Perhaps the reason of smaller quantity of radicals of coarse particles is an agglomeration of particles by radical mechanism.

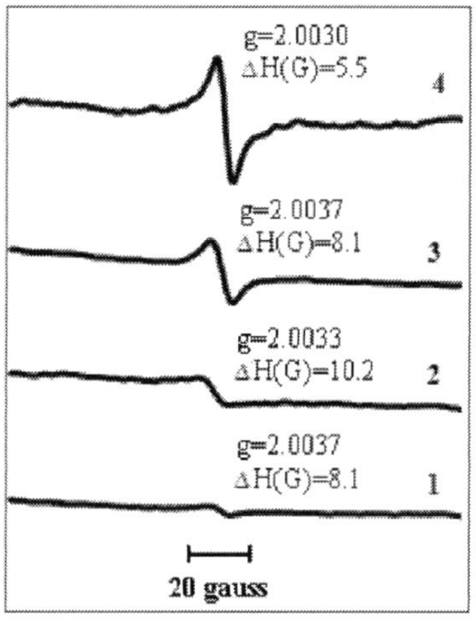

Figure 4. EPR spectra of airborne particulate matter (1 and 3 - PM_{10} and $PM_{2.5}$ from atmosphere of city respectively, 2 – from kerosene combustion, 4 – from paraffin candle burning)

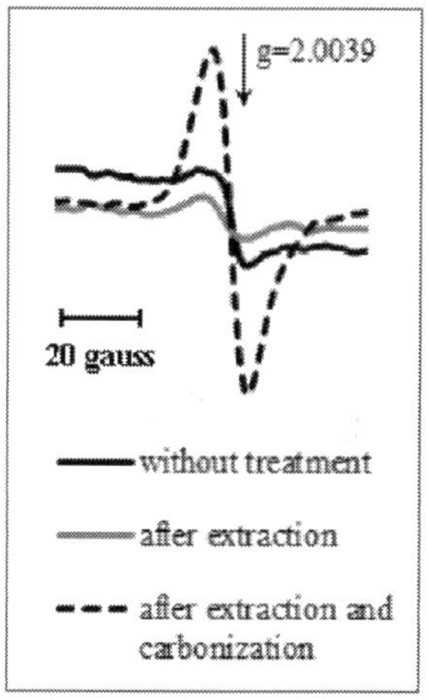

Figure 5. EPR spectra of particulate matter from kerosene combustion depending on treatment

EPR spectra of particulate matter from kerosene combustion after treatment are shown in Figure 5. After acetone and benzene extraction quantity of radicals was a little bit decreased (from $1.1 \cdot 10^{-5}$ % to $0.6 \cdot 10^{-5}$ %). Products of tar contained very low quantity of organic radicals. Extraction and carbonization in N_2 flow removed tar products and volatile compounds from particulate matter. In the EPR spectra we can see, that quantity of radicals is highly increased after carbonization (from $0.6 \cdot 10^{-5}$ % to $2.5 \cdot 10^{-4}$ %), which concurs with increase of intensity at 1055 cm^{-1} in the FTIR spectra (Figure 3, D).

4. Conclusions

Particulate matter obtained from candles and kerosene burnt in the closed chamber and from atmosphere of city consists of organic compounds with aromatic, aliphatic and hydroxyl functional groups. Particulate matter from paraffin candle burning and from atmosphere of city also contains ammonium ions. Particulate matter of various sources contains about $1 \cdot 10^{-4}$ % organic radicals, which are not identified for the time being. The smaller is particles the higher concentration of radicals was observed.

Acknowledgments
This work has been supported by the European Social Fund within the project «Support for Master Studies at University of Latvia». Margarita Baitimirova is grateful for financial support from the European Social Fund (ESF).

References

[1] Lighty J S, Veranth J H, Sarofim A F 2000 *Air & Waste Manage. Assoc.* **50** 1565-1618
[2] Baitimirova M, Osite A, Katkevics J, Viksna A 2012 *IOP Conf. Ser.: Mater. Sci. Eng.* **38** 012056
[3] Shooto D N, Dikio E D 2011 *Int. J. Electochem. Sci.* **6** 1269-76
[4] Maria S F, Russell L M, Turpin B J, Porcja R J 2002 *Atmospheric Enviroment* **36** 5185-96
[5] Tiwary A, Reff A, Colls J J 2008 *Aerosol Science* **39** 40-47
[6] Oberdörster G, Sharp Z, Atudorei V, Elder A, Gelein R, Kreyling W, Cox C 2004 *Inhal Toxicol.* **16** 437-45
[7] Лавров Н В 1971 *Физико-химические основы процесса горения топлива* (Москва: Наука) 272
[8] Maria S F, Russell L M, Turpin B J, Porcja R J, Campos T L, Weber R J, Huebert B J 2003 *J.Geophys. Res.* **108** 8637
[9] Silverstein R M, Bassler G C, Morill T C 1991 *Spectrometric identification of organic compounds* (New York: John Wiley & Sons, Inc.) 91-164

IOP Conf. Series: Materials Science and Engineering **49** (2013) 012026 doi:10.1088/1757-899X/49/1/012026

Effect of porosity on the electrical properties of PZT ceramics

E V Barabanova[1], O V Malyshkina[1], A I Ivanova[1], E M Posadova[1], K M Zaborovskiy[1] and A V Daineko [2]

[1]Tver State University, Tver, Russia,
[2]Research Institute "ELPA", Zelenograd, Russia

E-mail: pechenkin_kat@mail.ru

Abstract. In the present work the effect of porosity ceramics PZT on the dielectric properties was studied. It is shown that the presence of pores leads to decrease of the real part of the dielectric permittivity in the frequency range of 25 Hz – 1 MHz also as to non-stability of the dielectric characteristics in the phase transition region.

1. Introduction

Ferro-piezoelectric ceramics on the basis of lead zirconate titanate perovskite solid solutions PbZr1−xTixO3 (PZT) continue to be the most important ferroelectric and piezoelectric materials in commercial devices [1]. Curie temperature T_c of these materials more than 250 °C. Above this temperature ceramics is in the paraelectric cubic phase. Below the Curie temperature the PZT components have a ferroelectric ($PbTiO_3$, $PbZrO_3$) or antiferroelectric ($PbZrO_3$) ordering. This is confirmed by experimental data on crystals and ceramics of similar composition [2-4]. PZT materials systems have high and stable piezoelectric properties in a broad temperature range. For improving the piezoelectric properties of ceramics PZT it introduce various modifying agents.

The properties of PZT-based materials as well as of other solid solutions depend strongly on the preparation conditions, in particular on the porosity formed during sintering [5,6]. As a consequence both new techniques are developed and old technologies are improved to obtain materials with desired properties.

2. Experimental

The samples of PZT ceramics (PZT system with the addition of niobium pentoxide), which were produced with a given percentage (25%) of the pores were investigated. The PZT ceramics without pores for comparison electrical properties was used.

The dispersion of capacitance and dielectric loss tangent were measured at room temperature in the frequency range 25 Hz - 1 M Hz using immittance meter E7-20. Also the temperature dependence of the dielectric permittivity were measured in the temperature range 25-400°C at 100 Hz, 1 kHz, 10 kHz, 100 kHz and 1 MHz. The heating rate was 2K/min, cooling – 3.4 K/min.

To confirm percent of porosity the grain size and porosity were examined with the aid of scanning electron microscopy (JEOL JSM 6510LV). The images of the cross sections of the PZT samples obtained in secondary electrons on the SEM are shown in Figure 1. As can be seen from the pictures, samples greatly differ in the number and size of pores. In addition, the porosity ceramics have larger grains (about 3-4 microns) compared with normal (non–porosity), ceramics (about 1-2 microns). The results for the porosity sample coincide with a number of published data for PZT ceramics [7].

IOP Conf. Series: Materials Science and Engineering **49** (2013) 012026 doi:10.1088/1757-899X/49/1/012026

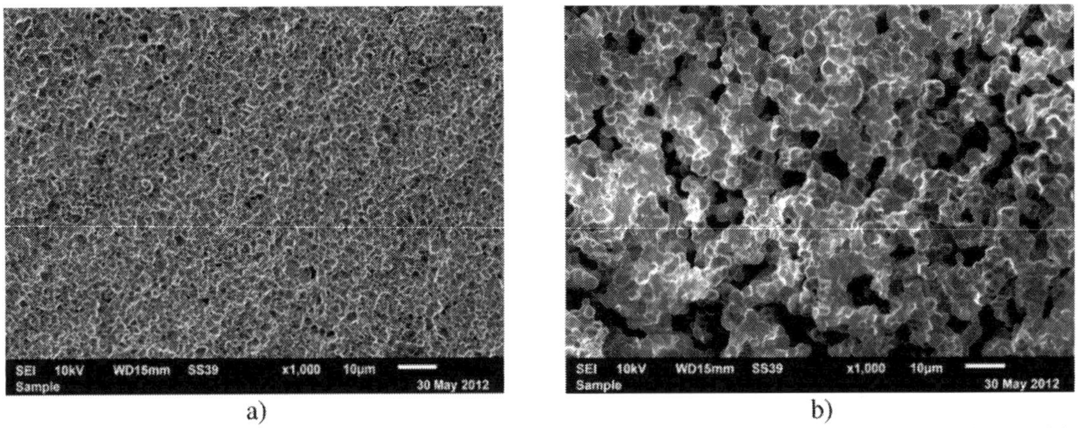

Figure 1. SEM images of the cross sections of the PZT samples pore-free ceramics (a) and that with porosity of 25% (b). Scale mark length 10 μm.

3. Results and Discussion

The frequency dependence of the complex dielectric characteristics for PZT ceramics were obtained from experimental data (Fig. 2). The real part the dielectric permittivity ε'(f) PZT ceramics, as may be seen from the graphs, decreases with increasing percentage content of pores in the sample [8]. Thus there are the dispersion in the frequency range under investigated is the same for both samples. The dielectric loss ε″(f) ceramics without pores are weakly dependent on the frequency, except for the high-frequency range (10^5–10^6 Hz). The absence of maximum of dielectric loss for both samples is probably related to the fact that the process of relaxation polarization corresponds to frequencies more than 10^6 Hz. Therefore for the two samples we do not see the arcs of semicircles on the diagrams of the dielectric dispersion (Fig.3).

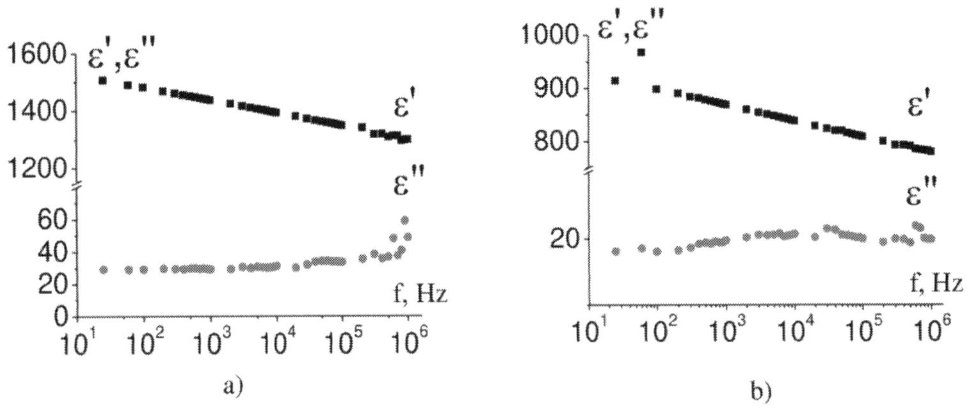

Figure 2. Frequency dependence of the real and imaginary parts of the dielectric permittivity of PZT-19 pore-free ceramics (a) and that with porosity of 25% (b)

Behaviors of the temperature dependence of the dielectric permittivity for the heating and cooling processes are different (Fig.4). The temperature of the maximum dielectric permittivity was observed during the cooling is less than during heating. The difference in temperature ranges is from 3 to 13 degrees. While that in all experiments the temperature change rate were the same both for heating and cooling, depending on given quantity on the frequency was not observed. It should be noted that just as at room temperature the magnitude ε' at the maximum for the porosity sample is less than the pore-free ceramics.

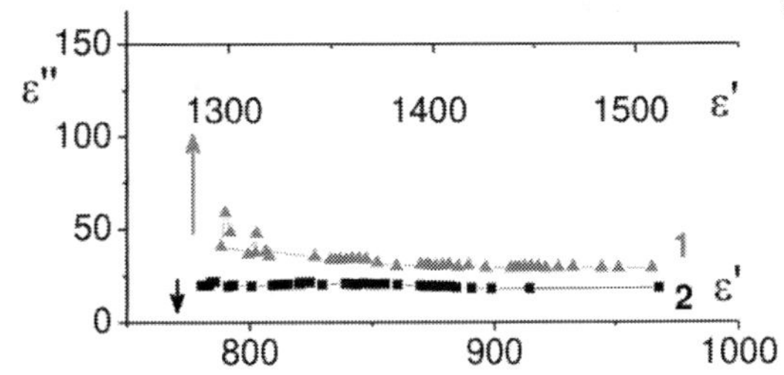

Figure 3. The diagrams of the dielectric dispersion $\varepsilon''(\varepsilon')$ of PZT pore-free ceramics (1) and that with porosity of 25% (2).

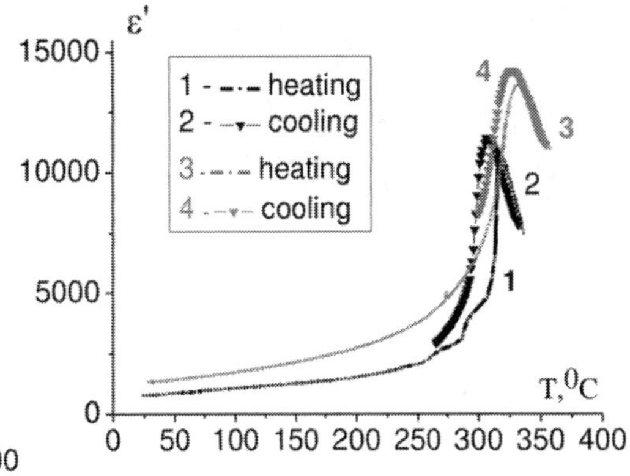

Figure 4. The temperature dependence of the permittivity PZT ceramic, pore-free (3, 4) and with porosity of 25% (1, 2) by heating (1, 3) and cooling (2, 4) at the frequency 1kHz.

Temperature dependences of the real part of the dielectric permittivity that occur during cooling at different frequencies are shown in Fig.5, a. In pore-free ceramic the positions of dielectric permittivity maximum, which corresponding to the phase transition, does not depend on the frequency measurement. It occurs at a temperature of 330 °C. In contrast, the maximum on the temperature dependence of the dielectric permittivity for ceramics containing pores of 25% (Fig.5, b) varies with the measuring frequency. At the same time, depending on the direction of displacement of these maximum from the frequency as it is proper to relaxors not observed. Such a difference in the behavior of the temperature dependence of the permittivity for a sample with porosity 25 % and pore-free ones is observed during heating.

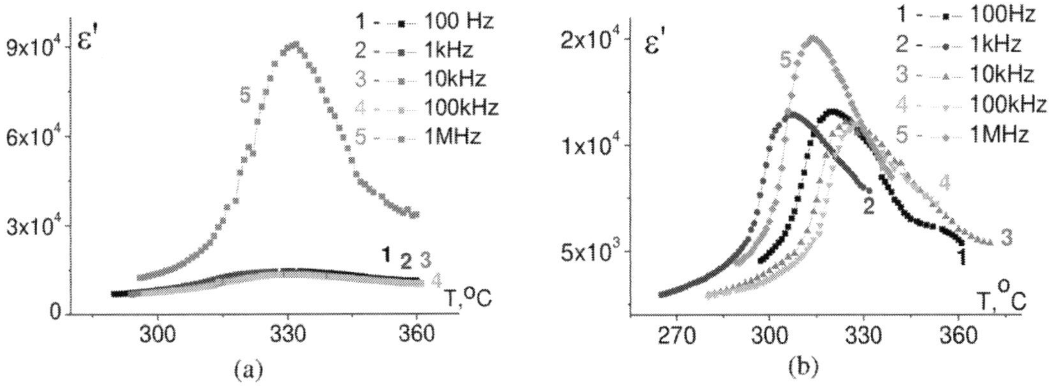

Figure 5. Temperature dependences were measured during cooling for the real part of the dielectric permittivity of PZT pore-free ceramics (a) and that with porosity of 25% (b)

The study of the dielectric hysteresis loops by the Sawyer-Tower method also showed the existence of differences between the properties of pore-free and porosity (25 %) ceramics (Fig.6). Calculated according the dielectric hysteresis loops were obtained remanent polarization of the pore-free sample is 16 μC/cm^2. For porosity (25 %) ceramics this value is about two times less (7 μC/cm^2). The values of the coercive fields are 4.4 kV/cm and 6 kV/cm for pore-free ceramics and that with porosity of 25% correspondingly.

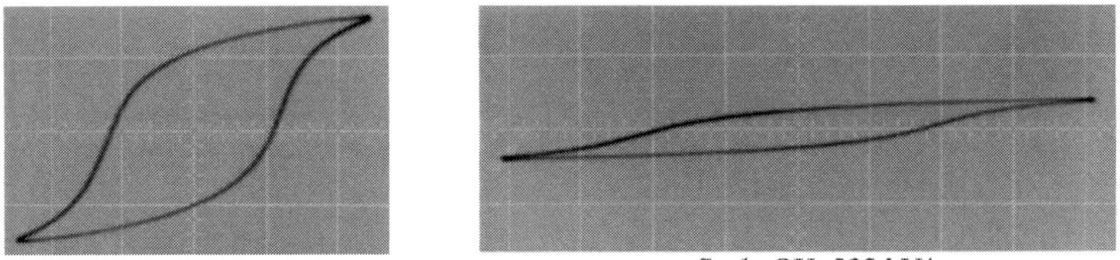

Scale OX: 400 kV/cm;
OY: 15,6 $\mu C/cm^2$.
(a)

Scale OX: 333 kV/cm;
OY: 28,6 $\mu C/cm^2$
(b)

Figure 6. Dielectric loops of PZT pore-free ceramics (a) and that with porosity of 25% (b).

4. Conclusions

Thus, the porosity of ceramics directly affects the dielectric characteristics such as permittivity and spontaneous polarization. Whereas the magnitudes of the above parameters decrease with the increase of porosity, the types of their frequency dependence remain the same. Therefore, specifying certain percent of porosity for the ceramic during the production process, it is possible to obtain samples with the dielectric properties that are necessary for a specific application. This is particularly important when ferroelectrics ceramics are used at high frequencies. In that frequency range, as shown above, increasing the porosity reduces dielectric loss.

The presence of porosity in the ceramic is not promoted to the arisen the relaxor properties, although it leads to instability of the dielectric properties of the phase transition region.

References

[1] ET-102 Piezoelectric Operated Actuators and Motors – A Global Industry and Market Analysis (Web: www.innoresearch.net)
[2] Yaffe B, Cook WR, Yaffe R 1971 *Piezoelectric ceramics* (London–NY: Academic Press)
[3] Heywang W, Lubitz K, Wersing W 2008 *Piezoelectricity Evolution and Future of Technology* Springer – Verlag Berlin Heidelberg
[4] Guo R, Wang C A, Yang A K, Fu J T 2010 J. Appl. Physics 108 124112
[5] Praveenkumar B, Kumar H H, Kharat D K 2005 Bull. Mater. Sci. 28 453-455
[6] . Ting R Y 1985 *Ferroelectrics* **65** 11-20
[7] Mercadelli E, Sanson A and Galassi C 2010 Porous piezoelectric ceramics *Piezoelectric ceramics* ed E Suaste Gymez (Rijeka, Croatia – Sciyo) pp 111-128 (www.intechopen.com)
[8] Piazza D, Galassi C, Barzegar A, Damjanovic D 2010 *J Electroceram* 24 170–176

IOP Conf. Series: Materials Science and Engineering **49** (2013) 012027 doi:10.1088/1757-899X/49/1/012027

Effects of illumination on the dielectric response of barium-strontium niobate ceramics

K Bormanis[1], AI Burkhanov[2], Luu Thi Nhan[3], SV Mednikov[3] and M Antonova[1]
[1] Institute of Solid State Physics, University of Latvia, Riga, LV-1063, Latvia;
[2] Volgograd State Architectural and Engineering University, Volgograd, Russia
[3] Volgograd State Technical University, Volgograd, Russia

E-mail: bormanis@cfi.lu.lv

Abstract. A study of the effects of white light on the low and infra-low frequency relaxation of polarization in the barium-strontium niobate (SBN) ceramics is reported. The light is found considerably decreasing the contribution of space charge at temperatures corresponding to the range of the relaxor phase (it is, around the T_m).

1. Introduction

Presently the studies of photosensitive materials are of high importance, for instance, the photo-response of submicron capacitor structures such as M/Pb(ZrTi)O_3(PZT)/M to visible or UV irradiation in relation to possible use of photovoltaic currents [1,2] for non-destructive reading of information. However, toxicity (evaporation of lead) during production of the PZT materials used in a variety of devices is urging to look for alternative materials without lead one of which being barium-strontium niobate (SBN) [3]. Since long the ferroelectric (FE) $Sr_{0.75}Ba_{0.25}Nb_2O_6$ (SBN-75) attracts attention of researchers as a photosensitive ceramic material [1,4,5].

The present study is focused on the effect the illumination by light at low intensity has on the dielectric response at low and infra-low frequencies in the SBN-75 ceramics.

2. Samples and methods

The sample of barium-strontium niobate $Sr_{0.75}Ba_{0.25}Nb_2O_6$ (SBN-75) used for the study was obtained by conventional ceramics technology. Dielectric measurements were performed on a plate of the size of 6mm×5mm×0.88mm supplied with silver paste electrodes. One electrode had several series of perforations 0.5 mm in diameter. A 5034W2C-DSA-A LED providing "cold" visible radiation at the intensity of 0,15 mW/cm^2 was used as the source of light directed at right angle to the perforated electrode parallel to the measuring field E. Since depth of penetration by visible light of that intensity into opaque ceramics is negligible, the effects were assumed to take place mainly in the layer adjacent to the perforated electrode, and surface area of the electrode was used in calculations of the dielectric response.

Measurements of polarization loops in dark and under light were performed sequentially heating up the sample beforehand cooled below the room temperature to a series of selected fixed thermostatic temperatures. Polarization loops were obtained under continuous illumination at different field amplitudes and frequencies of 0.1, 1, and 10 Hz. After that light was switched off and the next temperature set up to repeat the cycle of measurements.

Effective dielectric permittivity $\varepsilon'_{eff} = P_E / \varepsilon_0 E_m$ and dielectric loss ($\varepsilon''_{eff} = tg\delta \cdot \varepsilon'_{eff}$, $tg\delta = \dfrac{S}{\pi E_m P_E}$) where E_m is the amplitude of the sinusoidal field at measuring polarization loops, P_E – polarization at E_m, ε_0 – the dielectric constant, S – the polarization loop area in the *P-E* coordinates, were calculated from data obtained on a Sawyer-Tower circuit modified to be used connected to a computer.

3. Results and discussion

The shapes of polarization loops (PL) at frequency 0.1 Hz and temperatures of 14 °C, 20 °C, and 62 °C of dark (a) and illuminated (b) sample are shown in figure 1. The selected temperatures cover the range of the broad phase transition in SBN-75 ceramics as found [6] from $\varepsilon'(T)$ and $\varepsilon''(T)$ measurements at low field intensities.

As seen from figure 1, a considerable relaxation of polarization is displayed by the rounded PL at 62 °C in agreement with the conclusion [6] of an essential effect of conductivity around T_m of the material. The same general pattern of PL is maintained at 23 °C; only at $T < T_{room}$ (14 °C) a typical response of a FE material is displayed by the Rayleigh pattern of PL at amplitude $E<E_C$ (E_C – coercive field) [6]. Comparing the loops in dark and illuminated sample at 62 °C reveals a decrease of the loop area – narrowing of the loops caused by illumination. Effective parameters – effective dielectric permittivity $\varepsilon_{e\square ff}$, effective dielectric loss ε''_{eff} and $tg\delta_{eff}$, obtained from PL are used hereafter for quantitative assessment of the effects of illumination on the polarization process.

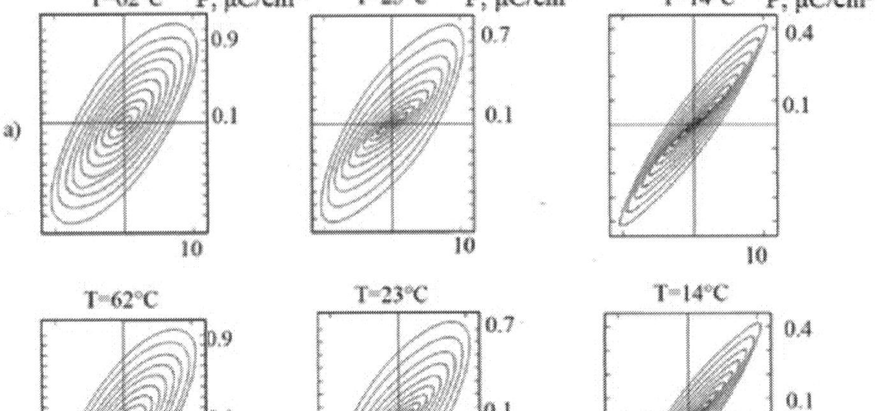

Figure 1. Dependence of polarization loops in SBN-75 ceramics on field intensity at frequency 0.1 Hz: a – in dark, b – at illumination.

Dependence on field intensity of $\varepsilon_{e\square ff}(E)$ and $\varepsilon''_{eff}(E)$ at frequency of 0.1 Hz at different temperatures in dark and illuminated sample (figure 2) shows anomalies on both curves at 62 °C. In the first case (a) – as a local maximum on $\varepsilon_{e\square ff}(E)$ at E ≈ 8–8.5 kV/cm and a flattening on the $\varepsilon''_{eff}(E)$ curve in the 6–8 kV/cm range (b). In the other case – as flattening of $\varepsilon_{e\square ff}(E)$ at E ≈ 8–10 kV/cm (c) and as a bending point on the $\varepsilon''_{eff}(E)$ curve in the 7–8 kV/cm amplitude range (d). Behaviour of the parameters of SBN-75 at 62 °C, which is below T_m ≈ 110 °C [6], might point to saturation of polarization P(E) as the cause of the additional maximum of $\varepsilon_{e\square ff}(E)$. The field intensity at which the broad maximum appears on $\varepsilon''_{eff}(E)$ would correspond to coercive field E_c when the material in the FE state starts "switching" under applied field exceeding E_c. However, the oval pattern of PL at 62 °C (figure 1) remaining the same at all field intensities does not allow relating the anomalies to the FE nature of switching polarization.

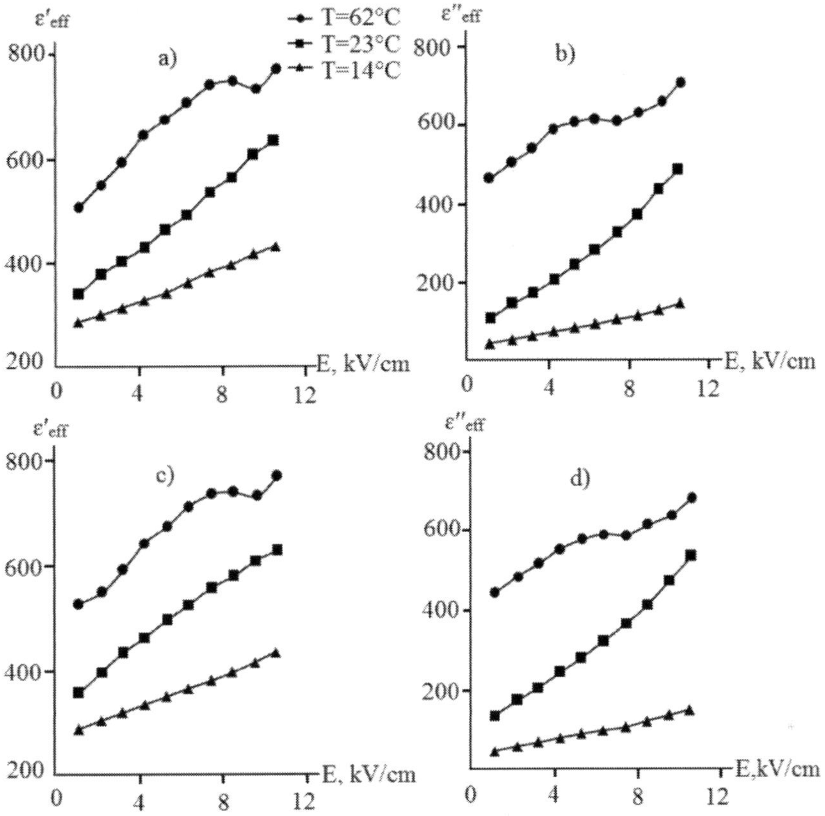

Figure 2. Dependence of ε'_{eff} *(E)* and ε''_{eff} *(E)* on field intensity at 0.1 Hz frequency at temperatures T = 14°C, T = 23°C and T = 62°C in dark (a, b) and illuminated (c, d) sample.

Most likely the anomalies are due to a substantial relaxation of space charge at 62 °C and frequency 0.1 Hz the maximum contribution made by field intensity amplitudes between 4 and 8 kV/cm. An evidence of it can be provided by $tg\delta_{eff}$ (figure 3) the magnitude of which is proportional to the area of the polarization loop.

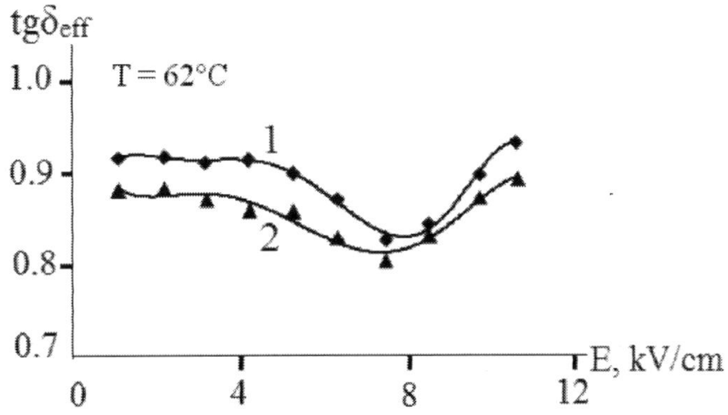

Figure 3. Field dependence of $tg\delta_{eff}(E)$ at frequency 0.1 Hz in dark (curve 1) and illuminated (curve 2) sample.

As follows from comparison of effective parameter values of dark and illuminated sample, the light mainly affects the behaviour of $tg\delta_{eff}$ and ε''_{eff}. Illumination decreases the values of $tg\delta_{eff}$ and ε''_{eff} while shifting the anomalies of $\varepsilon_{e\square ff}(E)$ and $\varepsilon''_{eff}(E)$ to a stronger field intensity E.

Since the main contribution to relaxation of polarization at infra-low frequencies and $T > T_{room}$ is provided by space charge, the decrease of dielectric loss observed at illumination of the sample might be related to the effect of it on conductivity of the material. In the present case, unbalanced carriers emerging in ferroelectrics under illumination [5] would reduce the space charge and, consequently, reduce the effect of it on the relaxation of polarization. Under conditions of through-flow conductivity starting at higher field amplitudes (the sharp increase of $tg\delta_{eff}$ at $E > 8$ kV/cm, figure 3) the light-induced charge carriers have no effect on the features of the dielectric response of the material. Behaviour of $\varepsilon_{e\square ff}(E)$ and $\varepsilon''_{eff}(E)$ at lower temperatures and higher frequencies when the contribution of conductivity to dielectric response is substantially smaller point to conductivity of the ceramics being mainly affected by illumination. As it follows from figure 1, the dielectric response at 14 °C and 0.1 Hz is typical to FE ceramics the effect of illumination being around 1 % – practically of the order of accuracy of the PL measurements.

The values of $\varepsilon_{e\square ff}$ and ε''_{eff} measured at frequency of 10 Hz in dark and illuminated sample are practically the same at 14 °C and 23 °C, only at temperatures closer to T_m the light-induced increase of $\varepsilon_{e\square ff}$ is detectable. In this case the increment of $\varepsilon_{e\square ff}$ can be related to the effect of unbalanced carriers on the intrinsic field of defects as shown by the study of dielectric response in SBN-75 single crystal [7]. The increase of $\varepsilon_{e\square ff}$ is achieved by partial compensation of the intrinsic field facilitating mobility of domain (phase) boundaries under applied field. Decrease of the coercive field E_c under illumination has been reported in studies of light-induced effects on polarization switching in SBN-61 crystals [8].

4. Conclusions

By comparing the values of effective parameters $tg\delta_{eff}$, $\varepsilon_{e\square ff}$ and ε''_{eff} obtained from polarization loops of dark and illuminated SBN-75 ceramic samples within the range of temperatures of the relaxor phase at frequencies below 10 Hz and characterising relaxation of polarization the effect of light is found to present a decrease of the $tg\delta_{eff}$ and ε''_{eff} values.

Unbalanced carriers induced by light in the SBN-75 ceramics decreasing the space charge and its contribution to relaxation of polarization most likely produce the decrease of the effective dielectric loss in illuminated samples.

5. References

[1] Glass A M, Von der Linbe D, Nerren T J 1974 *Appl. Phys. Lett.* **4** 233
[2] Delimova L A, Yuferev V S, Grehov I V et al 2009 *Solid State Physics* **51** 1149 (in Russian)
[3] Venet M, Santos I A, Eiras J A, Garcia D 2006 *Solid State Ionics* **177** 589
[4] Lines M E and Glass A M 1977 Principles and Application of Ferroelectrics and Related Materials (Oxford Clarendon Press)
[5] Fridkin V M 1979 Photoferroelectrics (Moscow Science in Russian)
[6] Bormanis K, Burkhanov A I, Mednhikov S V, Nhan L T, Kalvane A, Antonova M 2011 *Ferroelectrics* 417 58
[7] Burkhanov A I, Guzhakovsksya K P, Ivleva L I 2010 *News of RAS ser. Phys.* **74** 1292
[8] Smith P G R and Eason R W 1996 *Appl. Phys. Lett.* **69** 1509

The effect of annealing on spectra and decay time of X-ray luminescence of zinc oxide powders

K A Chernenko[1], O G Klimova[1], E I Gorokhova[2], G G Klimov[1], A V Semencha[1] and P A Rodnyi[1]
[1]St.Petersburg State Polytechnical University,
Politekhnicheskaya ul.29, St.Petersburg, 195251 Russia
[2]Research and technological institute of optical materials all-russia scientific center "S.I.Vavilov State Optical Institute",
ul. Babushkina 36, St.Petersburg, 190164 Russia

E-mail: nuclearphys@yandex.ru

Abstract. The annealing effects on zinc oxide powders have been investigated. Annealing was performed at 800, 1050°C in vacuum and air for 2 h. After that, the powders were annealed in gas mixture consisted of 15% H_2 and 85% Ar for 30 min at 800°C. After vacuum annealing the powders have green luminescence band located at 515 nm and decay time curve of the luminescence has three exponential decays with time constants: 10-50 ns, 100-200 ns, and 1100 ns. After air annealing powders have green luminescence band with peak at 525 nm and decay time curve of the luminescence has three exponential decays with time constants: 5 ns, 100-200 ns, and 4000-5000 ns. Annealing in hydrogenous ambience has no effect on peak position or intensity for powders previously annealed vacuum. In air annealed powders intensity of green luminescence decreases significantly after annealing in hydrogenous ambience.

1. Introduction

Zinc oxide (ZnO) is a direct wide band gap (3.37 eV) semiconductor material with a large exciton binding energy (60 meV). It gained significant attention motivated by potential applications as an oxide electronic material, and in optoelectronic and lighting applications. Luminescence of ZnO, generally, has two bands: near band edge luminescence (NBEL), ascribed to excitons [1], and green luminescence (GL). The origin of the GL is still a matter of debates, despite a great number of studies. As a rule, GL is ascribed to native defects. Lack of reliable theoretical calculations of defect energy levels and uncertainty of interpretation of EPR data makes difficult to analyze experimental results [2]. In most cases GL is attributed to zinc [3, 4] or oxygen [5, 6] vacancies. Some authors assume that GL can be connected to both defects [7, 8].

By applying different annealing atmospheres one can indicate if an increase/decrease in the concentration of a specific intrinsic defect is correlated with the GL properties.

In this work, we present the effect of vacuum or air and H_2-Ar annealing on the shift of the emission, intensity and decay time changes of ZnO powders.

2. Experimental

Chemical purity (powder #1, 99,9%, Russia) and high purity (powder #2 99,999%, USA) powders were used. Annealing was performed at 800, 1050°C in vacuum and air for 2 h. After that, powders were annealed in gas mixture of 15% H_2 and 85% Ar for 30 min at 800°C.

The X-ray excited luminescence (XRL) spectra were measured under steady state X-ray excitation (40 kV, 10 mA) with a FEU-106 PMT and a MDR-2 grating monochromator with 1200 grooves/mm. The spectra were corrected for the quantum efficiency of the PMT and for the monochromator transmission.

The decay time curves were recorded with the experimental system described in detail in [9]. A pulsed X-ray source with 1 ns pulse duration operates at 30 kV and maximum current amplitude of 500 mA. The detection system consists of a FEU-71 PMT, operating in single photon counting mode and related electronics. The system allows measurements with time resolution better than 400 ps. The XRL spectra and decay time curves were measured in the reflection mode: the angle between X-ray beam and the PMT was 90 deg. All measurements were performed at room temperature.

3. Results

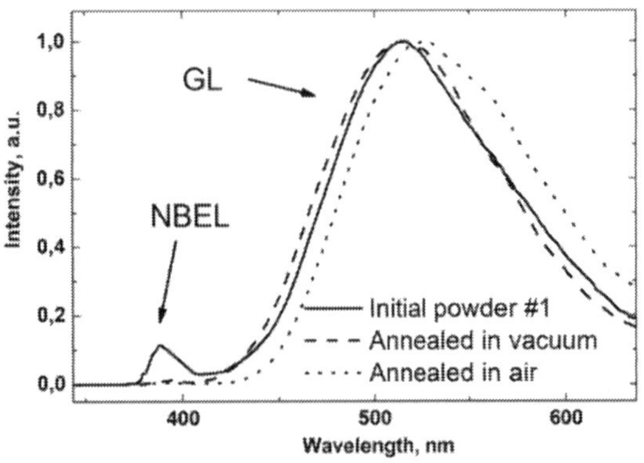

Figure 1. X-ray excited luminescence spectra of powder #1 after annealing in vacuum and air. All intensities are normalized.

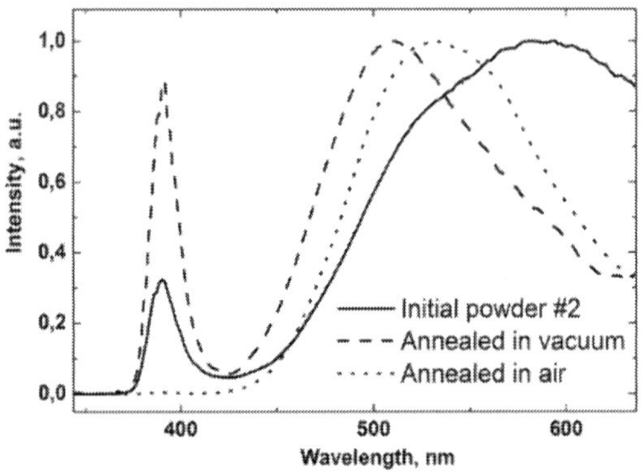

Figure 2. X-ray excited luminescence spectra of powder #2 after annealing in vacuum and air. All intensities are normalized.

The XRL spectra of powder #1 are shown in Fig. 1. Spectrum of initial powder has broad GL band at 513 nm and peak of NBEL at 389 nm with low intensity. After vacuum annealing GL maximum of powder #1 is around 513 nm (same as initial powder) and intensities of GL are about 10 (for 800°C) and 20 (for 1050°C) times greater than the initial one. GL maximum of both air annealed powders is around 525 nm and intensity is 5 times greater than the initial one.

The XRL spectra of powder #2 differ from that of powder #1 (Fig.2). Initial powder #2 has peak of GL intensity at 585 nm. After vacuum annealing NBEL intensity grows up towards to GL ones, but complete intensity of luminescence didn't change significantly. GL maximum of vacuum annealed powder is at 510 nm. Air annealing provides result completely the same to powder #1: rise of GL intensity and its maximum at 525 nm. Powder #2 had very weak luminescence intensity, so for further research we used powder #1.

Decay time curves of air and vacuum annealed powders are presented in Fig. 3. All the powders have complex decay time curves that consist of several exponential decays. Decay curves of air annealed powders are identical. One can see that slow component becomes faster for vacuum annealed powders. Decay time curve of air annealed powders have three clearly separable components. The estimated decay time constants are 5 ns, 100-200 ns, and 4000-5000 ns. Decay time curves of vacuum annealed powders can be approximated by two exponential decays but if we take into account the case of air

IOP Conf. Series: Materials Science and Engineering **49** (2013) 012028 doi:10.1088/1757-899X/49/1/012028

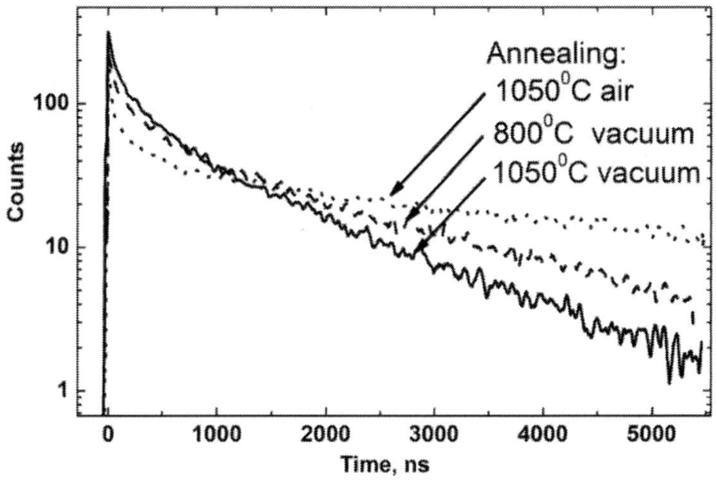

Figure 3. Decay time curves of powder #1 after annealing in vacuum and air.

annealed powders we should expect at least three decay times. Approximation by three exponential decays gives time constants of: 10-50ns, 100-200ns, and a slow decay time constant. For powder annealed at 800^{0}C decay time of slow component is about 1600 ns and for powder annealed at 1050^{0}C decay time of slow component is about 1100 ns.

Results of annealing in hydrogenous ambience are completely different for powders previously annealed in vacuum and air (Fig.4). For vacuum annealed powders annealing in hydrogenous ambience has no effect on peak position or intensity. Intensities of GL decrease significantly in air annealed powders and at the same time NBEL intensities grow up to GL ones.

Comparing decay time curves before and after annealing in hydrogenous ambience (Fig. 5) it can be seen that fast components of decay time curves have some differences and slow components are the same. So it can be clearly seen that decay time curves of vacuum annealed powders contained fast component which is quenched by annealing in hydrogenous ambience. Obviously, this component can produce inaccuracy in estimating of decay times for other components.

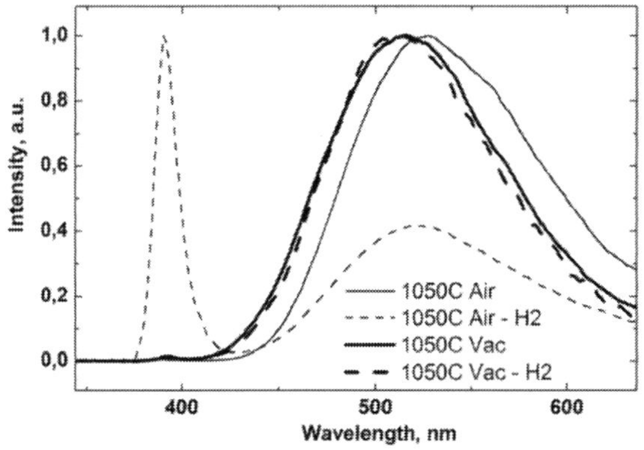

Figure 4. X-ray excited luminescence spectra of air and vacuum annealed powders #1 after annealing in hydrogenous ambience. All intensities are normalized.

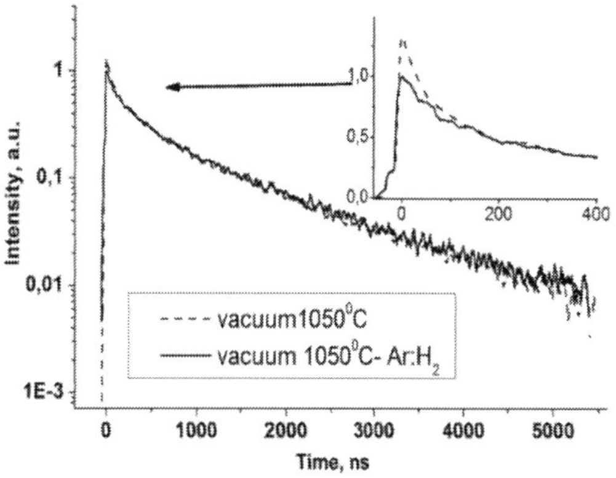

Figure 5. Comparison of the decay time curves of vacuum annealed powder #1 before and after annealing in hydrogenous ambience.

4. Discussion

The effects of annealing in vacuum and air should be caused mainly by creation of native defects: oxygen vacancies and zinc vacancies, respectively. For certain, there are a lot of other types of native defects in the zinc oxide but they have high formation energy [10] and cannot have a great influence except of interstitial zinc. Interstitial zinc has low formation energy [10] and acts as a shallow donor, so it can participate in energy transfer processes.

Hydrogen can extract an excess of oxygen; as it comes from DFT calculations [11], it can passivate zinc vacancies by creation of $(V_{Zn}-H_2)^0$ centers and it can act as shallow donor.

Changes in X-ray luminescence spectra that come after annealing can be explained by assuming that GL with peak at 510-515 nm is related to oxygen vacancies and GL with peak at 520-530 nm is related to zinc vacancies.

Annealing in vacuum creates oxygen vacancies therefore intensity of GL increases. After that, annealing in hydrogenous ambience can't produce significant effect.

Annealing in air reduces number of oxygen vacancies and it creates zinc vacancies. Formation energy of zinc vacancies is higher [10, 12], so the GL has lower intensity. The following annealing in hydrogenous ambience reduces the excess of oxygen and passivates zinc vacancies. As a result, GL intensity decreases dramatically.

Powders #1 and #2 had great difference in GL intensity after vacuum annealing. Oxygen vacancies have to have specific charge state to take part in the luminescence process. As shown in [13], they should be singly or doubly ionized vacancies. Charge state of the vacancies is determined by the Fermi level [10], which is dependent on presence of donor or acceptor impurities. We suppose that powder #1 had acceptor impurities that led to creation of singly ionized vacancies and strong GL.

Decay time curves are not clearly understood yet. They have complex structure that consists of several components with similar decay times, so it is hard to clearly resolve them. In addition, it is unclear why the change of the slow component decay time occurs after annealing. As a hypothesis, an appearance of a new nonradiative decay channel or changes in the properties of a complex luminescence center can be assumed. It should be noted that if donor-acceptor pairs had taken part in the luminescence then decay time curves would have had hyperbolical decay.

5. Conclusions

Vacuum and air annealing of ZnO powders causes rise of green luminescence with peaks at 510-515 nm and 525-530 nm, respectively. Despite the fact that peaks have very small difference in position, luminescence has different origins. Luminescence band with maximum at 510-515 nm is connected with oxygen vacancies and luminescence band with maximum at 525-530 nm is connected with zinc vacancy. However, complex decay time curves show that luminescence cannot be associated with single vacancies.

6. References

[1] Meyer B, Alves H and Hofmann D 2004 *Phys. stat. sol.* B **241** 231
[2] Janotti A and Van de Walle C G 2009 *Rep. Prog. Phys.* **72** 126501
[3] Kohan A, Ceder G, Morgan D and Van de Walle C G 2000 *Phys. Rev.* B **61** 15019
[4] Chen H, Gu S and Tang K 2011 *J. Luminesc.* **131** 1189
[5] Leiter F, Alves H, Hofstaetter A, Hofmann D and Meyer B 2001 *Phys. Status Solidi* B **226** R4-R5
[6] Cao B, Cai W and Zeng H 2006 *Appl. Phys. Lett.* **88** 161101
[7] Berseth M, Swensson B, Kuznetsov A, Klason P, Zhao O and Willander M *Appl. Phys. Lett.* **89** 262112
[8] Ton-That C, Weston L and Phillips M 2012 *Phys. Rev.* B **86** 115205
[9] Potapov A, Rodnyi P. and Mikhrin S 2004 *Rad. Measurements* **38** 839
[10] Janotti A and Van de Walle C G 2007 *Phys. Rev.* B **76** 165202
[11] Lavrov E, Weber J and Bornnert F 2002 *Phys. Rev.* B **66** 165205
[12] Richard C, Catlow A, Sokol A and Walsh A 2011 *Chem. Commun.*. **47** 3386-3388
[13] Manoranjan G, Ningthoujam R and Vatsa R. 2011 *J. Appl. Phys.* **110** 054309.

IOP Conf. Series: Materials Science and Engineering **49** (2013) 012029 doi:10.1088/1757-899X/49/1/012029

On the Possibilities for Improving the Efficiency of Radiation in Heterostructures Based on IV-VI Semiconductors

A. M. Pashaev[1], O. I. Davarashvili[2], M. I. Enukashvili[2], Z. G. Akhvlediani[2, 3], L. P. Bychkova[2] and M. A Dzagania[2]

[1] National Aviation Academy, Baku, Azerbaijan
[2] Iv. Javakhishvili Tbilisi State University, Tbilisi, Georgia
[3] E. Andronikashvili Institute of Physics, Tbilisi, Georgia

E-mail: omardavar@yahoo.com

Abstract. It is shown that, even in isoperiodic heterostructures, significant mismatch arises during the process of epitaxial growth because of interdiffusion of the components at the heteroboundary. As a result, nonradiative recombination at the formed defects increases, which significantly decreases the quantum yield of radiation. The heterostructures with active layers of pre- and post-inverse compositions are discussed. For overcoming the adverse effect of diffusion mismatch, it is proposed to bring the composition of the emitter closer to that of the active layer or to dope this layer with the impurities increasing the elastic limit.

1. Introduction

The search for new ways of increasing the operating temperatures of IR semiconductor lasers based on the IV-VI compounds continues [1]. It was noticed that, during the process of epitaxial growth of heterostructure layers, because of interdiffusion of the components and in the case of initially isoperiodic layers, a transition region of a variable composition was formed. The halfwidth of this region is larger than the critical thickness, while the maximum epitaxial deformation in it approximates the limiting value of the elastic state [2, 3]. If we consider the heterostructure as a radiative one, i.e. as an emitter-active layer pair, the approximation of elastic deformations to their critical values (formation of a network of dislocations) can cause decreases in the current carrier lifetime and in the quantum yield of radiation in the active layer.

This work deals with modeling of diffusion mismatch in the IV-VI heterostructures and the investigation of the ways of overcoming its adverse effect on the properties, in particular, of radiating devices (lasers).

2. Diffusion Mismatch Model and its Topicality

Behind the diffusion mismatch model are the experimental and theoretical investigations where the heteroboundary smearing mainly in isoperiodic structures was revealed and assessed [4]. The structures with active layers of pre- and post-inverse compositions were considered. As is well known, in solid solutions of chalcogens with lead and tin, with the increasing tin concentration the forbidden gap width decreases to zero at a certain tin concentration and then increases again in post-inverse compositions. In this case, the derivatives of the forbidden gap width, for instance, by composition and temperature reverse the sign. Table 1 gives the maximum relative mismatch (deformation) $(\Delta a/a_0)_{max}$ and the values of critical deformations calculated by the interpolation formula. Because of the difference in the tabular values of components in heterostructure layers at typical epitaxy temperature 500–600 °C, counterflows emerge at the heteroboundary. Owing to high concentration of nonstoichiometric defects $>10^{18}$ cm^{-3}, the diffusion coefficients of tin and tellurium are taken equal to 10^{-14} and 10^{-12} cm^2/s, respectively. The diffusion distribution by composition (x, y) along the active layer was determined by the probability integral $\mathrm{erf}\left(\frac{d}{2\sqrt{D_{Sn,Te}\,t}}\right)$ depending on d, diffusion coefficients D_{Sn} and D_{Te} and epitaxy time t. For definite time t, local compositions (x, y) at corresponding values of d, and through them – function $a(d)$ were determined by the relation :

$$a = 6.126 - (0.01y + 0.123)x + 0.334y \tag{1}$$

Table 1. Maximum relative and critical deformations for active layers of pre-inverse compositions $Pb_{1-x}Sn_xSe_{1-y}Te_y$

No.	Emitter composition		Active layer composition		$\Delta y = y_a - y_e$	$(\Delta a/a_0)_{max}\cdot 10^{-3}$	$\varepsilon_{cr}\cdot 10^{-3}$
	x	y	x	y			
1	0	0.734	0.231	0.825	0.091	1.87	1.7
2	0	0.213	0.178	0.280	0.067	1.42	2.11

Taking the plane with the maximum value $\Delta a/a_0$ as a new heteroboundary, it makes sense to compare the obtained values with critical deformations ε_{cr} obtained when measuring the internal friction [5]. The following ε_{cr} values were obtained for binary compounds $\varepsilon_{cr}\cdot 10^{-3}$: 2.1 (PbSe), 3.7 (SnSe), 1.8 (PbTe), 3.5 (SnTe). In the first and second cases (Table 1), the ε_{cr} values were calculated by the interpolation formula: $\varepsilon_{cr} = (x, y) = \varepsilon_{PbSe}(1-x)(1-y) + \varepsilon_{SnSe}\,x\,(1-y) + \varepsilon_{PbTe}(1-x)\,y + \varepsilon_{SnTe}\,x\cdot y$ (2)

The ε_{cr} values given in Table 1 correspond to the elastic limit deformations under torsional vibrations of block monocrystals. Though, for the layers, ε_{cr} values should have been smaller, the maximum relative deformations had some spread of values even without consideration for additional defects by the dislocation density in individual cases and approximated the calculated ε_{cr} values. A similar analysis could be performed by comparing the critical thickness h_{cr} with the halfwidth of mismatched regions Δ. However, because the mismatched region halfwidth appreciably depends on the growth time, while h_{cr} changes less in time, such a comparison would be time-dependent as well as in the first case. Although, just at the first consideration by the maximum relative deformations, the elastic state was revealed at the thickness less than h_{cr}. Thus, relation between $(\Delta a/a_0)_{max}$ and ε_{cr} and position $(\Delta a/a_0)_{max}$ near the heteroboundary depend from the growth time and thickness of the layers. Table 2 gives the values of maximum relative and critical deformations for active layers of post-inverse compositions. As is obvious from table 2, in this case the maximum deformations exceeded the elastic limit and plastic deformation occurred, which caused dislocation multiplication, a decrease in the current carrier lifetime and, accordingly, in the quantum yield of radiation. In this case, because of the growing rate of nonradiative recombination, threshold current in lasers increases.

According to [1]:

$$I_{th} \sim \left(\frac{\tau_{ef}}{\tau}\right)^{-1} = \frac{d + 2S\tau}{d} \tag{3}$$

where τ_{ef} is the current carrier lifetime in the active layer with account for recombination centers at mismatch, τ is the initial current carrier lifetime in the active region (corresponding to the current carrier concentration), d is the active region thickness, S is the surface recombination rate. In turn, $S = \sigma\cdot v\cdot N$, where $\sigma = a^2$ is the recombination cross-section; v is the thermal velocity of current carriers; $N = 1/a^2\cdot\Delta a/a_0$ is the density of recombination centers (two-dimensional density of dislocations).

Table 2. Maximum relative and critical deformations for active layers of post-inverse compositions $Pb_{1-x}Sn_xSe_{1-y}Te_y$

No.	Emitter composition		Active layer composition		$\Delta y = y_a - y_e$	$(\Delta a/a_0)_{max} \cdot 10^{-3}$	$\varepsilon_{cr} \cdot 10^{-3}$
	x	y	x	y			
1	0.052	0.720	0.506	0.900	0.18	3.68	2.67
2	0.020	0.177	0.427	0.375	0.20	3.26	2.68

Taking into consideration that $\Delta a/a_0$ are maximum values equal to $3 \cdot 10^{-3}$ in the mismatched layer, $v=3 \cdot 10^7$ cm/s, $\tau = 5 \cdot 10^{-9}$ s and $d = 0.6$ μm, because of diffusion mismatch, the threshold current in lasers increases ~ 8 times. Thus, it is important to find the ways of neutralization of the adverse effect of diffusion mismatch.

3. The Ways of Elimination of the Adverse Effect of Diffusion Mismatch on the Characteristics of Radiating Devices (Lasers)

From the above data, it is obvious that, if we decrease the concentration of Te between the active layer and the emitter by 0.024, the maximum mismatch will decrease by almost a fourth. It is possible to decrease significantly the maximum mismatch $(\Delta a/a_0)_{max}$ by bringing the composition of the emitter closer to that of the active layer with simultaneous doping of the emitter with calcium or strontium. Calcium and strontium have low diffusion coefficients, but their introduction allows avoiding the decrease in the forbidden gap width of the emitter (heterobarrier) when its composition is changed. Table 3 gives the values of maximum relative and critical deformations for post-inverse compositions of active layers.

Table 3. Maximum relative and critical deformations for active layers of post-inverse compositions $Pb_{1-x}Sn_xSe_{1-y}Te_y$ (with bringing of the composition of the emitter closer to that of the active layer)

No.	Emitter composition		Active layer composition		$\Delta y = y_a - y_e$	$(\Delta a/a_0)_{max} \cdot 10^{-3}$	$\varepsilon_{cr} \cdot 10^{-3}$
	x	y	x	y			
1	0.205	0.780	0.506	0.900	0.12	2.03	3.68
2	0.205	0.390	0.427	0.375	0.08	1.80	3.26

As is seen from table 3, in the result of bringing of the composition of the emitter closer to that of the active layer, the maximum mismatch $(\Delta a/a_0)_{max}$ in the heterostructure significantly decreases. In the general case the elastic limit was assessed by critical strain $\sigma = E/1-\nu \cdot \varepsilon_{cr}$, where E is the modulus of elasticity; ν is the Poisson ratio; ε_{cr} is the critical deformation. In the above cases, the modulus of elasticity was constant. However, as the investigation of internal friction showed [5], the modulus of elasticity increased 1.8–2 times at doping with such impurities as manganese or chrome (decreasing the lattice constant of the layers of IV-VI semiconductors). Then, with account for the 3-fold increase in the critical deformation at doping, the critical strain increased ~ 6 times. The possibility of doping of the active layer, especially of doping without decreasing the quantum yield of radiation, could be universal for expanding the elastic deformation range. This opens up possibilities for avoiding the dislocation multiplication, the decrease in the current carrier lifetime and, accordingly, the decrease in the radiation yield. As a result, actually measurable threshold current decreases by a factor of 5 to 10 and, accordingly, the operating temperature of lasers can be increased. As the lattice constants of considered binary compounds and triple solid solutions of IV-VI semiconductors are higher than those of fourfold ones, in these cases efficient "negative" pressure is realized which maybe used both in lasers and photodetectors [6, 7].

It is noteworthy that of interest are the heterostructures with "temperature-self-regulating" heterojunctions. In the structures of that kind, the emitter is designed on the base of fourfold solid solutions of the pre-inverse composition and active layers of the post-inverse composition and doped by manganese. At the same time, the compositions of the layers are selected so that, at liquid–nitrogen temperature, the forbidden gaps have similar width equal to 0.18 eV. However, because of sign-inverse temperature coefficients dE_g/dT in these layers, at room temperature E_g of the active layer decreases, while that of the emitter increases. With consideration for $dE_g/dT \approx 4 \cdot 10^{-4}$ eV/K , the energy barrier $\Delta E_g = 0.18$ eV at room temperature is provided.

4.Conclusions

It is obvious that, according to the model considered, complete matching of the lattice constants of conjugated lattices at the heteroboundary of the layers at technologically conditioned temperature of epitaxial growth yet does not provide the desired matching even with account for corrections made during heterostructure cooling to room temperature. The interdiffusion of components between the layers at the heteroboundary causes disturbances in isoperiodicity, appearance of a mismatched layer and emergence of strain. In most cases, the emerged deformations and, accordingly, strain exceed the elastic limit - the strain relaxes with formation of mismatch dislocations.

In the work, it is shown that bringing of the composition of the emitter closer to that of the active region and doping of this region with manganese impurity on the one hand decreases the maximum relative deformations and on the other hand increases the modulus of elasticity, critical deformations and the elastic limit at large. This allows preventing the strain relaxation, spread of nonradiative recombination centers, and, accordingly, suppressing the threshold current in lasers.

References

[1] Gegiadze G G, Davarashvili O I, Enukashvili M I, Kekelidze N P, Darsavelidze G S, Gabrichidze L L and Zlomanov V P 2005 A New Role of Impurities in the Improvement of Characteristics of Infrared Semiconductor Lasers and Photodetectors *Georgian Engineering News* **2** 54–56

[2] Pashaev A M, Davarashvili O I, Aliyev V A, Bychkova L P, Enukashvili M I and Zlomanov V P 2008 On Critical Thickness in A^{IV} B^{VI} Heterostructures at Diffusion Mismatch *Georgian Engineering News* **4** 7–10

[3] Davarashvili O I, Darsavelidze G S, Bychkova L P, Enukashvili M I et al. 2009 Diffusion Mismatch in Semiconductor Heterostructures *Bull. Georg. Nat. Acad. Sci.* **3** (35) 331–336

[4] Bychkova L P, Gegiadze G G, Davarashvili O I, Enukashvili M I et al. 2001 Strain in Isoperiodic Heterostructures PbSeTe-PbSnSeTe *Fizika* **2** (7) 21-24

[5] Pashaev A M, Davarashvili O I, Aliyev V A, Darsavelidze G S, Enukashvili M I and Zlomanov V P 2008 Experimental Modeling of the Elastic Properties of Binary A^{IV} B^{VI} Compounds *Trans. Acad. Sci. Azer.* **5** 96–100

[6] Pashaev A M, Davarashvili O I, Akhvlediani Z G, Enukashvili M I, Bychkova L P and Dzagania M A 2012 Study on the Forbidden Gap Width of Strained Epitaxial Lead Selenide Layers by Optical Transmission *Mat. Sci. Eng.* **2** 142–150

[7] Pashaev A M, Davarashvili O I, Enukashvili M I, Akhvlediani Z G, Gulyaev R G and Dzagania M A 2012 Supercritical Epitaxial Nanolayers of Lead Selenide In: *Book of Abstracts of the 3rd Int. Conf. on the Physics of Optical Materials and Devices (Belgrade, Serbia, 3–6 September 2012)* 217

IOP Conf. Series: Materials Science and Engineering **49** (2013) 012030 doi:10.1088/1757-899X/49/1/012030

Thermal properties of hemp fibre non-woven materials

Liga Freivalde[1], Silvija Kukle[2], Stephen Russell[3]
[1,2] Riga Technical University, Institute of Textile Technology and Design, Azenes Street 14, Riga, Latvia, LV-1048.
[3]University of Leeds, Centre for Technical Textiles, Leeds, LS2 9JT, UK.

E-mail: liga.freivalde@rtu.lv

Abstract. This review considers the thermal properties analysis of hemp fiber non-woven materials made by three different manufacturing technologies - thermal bonding, needle-punching and hydro-entanglement. For non-wovens development two hemp fibers cultivars grown in Latvia were used – Purini and Bialobrzeskie. Thermal resistance, conductivity and the effects of several parameters on thermal performance are revised.

1. Introduction

Reductions of energy consumption in the building sector are important measures needed, as guided by the new energy policy, EU has to achieve at least a 20 % reduction of energy and greenhouse gas emissions by 2020 compared to the year 1990. In Latvia high hemp straw yields are obtainable with excellent physical and mechanical fibers properties. [1,2] This study is concerned with the non-woven samples development from local hemp dioeciously genotype Purini (P) and the EU registered monoecious variety Bialobrzeskie (B) both farmed in climatic and soil conditions found in Latvia. Their physical and thermal properties testing and comparative analysis are carried out.

2. Materials and methods

Hemp fibres, used for this study were obtained from hemp stems harvested from a trial plot of Latgalian Agriculture Research Centre of Latvia. The harvested hemp stems of both varieties were left for dew retting on the field (4 weeks). Linear density for P is 23.4 Tex, for B is 36.2 Tex. [3]

2.1. Samples developed by thermal bonding technology

The hemp/polyethylene-polypropylene webs with fibers ratio 90:10% were produced by carding technique and non-woven samples were made from these webs by means of thermal bonding, where each sample stayed about 1 minute for each thickness cm of the material at 145° hot air regime.

2.2. Samples developed by needle-punching technology

Non-woven specimens were made from carded webs by means of needle-punching technology performed on 30 cm wide sample needle loom with 36 g regular 3 barb needles. The shank gauge of the needle is 15 – 18 – 42. Penetration depth of needles: 6.4 mm on both sides. Punch density: 75 punches/ cm^2.

2.3. Samples developed by hydro-entanglement technology

Hydro-entanglement was performed on a seven injector 0.5 m wide machine. Webs were pre-wetted and hydro-entangled using jet strips with a nozzle diameter of 150 μm; pressure 100 bar was applied in an alternating face and back profile. The conveyor speed was fixed at 5 m/min. During the samples production was observed that filters of hydro-entanglement machine blocked up fast as no special measures were taken to remove water-soluble components and impurities (dust) from hemp fibres. Jet strips nozzles blocking were not observed.

2.4. Thermal bonded, needle-punched and hydro-entangled non-woven samples testing methods

An abbreviation system has been used to describe the samples. 3 needle-punched (N), 3 thermal bonded (T) and 4 hydro-entangled (H) samples for each variety were used to produce the non-wovens corresponding to different thicknesses: N1, N2, N3; T1, T2, T3 and H1, H2, H3, H4. Samples have also been denoted according to the origin of the constituent hemp fiber, i.e. Purini (P) or Bialobrzeskie (B). All non-woven samples were conditioned under standard atmospheric condition EN 20139:1992. Standard test methods were utilized for the measurement of non-woven properties. Non-woven thickness was measured according to BS EN ISO 9073-2:1997, where method C was adopted with slight modifications: uniform pressure 0.02 kPa, i.e., 20 grams on 100 cm^2 for N and T groups samples, for H group samples: 10 grams on 50.2 cm^2. Surface density (m) was determined according to BS EN 29073-1:1992, modification: test area of the sample was 10 000 mm^2. The density of the non-woven samples has been calculated by dividing the non-woven surface density by the material volume. Both, thermal resistance ($m^2 \cdot K/W$) was measured by using two-plate method: fixed pressure procedure and thermal conductivity (λ), W/ (m·K), was calculated according to BS 4745:2005.

3. Results

Through all experiment thermal conductivity (λ) is in the range from 0.028 up to 0.04 W/ (m·K), overall difference is 44 %. It can be explained as in general, thermal conductivity depends on the thermal resistance, density, temperature, moisture content, as well as components and porosity [4]. Therefore, the material λ can vary within a relatively wide range.

Test results of hydro-entangled samples show lower λ values for cultivar P comparing with variety B in all cases except H4 is slightly higher. Differences between cultivars mostly are insignificant, for samples H4 it is only 0.2%, for samples H3 and H2 it is 2%. For samples H1 difference is higher, i.e., 22% (Fig.1).

By comparing hydro-entangled samples with thermal bonded and needle-punched sample groups it can be seen that changes in the thermal conductivity are relatively small, except sample NP2 is noticeably higher than sample TP2 (Fig. 1), compared to the changes in the thicknesses (Fig.2).

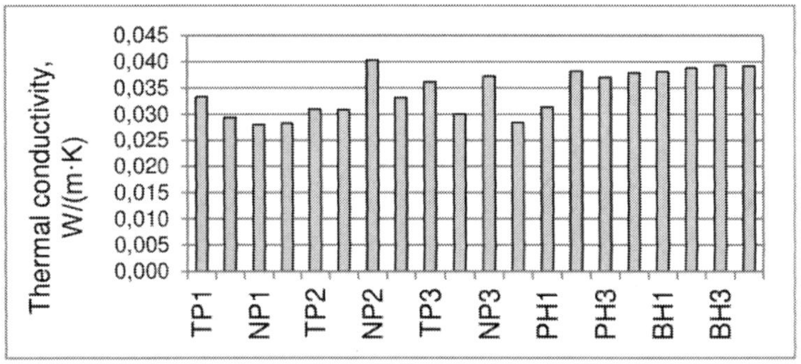

Figure 1. Thermal conductivity of T, N and H samples groups

Throughout the experiment analysis of covariance show strong relationship between thermal resistance and thickness as high correlation take place: r = 0.99, correlation between thermal resistance and surface density is slightly lower: r = 0.94. Results for relationship among non-woven thermal resistance and thickness (Fig. 2) can be described as linear relationships for thermal bonded $Y_{TP/B}$ as well as needle-punched $Y_{NP/B}$ samples groups (eq. 1-4).

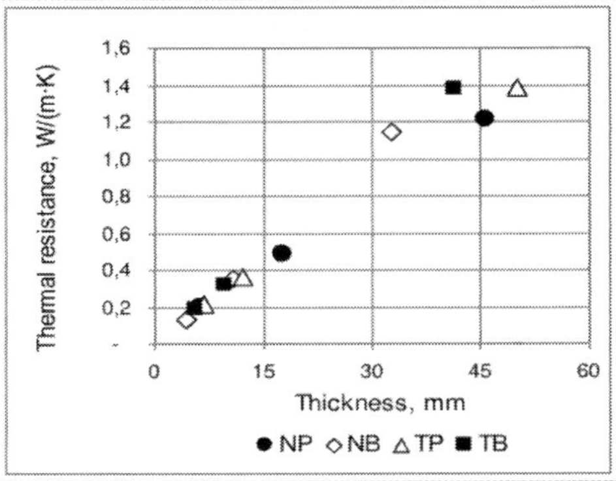

Figure 2. Thermal resistance as a function of thickness for cultivars P and B, made by T and N methods.

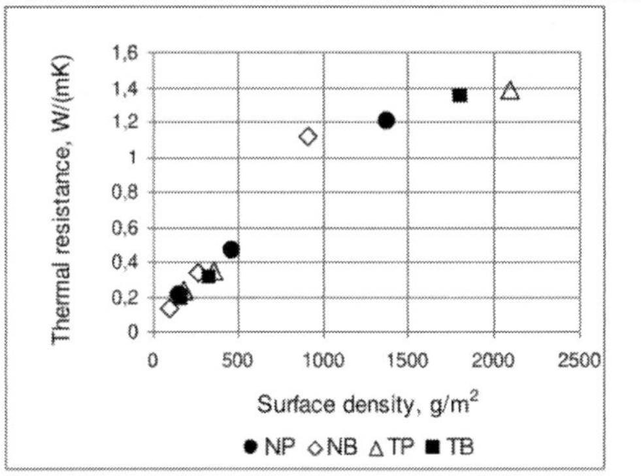

Figure 3. Thermal resistance as a function of surface density for cultivars P and B, made by T and N methods.

Empirical regression models are linear:

$$Y_{TPR} = 0.0388 + 0.0269d_1 , \quad (1)$$
$$Y_{TBR} = -0.0022 + 0.0333d_2, \quad (2)$$
$$Y_{NPR} = 0.0316 + 0.0258d_3, \quad (3)$$
$$Y_{NBR} = -0.027 + 0.0358\ d_4, \quad (4)$$
$$Y_{TPR} = 0.1308 + 0.0006m_1, \quad (7)$$
$$Y_{TBR} = 0.0708 + 0.0007m_2, \quad (8)$$
$$Y_{NPR} = 0.0873 + 0.0008m_3, \quad (9)$$
$$Y_{NBR} = 0.0255 + 0.0012m_4 \quad (10)$$

where design factors d_1, d_2, d_3, d_4 represent non-woven thickness (mm); m_1, m_2, m_3, m_4 - surface density (g/m^2), response variables Y_{TPR} – thermal resistance of thermal bonded (TP, TB) and needle-punched (NP, NB) samples of both varieties.

By comparing thicknesses of TP1 and NP1 samples, which are respectively 6.7 mm and 6.5 mm, it can be seen that changes between methods are minor, only 2%. Meanwhile differences of surface density (Fig.4) are more significant - 13 % (172 g/m^2 for TP1 and 153 g/m^2 for NP1). Density between methods of the same samples varies by 9 %, but between λ values by 19 %.

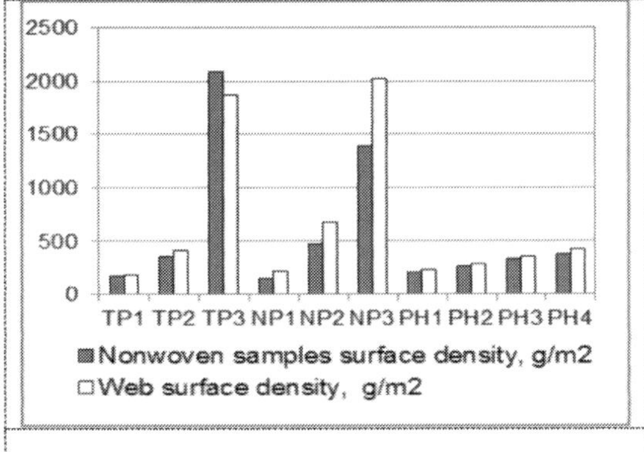

Figure 4. Surface density for carded web's and non-woven samples (T, N, H), g/m^2 of genotype P.

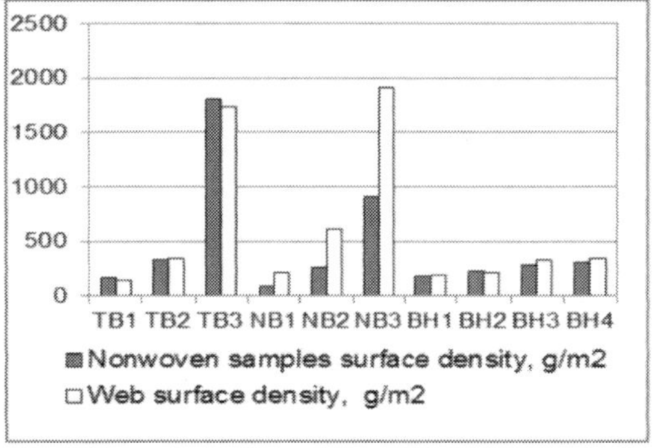

Figure 5. Surface density for carded web's and non-woven samples (T, N, H), g/m^2 of variety B.

Thickness for the samples PH2 and BH2 are equal: 2.7 mm. Difference between densities as well as surface densities is 16 %. It can be explained due fact that for webs processing by carding was used the same amount of entering fibers, but after carding the web weight and respectively also web surface density was higher for variety P (Fig. 4, 5). For variety B 29% more fibers went to the waste in the carding process, thence less fiber and non-woven surface density appeared (Fig.5). For the samples

BH1 and BH2 difference between surface densities and densities is 25.4 %. As the thickness 2.7 mm is equal for both samples it leads to different fiber congestion (Fig.5). Changes between thermal properties are insignificant: 0.8 %. Thermal resistance for samples PH1, PH2, BH1 and BH2 is 0.07 m^2K/W. Slightly higher values are for PH3 and PH4, i.e., 0.08 m^2K/W. For BH3 and BH4 it is 0.09 m^2K/W. For P group samples, by increasing thickness, thermal resistance values grows gradually, but for B group samples thermal resistance develop by leaps reaching a certain thickness (Fig.6, 7).

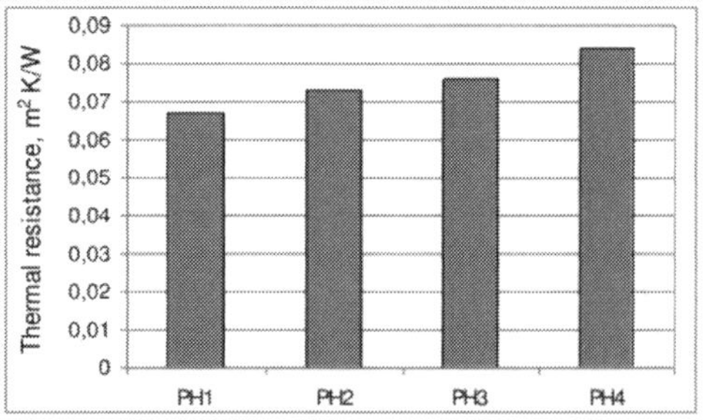

Figure 6. Thermal resistance for cultivars P and B made by hydro-entangling method.

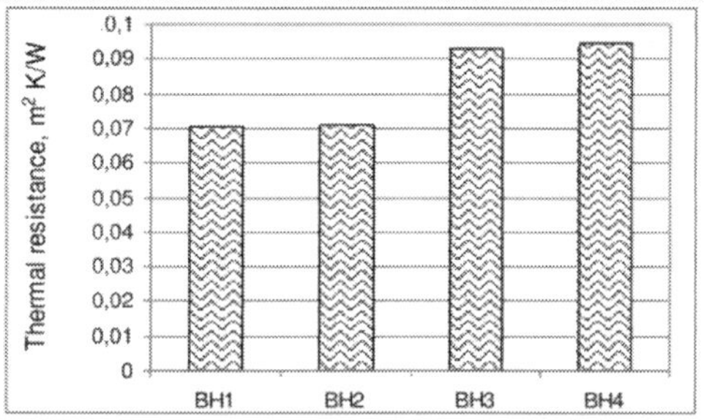

Figure 7. Thermal resistance for variety B made by hydro-entangling method.

The trend within experiment shows that thermal resistance of non-wovens of all technologies (T, N, H) and both varieties increased with non-woven thickness and surface density increase (Fig.2, 3, 6, 7).

4. Conclusions

Thermal properties of non-woven samples of technical hemp fibers made through the thermal bonding, needle-punching and hydro-entanglement technologies were explored in this study. For thermal bonded non-wovens thermal resistance is in the range: 0.20-1.39 m^2K/W, λ : 0.0294 - 0.0361 W/m·K, density: 25.7- 41.7 kg/m^3 for cultivar P and 0.19-1.38 m^2K/W; 0.029 – 0.031 W/m·K, 29.6 - 43.8 kg/m^3 for variety B. For needle punched non-wovens thermal resistance is 0.24 - 1.23 m^2K/W, λ is 0.028 - 0.040 W/m·K, density 23.6 - 30.3 kg/m^3 for variety P and for variety B: 0.16-1.15 m^2K/W, 0.028- 0.033 W/m·K, 20.6-27.8 kg/m^3. For hydro-entangled samples thermal resistance is 0.067-0.084 m^2K/W, λ 0.031-0.039 W/m·K, density 98.1-116.2 kg/m^3 for P; 0.071-0.094 m^2K/W, 0.0379-0.0392 W/m·K, 68.5-85.9 kg/m^3 for B. These results shows the same or better thermal properties than other commonly used thermal insulation materials, e.g., mineral wool with λ 0.030-0.045 W/m·K, if density is 13-180 kg/ m^3. By comparing hydro-entangled samples with thermal bonded and needle-punched, it can be seen that changes in the thermal conductivity are relatively small, except sample NP2 is noticeably higher than sample TP2, compared to the changes in the thicknesses.

Acknowledgements

This research has been supported by the European Social Fund within the project "Support for the Implementation of Doctoral Studies at Riga Technical University".

References

[1] Baltiņa I, Zamuska Z 2010 Effects of Late Hemp Straw Harvesting on Fiber Quality *Scientific Journal of RTU- Material science* **5** 194-198
[2] Freivalde L, Kukle S, Purina B and Stramkale V 2010 Durability properties of hemp fibers *Proceedings of the Inter.Textile Clothing and Design Conference ITC&DC*, 1847-7275
[3] Kukle S, Stramkale V, Kalniņa D, Soliženko R 2012 Comparision of Hemp Fibre Properties *Proceedings of the International Textile Clothing and Design Conference ITC&DC* 76-80
[4] Al-Ajlan S 2006 Measurements of thermal properties of insulation materials by using transient plane source technique *Applied Thermal Engineering* **26** 2184-2191.

IOP Conf. Series: Materials Science and Engineering **49** (2013) 012031 doi:10.1088/1757-899X/49/1/012031

Structural, microstructural and impedance spectroscopy study of functional ferroelectric ceramic materials based on barium titanate

B Garbarz-Glos[1], W Bąk[1], M Antonova[2] and M Pawlik[1]
[1]Institute of Physics, Pedagogical University of Cracow, Poland,
[2]Institute of Solid State Physics, University of Latvia

E-mail: barbaraglos@gmail.com

Abstract. The differences between the physical properties of barium titanate $BaTiO_3$ and newly obtained $BaHf_xTi_{1-x}O_3$ were identified. These ceramics were prepared by solid-phase reaction from simple oxides and carbonates using the conventional method. The structure and morphology of investigated samples were characterised by X-ray diffraction (XRD) and scanning electron microscopy (SEM). The application of impedance spectroscopy made possible to characterize of these materials in the terms of electrical properties.

1. Introduction

Ceramic ferroelectric materials due to the availability of advanced materials are now the increasingly important elements of active intelligent systems. The systems, in which there is a feedback loop between a sensor and a actuator and the response of the system is varied and flexible. Among these materials, ferroelectrics with perovskite ABO_3 structure type have a special place, barium titanate in particular. $BaTiO_3$ (BT) is the piezoelectric material of the greatest importance (module piezoelectric d_{33} is of the order 10^{-12}C/N and electromechanical coupling coefficient $k_p = 0.38$) [1-3]. Therefore it is used as mechanical-electrical sensors measuring and recording e.g. vibration, speed, acoustic emission signals, etc. The ceramics based on $BaTiO_3$ has a great range of physicochemical properties. In many recent works [4-9] the effects of isovalent substitutions such as Zr^{4+}, Sn^{4+} for Ti^{4+} and Ca^{2+}, Sr^{2+} for Ba^{2+} on the microstructure, dielectric properties and phase transition of BT ceramics have been extensively studied. The purpose of the work was the investigation of the role of Hf^{4+} substitution on the structural, microstructural and dielectric properties of the system $BaTiO_3$ - $BaHfO_3$ over a wide temperature and frequency range.

2. Experimental methods

The investigated samples $BaTiO_3$, $BaTi_{0.925}Hf_{0.075}O_3$, $BaTi_{0.85}Hf_{0.15}O_3$ (abbreviated to BT, BTH7.5, BTH15) were prepared by a solid state synthesis reported earlier [10]. The X-ray studies were performed by means of an X'Pert PRO (PANalytical) diffractometer with the Cu-Kα radiation (λ=1,5406 Å) and a graphite monochromator [9, 11]. A profile-fitting program FULLPROF [12] based on the Rietveld method was used to analyse and fit the spectra. In order to study the surface structure and morphology of the samples the high performance Hitachi S4700 electron scanning microscope with field emission and the Noran Vantage EDS (Energy-Dispersive X-Ray Spectroscopy) system was applied [9, 11]. The investigations of microstructure were performed on fractures and polished sections. Before the measurements the samples were sputtered with carbon. Dielectric measurements were carried out using an Alpha-AN modular measurement system together with cryogenic temperature control system - Quatro Cryosystem and WinDETA Novocontrol software, at frequency varying from 0.1 Hz to 10 MHz and temperature range from 130 K to 600 K.

3. Results and discussion

A powder X-ray diffraction pattern shows (figure 1) that the $BaHf_xTi_{1-x}O_3$ polycrystalline samples at room temperature are of single-phase perovskite structure. In figure 1, shifting the diffraction peaks towards lower angles with comparison to the pure barium titanate $BaTiO_3$ with increasing Hf content are visible. This behavior indicates, that the values of the lattice parameters in BTH7.5 and BTH15 increased. The merging of the splitting of some peaks indicates a transformation from tetragonal symmetry phase (P4mm) to a higher symmetry –rhombohedral (R3m). The values of the lattice constants at room temperature were calculated. The determined cells parameters are: for $BaTiO_3$ a=3.9943Å, $BaTi_{0.925}Hf_{0.075}O_3$ a=4.0215Å and for $BaTi_{0.85}Hf_{0.15}O_3$ a=4.0348Å.

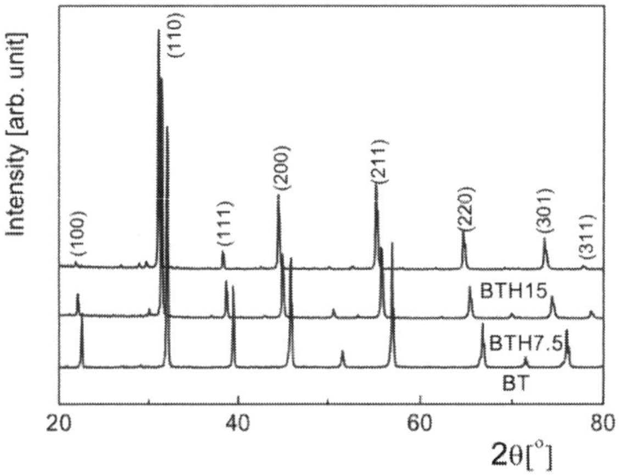

Figure 1. Room temperature X-ray diffraction profiles for BT, BTH7.5 and BTH15 samples.

The atomic composition of the ceramics BT, BTH7.5 and BTH15 was carried out by using the EDS. In an EDS survey spectrum (not shown here) nothing more than the expected elements are seen. They confirm the high purity and the expected qualitative composition. The images of fractures of investigated samples are shown in figure 2. The surface of the fracture goes along grains as well as along the boundaries between the grains. The fracture has a fragile character and a tendency to formation of crystalline structures can be observed in the grains. A good homogeneity of the microstructures and a small degree of porosity were observed. Pores have an irregular shape without curvings and narrowings. The growth terraces of grains are clearly seen. They indicate that the growth of grains took place according to a layer mechanism with a screw dislocation. Using the linear intercept method the grain size was determined from SEM photographs. The average grain size of BTH ceramics is included in the range from 8μm to 11μm.

Figure 2. The photograph of microstructure of BT, BTH7.5 and BTH15 samples (mag. 10000x).

An impedance spectroscopy is useful for attempting to separate intra-granular (bulk), inter-granular and sample electrode interface effects [13]. These effects usually are related to the sites of traps for oxygen vacancies and other defects. The impedance analysis is also helpful in establishing a space charge polarization (i.e. an accumulation of charges at the interface between the sample and electrodes) and its relaxational mechanism. In figures 3-6 the impedance data of the investigated samples are presented in the form of Nyquist diagrams (Z" vs. Z') for several representative temperatures (>570K).

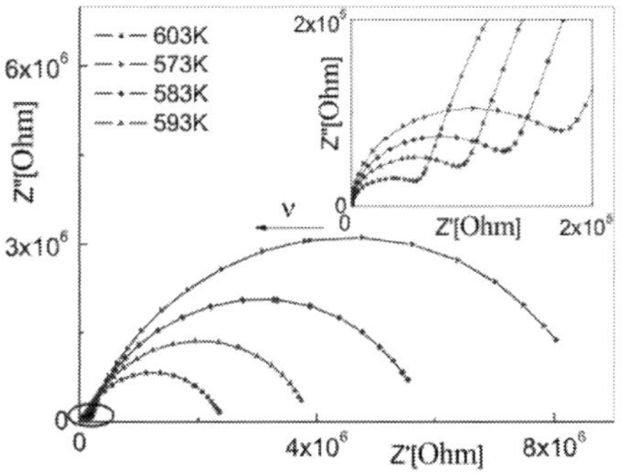

Figure 3. The imaginary versus the real part of impedance for BT sample.

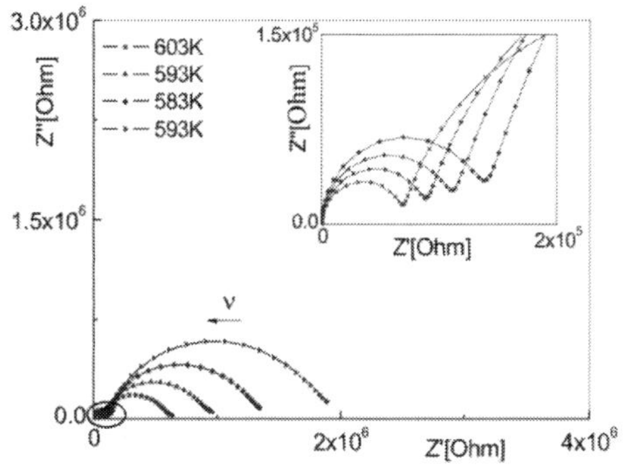

Figure 4. The imaginary versus the real part of impedance for BTH7.5 sample.

The impedance plots Z"(Z') are composed of two semicircle arcs which cannot be presented (and observed) simultaneously within the same scale in the one graph. The first process takes place within impedance regions from about 0 to $\sim 10^5\Omega$ and second one (stronger) reaches the maximum values of impedance even one order of magnitude higher ($\sim 10^6\Omega$). Thus the weaker process is presented as expanded insets with visible closer details. The weaker process can be prescribed to the grain (bulk) of impedance response. We postulate that the second one (stronger) originates from the free charge and the induced polarization occurring on grain boundaries. For BTH15 ceramics in the low frequency range, there is an additional short inclined straight line which may be attributed to the interface between the sample and the electrode relaxation (space charge polarization). The two semicircular arcs of the impedance plot can be expressed as an equivalent circuit consisting of two parallel R (resistance) and C (capacitance) elements connected in series; one RC element for the bulk and the other for the grain boundary response [14]. The relaxation time for the grain and grain boundary can be calculated from the relation $\omega\tau = 2\pi\nu_m RC = 1$, where the ν_m is the frequency of the maximum of the semicircle, $\omega = 2\pi\nu_m$ and $\tau = RC$ is a relaxation time. Both processes are represented by not perfect but flattened semicircles. This flattening of the electrical response of the grain and grain boundary might be due to a nonexponential process, such as a diffusion or nonuniformities in the microstructure of material, leading to the distribution of relaxation times, characteristic for this ceramics type. It is frequently described as the semicircles with centres shifted below the real axis Z'. In the case of BTH7.5 sample the smallest "diffuse" of semicircle arcs for bulk (grain) are observed. The temperature increase causes the decrease of the experimentally observed semicircle radiuses for all investigated samples. This is attributed to a cationic disorder due to the random distribution of B-site cations (Hf^{4+}, Ti^{4+}) having a different ionic radius. It can have an influence on a diffusion of Hf ions during the synthesis. In these materials the partial substitution of Ti by Hf takes place, and therefore the formation of new crystalline structures depending on the level of substitution occurs. Due to the fact that the ionic radii of Ti^{4+} and Hf^{4+} and the covalence degree of the bonds are different, it should be expected that the competition between Ti-O and Hf-O bonds exists. The variation of the relaxation times as a function of reciprocal temperature shows the Arrhenius behavior in the paraelectric phase (figure 6). The activation energy was determined from the equation $\tau = \tau_0\exp(-E_a/kT)$, where τ_0 is a pre-exponential factor, and k and T have their usual significance. The grain (bulk) and grain boundary relaxation activation energy were evaluated and were given in figure 6. The similar values of activation energies obtained from the impedance and electrical plots modulus (not shown here) indicate the localized conduction (dielectric relaxation) of charge species in the boundary of sample.

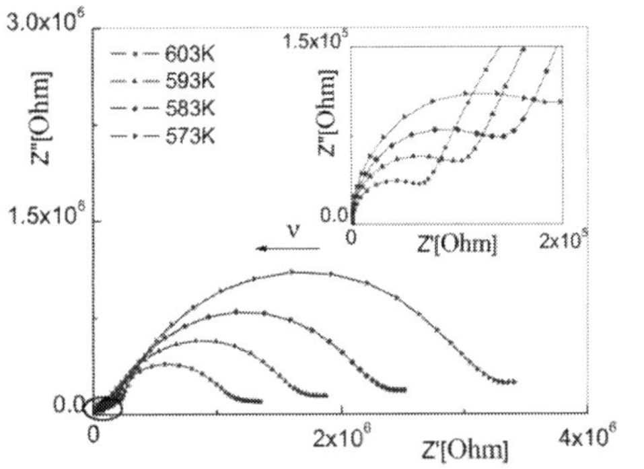

Figure 5. The imaginary versus the real part of impedance for BTH15 sample.

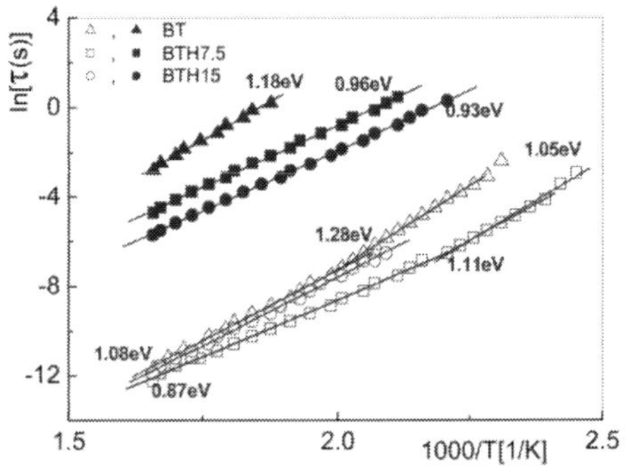

Figure 6. Relaxations times for grains (open points) and grain boundaries (solid points) as a function of reciprocal temperature.

4. Conclusions

The polycrystalline samples of $BaHf_xTi_{1-x}O_3$ were obtained by means of a conventional ceramic technology. The synthesized material was determined by the X-ray diffraction and scanning electron microscopy. The analysis of the X-ray diffraction pattern revealed the formation of the single perovskite phase at room temperature. From the obtained results it was stated that the values of the lattice parameters in BTH7.5 and BTH15 increase in comparison with the values for pure barium titanate. The performed SEM and EDS studies revealed that the samples were of good quality and chemically homogeneous. The investigated material was dense and possessed a compact structure. It is found that hafnium concentration has significant influence on dielectric properties of BT. The grain and grain boundary relaxation activation energies were evaluated. The obtained results showed, that the conduction process occurs both in the grains and grain boundaries.

References

[1] Kong L B, Zhang T S, Ma J, Boey F 2008 Progress in synthesis of ferroelectric ceramic materials via high-energy mechanochemical technique *Prog. Mater. Sci.* **53** 207-322
[2] Rehrig P W, Park Seung-Eek, Trolier-McKinstry S, Messing G L, Jones B and Shrout T R 1999 Piezoelectric properties of zirconium-doped barium titanate single crystals grown by templated grain growth *Journal of Applied Physics* **86** 1657-1661
[3] Chung Deborah D. L 2010 *Functional materials* (Engineering Materials for Technological Needs vol 2) ed Deborah D L Chung (Singapore: World Scientific Publishing Co. Pte. Ltd.) chapter 3 149-150
[4] Yu Z, Ang C, Guo R, Bhalla A S 2007 Dielectric properties of $Ba(Ti_{1-x}Zr_x)O_3$ solid solution *Mat. Lett.* **61** 326-329
[5] Kajtoch C 2011 Influence of Zr-substitution on phase transitions character in polycrystalline $Ba(Ti_{1-x}Zr_x)O_3$ *J Mater Sci* **46** 1469–1473
[6] Shvartsman V V, Dec J, Xu Z K, Banys J, Keburis P, Klemann W 2008 Crossover from ferroelectric to relaxor behavior in $BaTi_{1-x}Sn_xO_3$ solid solution *Phas. Trans.* **81** 1013-1021
[7] Kajtoch C 2011 Dielectric properties of $BaTi_{1-x}Sn_xO_3$ in the paraelectric phase *Ceram. Inter.* **37** 387-391
[8] Varatharajana R, Samanta S B, Jayavel R, Subramaniana C, Narlikar A V, Ramasamya P 2000 Ferroelectric characterization studies on barium calcium titanate single crystals *Mat. Charact.* **45** 89-93

[9] Śmiga W, Garbarz-Glos B, Kalvane A, Antonova M, Suchanicz W, Sternberg A 2009 Strontium Concentration Dependence of Selected Structural and Mechanical Properties of Polycrystalline $Ba_{1-x}Sr_xTiO_3$ *Integrated Ferroelectrics* **108** 77-88

[10] Garbarz-Glos B, Piekarczyk W, Smeltere I, Śmiga W and Antonova M 2012 Ultrasonication as a Method of Investigation of the Mechanical Properties of Doped Hafnium Barium Titanate *Ferroelectrics* **436** 87-95

[11] Garbarz-Glos B, Śmiga W, Bujakiewicz-Korońska R, Suchanicz W, Dambekalne M, Livinsh M and Sternberg A 2009 Microstructural and Ferroelectric Properties of $BaZr_{0.20}Ti_{0.80}O_3$ Ceramic *Integrated Ferroelectrics* **108** 67-76

[12] Rodriguez-Carvajal J 1990 Abstract Satelite Meeting on Powder Diffraction, Congr. Int. Union of Crystallography. Toulouse. France

[13] Bauerle J E 1969 Study of solid electrolyte polarization by a complex admittance method *Journal of Physics and Chemistry of Solids* **30** 2657- 2670

[14] Sinclair D C and West A R 1989 Impedance and modulus spectroscopy of semiconducting $BaTiO_3$ showing positive temperature coefficient of resistance *Journal of Applied Physics* **66** 3850-3863

IOP Conf. Series: Materials Science and Engineering **49** (2013) 012032 doi:10.1088/1757-899X/49/1/012032

Photosensitive properties of composite films based on copper chloride in polymer matrix

A Gerbreders [1,2], **A Bulanovs** [2], **E Sledevskis** [2], **V Gerbreders**[2], **J Teteris**[1]
[1]Institute of Solid State Physics, University of Latvia, Latvia
[2]G.Liberts' Innovative Microscopy Centre, Daugavpils University, Latvia

E-mail: andrejmah@gmail.com

Abstract. The composite films based on PMMA and $CuCl_2$ were prepared. Transmission spectra of the films before and after illumination by different laser lines (405, 448, 456 nm) were measured. The reversible character of the composite photosensitivity was fixed. The phenomenon of direct surface relief formation after laser beam illumination of composite films was described. The holographic gratings were recorded by different exposure (200 – 500 J/cm^2) in the films with different concentration (8 and 20 wt. %) of copper salt. The film sensitivity dependence on polymer molecular weight was studied. Surface relief of the gratings was measured by AFM.

1. Introduction

Optical composite materials are wider used in modern technologies [1]. Combination of more than one component in one material lets to change the properties of device in wide ranges. Using polymer materials as a matrix in composites gives the possibility to refuse from complex and expensive installations of vacuum or magnetronic sputtering.

Recently, we already have described the composite materials, containing from organic polymer matrix and non-organic photosensitive filler [2]. We have showed the possibility of surface relief formation by laser illumination with light wavelength which interacts only with filler. Composite films were obtained from a common solution by evaporation of the solvent after the coating liquid layer on the substrate. The high viscosity of the polymer matrix does not allow the filler particles to coagulate. However, the system is often found in non-equilibrium thermodynamic state, if the concentration of filler is relatively high. Under the action of the corresponding radiation, the particles of photosensitive filler change their thermodynamic state, which in some cases can lead to moving of the polymer matrix macromolecules and, consequently, lead to the formation of surface relief. Over the past few years in many papers the direct surface relief formation on polymer or amorphous materials under the action of polarized light was described [2-4]. In our opinion, composite materials have to be a great potential in this technology.

In this paper we present a system based on polymethylmetacrylate and copper (II) chloride. To improve the simplicity of the system, a challenge is to find a common solvent for the components. As such, we used 1-methoxy-2-propanol which appeared as well as the polymer dissolves and copper (II) chloride dihydrate, to 20 wt. % concentration. However, previously the composites of copper (II)

---

[1] Institute of Solid State Physics, University of Latvia,
8 Kengaraga str., Riga LV 1063, LATVIA

chloride and other polymers were described. In particular, in [5] the composite of copper (II) chloride and polyethylene oxide were studied. In [6], the complexes of copper (II) chloride and polyvinylpyrrolidone have been studied. In the papers, the photosensitivity of such composites have been marked and showed that changes of the optical properties is due to the reduction of Cu^{2+} to Cu^{+} and the excretion of chlorine, which, because of their activity react with a polymer macromolecule.

2. Experimental

Polymethylmetacrylate (PMMA) (Aldrich, CAS# 9011-14-7, analytical standard M_w=97 000) and copper (II) chloride dihydrate (Scharlab s.l., min 99%, CAS# 10125-13-0) were dissolved in 1-methoxy-2-propanol (ROTH, CAS# 107-98-2). Concentration of both solutions was 20 wt. % (copper (II) chloride solution concentration was calculated to 20 wt. % of dry $CuCl_2$). The solutions were mixed in such correlations, that after films drying on substrate a mass part of copper (II) chloride was 5, 8, 10, 15, 20, 25 and 35 % (wt).

Fabrication of the films from solutions were performed by company BYK Gardner GmbH applicator, the thickness of liquid layer was 30, 60, 90 μm (the thickness of dry films was 4, 7, and 10 μm respectively). A glass and quartz plates were used as substrate. The films were dried in dark place at the room temperature 12 h. The thickness of dry films and surface relief were determined by the Veeco Dektak 150 surface profiler.

Illumination of the composite films was performed by semiconductor lasers with the wavelengths of 405 and 456 nm at exposure 50 – 2000 J/cm^2.

For all films the transmission spectra were measured by the Ocean Optic HR4000CG spectrometer.

The holographic gratings with a period of Λ = 0.88 μm were recorded by two symmetrically laser beams (wavelength 456 nm) of equal intensity (15 mW/cm^2) with linear horizontal (p-p) or vertical (s-s) polarization. The readout of transmission diffraction efficiency was made by the second order maximum of recording beam.

3. Result and discussion

The experiments showed that the composite films with concentration of copper (II) chloride more than 20 wt. % of PMMA are relatively unstable. The instability is expressed as a crystal growth of filler on the film surface during the drying process after application, as well as in film cracking at mechanical damage. Obviously, this instability is increased with an increasing of layer thickness.

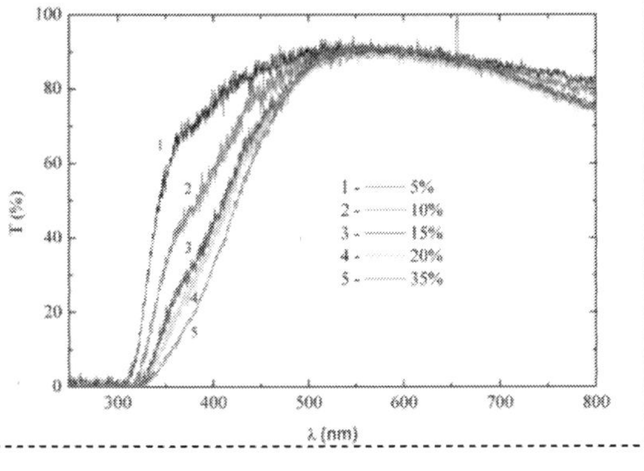

Figure 1. Transmission spectra of $CuCl_2$-PMMA films in glass substrate. $CuCl_2$ concentration is 5-35%, film thickness - 5 μm.

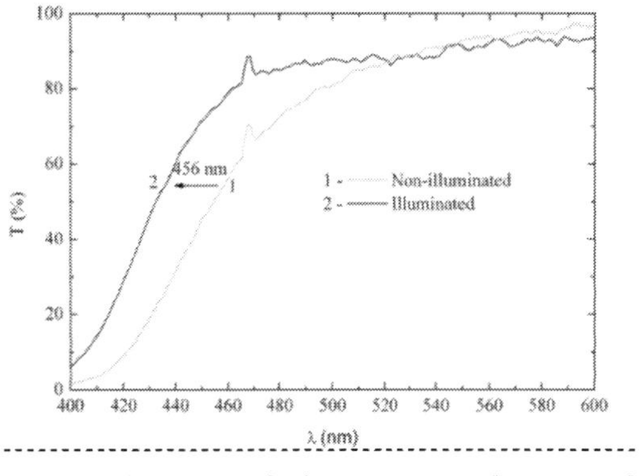

Figure 2. Transmission spectra changes of $CuCl_2$-PMMA films after illumination by laser 456 nm on glass substrate. $CuCl_2$ concentration is 8%, film thickness is 10 μm. Exposure dose is 267 J/cm^2

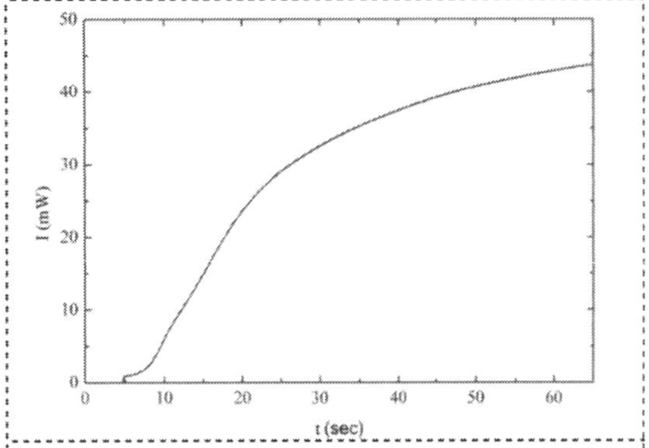

Figure 3. Dependence of 405 nm laser line transmission of $CuCl_2$-PMMA film on illumination time. $CuCl_2$ concentration is 8%, film thickness is 10 μm. Intensity is 10 W/cm^2.

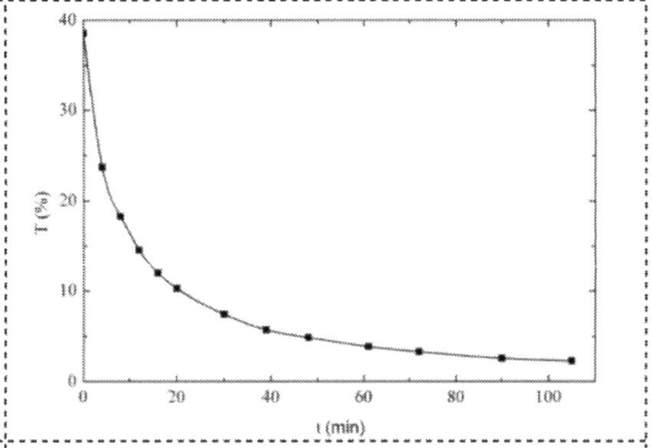

Figure 4. Relaxation of 405 nm laser line transmission of $CuCl_2$-PMMA film after illumination with the same spectral line at exposure dose of 600 J/cm^2. $CuCl_2$ concentration is 8%, film thickness is 10 μm.

The optical transmission spectra and their changes by irradiation for composite films are shown in Fig. 1 and 2. It can be seen quite regular increase of optical density at increasing of the copper chloride concentration in the composite. The optical density of the films increases with the film thickness. The absorption edge of the material is in the blue region of the spectrum, and, as it was proved the composite absorbance decreases by the corresponding wavelength of laser irradiation.

It is interesting to note the fact of reversible spectral changes in the films with copper (II) chloride concentration less then 15%. Figures 3 and 4 illustrate the changes of the beam of wavelength 405 nm transmissions and it relaxation. The relaxation time is increased at increasing of the copper chloride concentration in the composite.

In the films with $CuCl_2$ concentrations more than 10 wt % the formation of the surface relief under irradiation of blue laser spectrum was fixed. Figure 5 shows the surface topography image after laser irradiation at 405 nm with an exposure 73 J/cm^2. We can note the cavity formation in the place of irradiation. The narrow peaks in the graph, probably, are displaying the dust particles situated in the surface of the film. The pattern of surface topography changes during the relaxation process has not yet been identified and it is the subject of future research.

For holographic record we have chosen the composite film with the concentration of $CuCl_2$ 25 wt. %. Figure 6 shows the sensitivity of the composite film to different polarization of recording beams, in other words it is demonstrate the material sensitivity to the direction of the electric field vector. A substantial increasing of the diffraction efficiency as well as the occurrence of a surface relief in the case of p-p polarization of the recording beams is fixed.

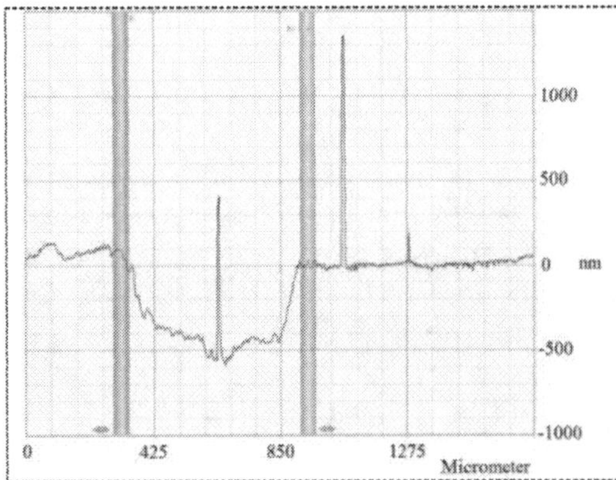

Figure 5. Surface relief of $CuCl_2$-PMMA film after illumination by laser 405 nm. $CuCl_2$ concentration is 10%, film thickness is 5 μm. Exposure is 72 J/cm^2.

IOP Conf. Series: Materials Science and Engineering **49** (2013) 012032 doi:10.1088/1757-899X/49/1/012032

Figure 7 shows the topography of recorded grating, made by AFM. The depth of the surface relief is up to 30 nm. It can be seen that the relief is formed by structures with a diameter of 200 nm, it is likely that this is the filler particles.

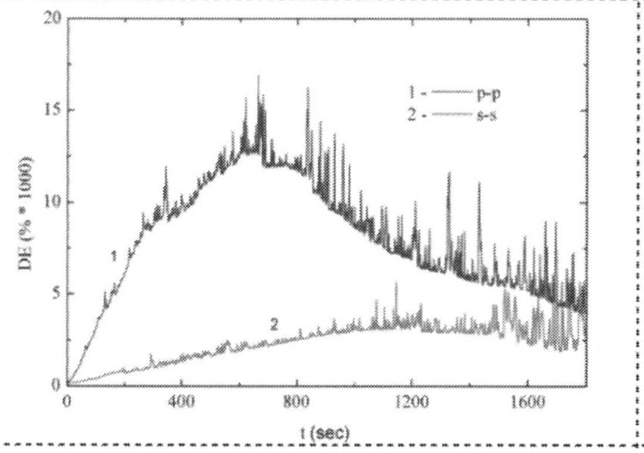

Figure 6. Dependence of difraction efficiency in the second diffraction maximum on recording time and beam polarization. λ=456 nm, intensity of each beam is 15 W/cm^2, grating period is 0.88 μm. $CuCl_2$ concentration is 25%, film thickness is 6 μm.

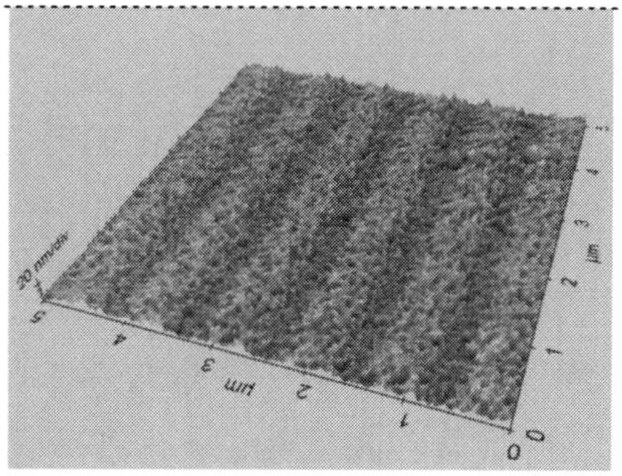

Figure 7. Surface relief of holographic grating, recording by laser 456 nm, intensity of each beam is 15 W/cm^2, time of record is 600 sec. Relief depth is 20-30 nm. $CuCl_2$ concentration is 25%, film thickness is 6 μm.

4. Conclusions

Composite films, based on copper (II) chloride (5-35 wt %) and polymethylmetacrylate are photosensitive materials with absorption edge located in blue part of visible spectra. The films can be prepared from solution in 1-methoxy-2-propanol with concentration of $CuCl_2$ less then 20 wt % according PMMA, the films with higher concentration of $CuCl_2$ are unstable. The composite with a concentration of copper (II) chloride less than 15 wt % shows the reversible changes in transmission spectra.

The direct relief on the composite films surface can be formed by illumination of lasers at 405 and 456 nm in exposure range of more than 70 J/cm^2. In the films one can record holographic gratings and surface relief gratings with depth of relief up to 30 nm.

Acknowledgment

This work has been supported by the European Regional Development Fund within the project Nr. 2010/0275/2DP/2.1.1.1.0/10/APIA/VIAA/124.

5. References

[1] A. Сидоров et al. "Материалы и технологии интегральной оптики", СПГУ, 2009, 44-48.
[2] A. Gerbreders et al. *Proc. SPIE* 7142, 714212
[3] A. Ozols et al. 2008 *Thin Solid Films* **516,** 8887–8892
[4] K.Yang et al. 2006 *Phys. Rev.* B **73**, 1652042
[5] A. Lyons et al. 1988 *Macromolecules*, **21** No 11, 3125-3134
[6] H. Kaczmarek et al. 1996 *Polymer*, **37** No 18, 4061-4068

IOP Conf. Series: Materials Science and Engineering **49** (2013) 012033 doi:10.1088/1757-899X/49/1/012033

Engineering aspects of multilayer piezoceramic actuators

V.A. Golovnin[1], I.A. Kaplunov[2], A.I. Ivanova[2], R.M. Grechishkin[2]

[1]Research Institute ELPA, JSC, Zelenograd, Russia
[2]Tver State University, Tver, Russia

e-mail: rmgrech@yandex.ru

Abstract. With the increasing demand for multilayer ceramic chip components a full understanding of the co-firing of ceramics with metal electrodes becomes important. In the present work the processing of a piezoelectric monolithic actuator by stacking and cofiring Ag-Pd electroded tape cast layers was studied. The inter-diffusion and microstructure of the co-fired interface of PZT ferroelectrics and Ag-Pd metal electrode were examined by scanning electron microscopy (SEM) and energy-dispersive microanalysis. No strong structural distortions and interdiffusion were observed at the co-fired ceramic-electrode interface.

1. Introduction

Piezoelectric multilayer monolithic stacks (MLMS) are increasingly used in a great variety of high-precision actuators and microelectromechanical systems (MEMS) [1-3]. The application-specific adjustment of the working parameters of the MLMS is achieved by the choice of the piezoelectric material (mainly high-strain PZT ceramics), MLMS geometry, electrode material, etc. The microstructure of the MLMS is defined generally by the sintering process, though the result depends on the composition of PZT and electrode [4-8]. The volumetric ratio of the PZT/electrode interfaces increases inevitably with the miniaturization trend of the chip devices, so it is imperative to investigate the co-fired interfaces in greater detail. In the present work we study the cofiring process with the aid of scanning electron microscope (JEOL JSM 6510LV) and X-ray microanalysis (Inca+ , Oxford Instruments.

2. Experimental

Fabrication of the samples included paste preparation, multiple printing and drying, isostatic compression, co-firing, cutting, electrical connection, poling and testing. 70%Ag-30%Pd powders with a liqiudus temperature of ~1150°C served to form the inner electrodes.

Before firing green multilayer piezoelements were formed by ceramic powder layers (91.7±1 and 8.3±1 wt% (40 vol% of organic binder) with a thickness of ~50 μm interlaced with Ag-Pd paste with both polished layers ~5 μm thick. The specific surface of the PZT-46 powder was in the limits of 10000…11000 cm^2/g.

At 160 °C the butyral resin decomposes releasing water and oil aldehyde, while butyl benzyl phthalate (SANTICIZER 160) flares at 199 °C. The weight loss during heat treatment at 380 °C amounts to 7.8 wt%; the weight loss process starts at ~90 °C (warm-up time 1000 s) and finishes at ~400 °C (warm-up time 4500 s).

The sintered microstructure was investigated with both mechanically polished or fractured perpendicular sections of the samples. The flexural deformation and hysteresis characteristics of prepared actuators were measured in a quasistatic mode with the aid of a microscope. The thickness of the multilayer samples is defined by the number (typically 20) and thickness of the individual layers (~50 μm).

3. Results and discussion

The thermogravimetric analysis (TGA) (figure 1) shows that the organic components of green multilayers start to burn at temperatures as low as 50 °C, while the main burning out of the plasticizer and binder proceeds in the 200 to 400 °C range of temperatures. The differential thermal analysis (DTA) data (figure 1) are indicative of intense organic compound decomposition processes at temperatures up to 500 °C. In this view great care should be taken in the proper choice of slow enough heating rates of the green multilayer composite.

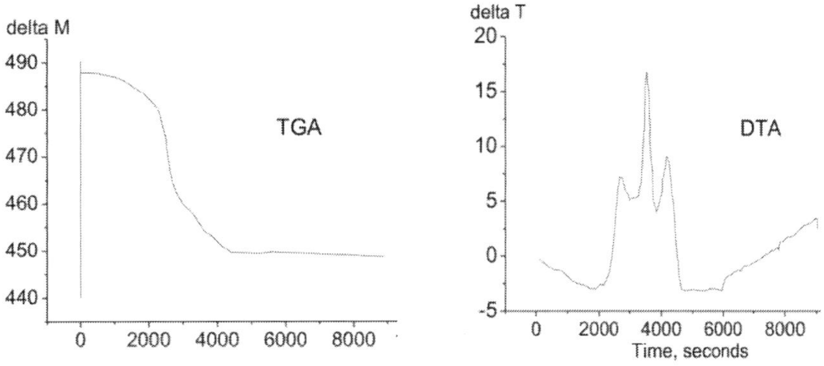

Figure 1. TGA and DTA curves of the green workpiece under temperature ramp of 800 °C / 2.5 hrs

Figure 2 shows a SEM micrograph of the multilayer sample surface in the region of partial delamination of the outer Ag electrode. It is seen that the underneath ceramic is dense and there are hardly any pores in the field of view. It should be noted that the porosity of the outer electrode is markedly higher than that of ceramics. The typical size of the pores is about 3 μm and they propagate deep into the metal (figure 3).

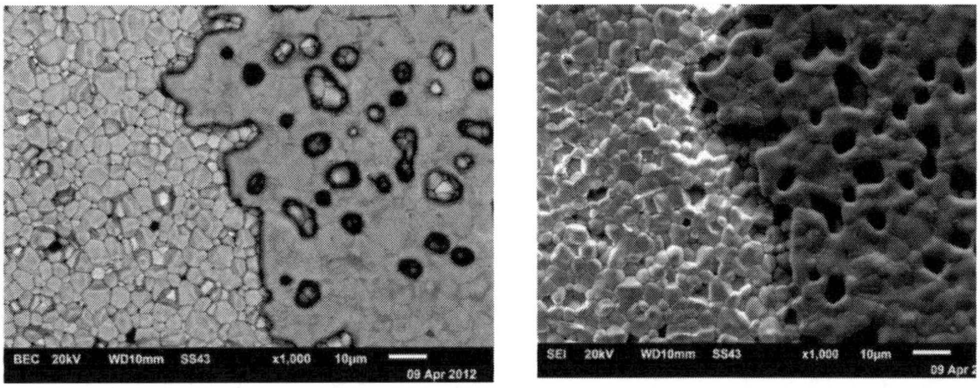

Figure 2. Multilayer sample surface in the region of outer Ag electrode partial delamination observed in back-scattered (BEC) and secondary electrons (SEI). Left parts of the images – exposed ceramics, right parts – silver electrodes

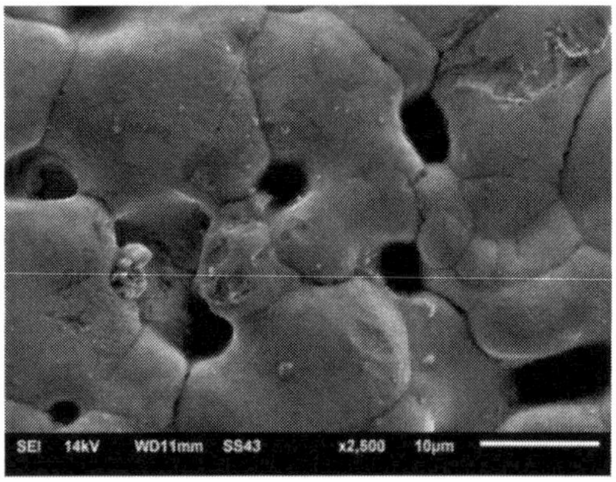

Figure 3. Pore and crystallite structure of the outer Ag electrode

Figure 4 shows the fractured profile of the multilayer sample in the region of the inner Ag/Pd/70/30 electrode. This figure reveals that the internal electrode is well defined. The metal is closely bonded to the ceramic and practically no pores are observed along the interface.

In the lower parts of figure 4 are given the results of local microanalysis of the composition of the composite in the regions of inner electrode and the interelectrode area.

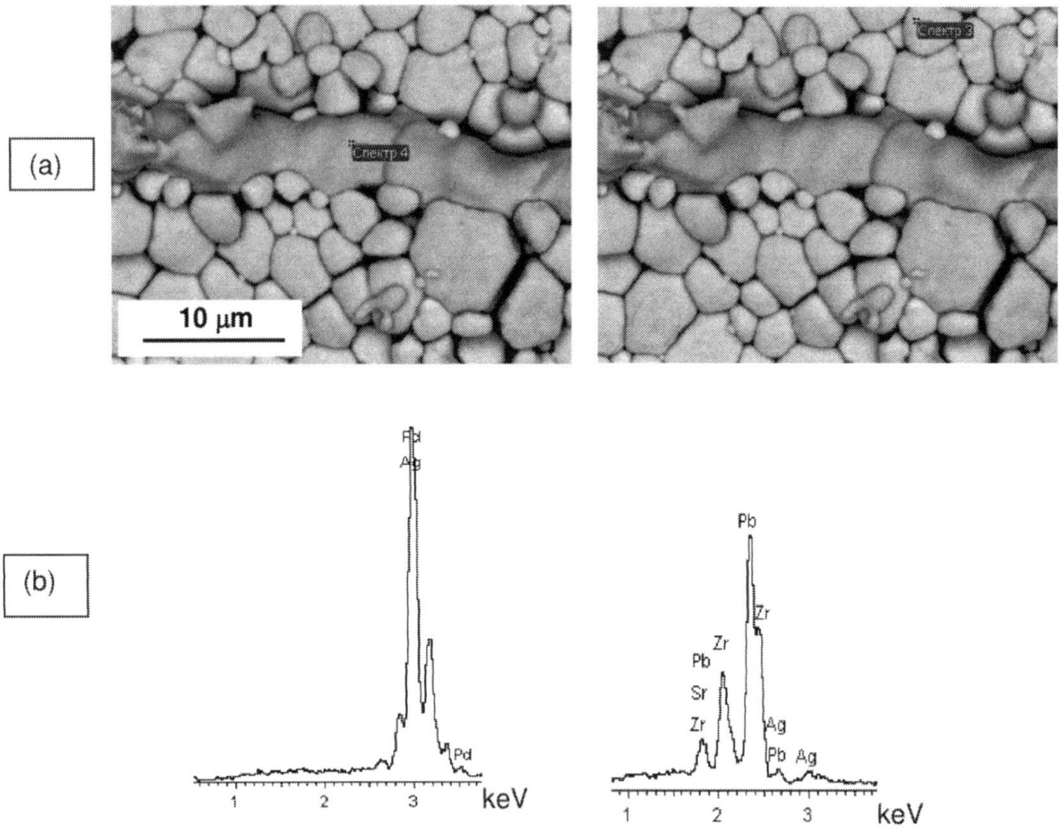

Figure 4. SEM microstructure morphology of the internal electrode area (*a*) and EDS analysis (*b*) of the Ag-Pd electrode and of bulk material. Areas of interest are marked with rectangles

The microanalysis has shown that the atomic composition of the electrode after co-firing coincides with that of the initial Ag-Pd powder except for Pb traces observed in the electrode at distances ≤ 500 nm from the PZT/Ag-Pd interface.

Strain of the finished multilayer sample as a function of electric field is presented in figure 5.

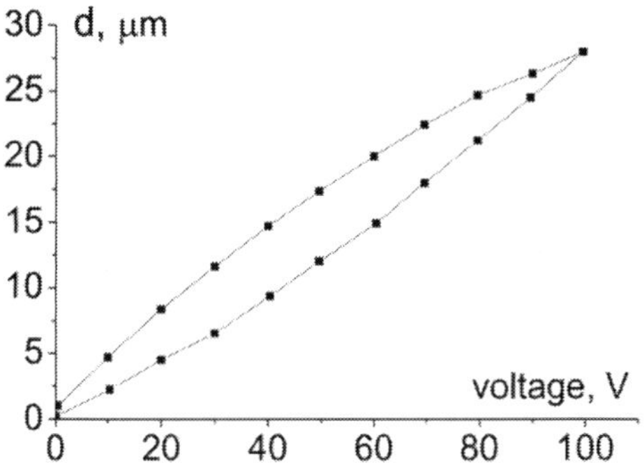

Figure 5. Strain vs voltage hysteresis curve of a finished multilayer actuator. Fitting equations for the descending and ascending parts of the loop are
$y_1 = - 0.0012x^2 - 0.3928x + 0.7282$
and
$y_2 = 0.0008x^2 + 0.2005x - 0.0082$

4. Conclusion

The process for the preparation of multilayer actuators was described starting from PZT and Ag-Pd individual layers produced by by tape casting. Uniform layers with minimal porosity well interconnected with the co-fired internal 90%Ag-10%Pd electrodes were obtained. The microanalysis has shown that the atomic composition of the Ag-Pd electrode after co-firing coincides with that of the initial powder except for Pb traces observed in the electrode at distances ≤ 500 nm from the PZT/Ag-Pd interface. Low-hysteresis multilayer actuator with a strain of ~30 μm with applied voltage of ~100 V is demonstrated.

References

[1] Heywang W, Lubitz K, Wersing W 2008 *Piezoelectricity – Evolution and Future of a Technology* (Berlin/Heidelberg: Springer)
[2] Kushnir U, Rabinovitch O Advanced piezoelectric-ferroelectric stack actuator 2009 *Sens. & Act.* A **150** 102-9
[3] Zhu D, Beeby S, Tudor J, White N, and Harris N 2011 Improving output power of piezoelectric energy harvesters using multilayer structures *Proc. Eng.* **25** 199-202
[4] Kühnlein T, Stiegelschmitt A, Roosen A, Rauscher M 2013 Development of a model for the sintering of PZT multilayer ceramics and their dielectric properties *J. Eur. Ceram. Soc.* **33** 991-1000
[5] Galassi C, Camporesi G, Fabbri G, Costa A L, Roncari E 2001 Processing of a multilayer bender type actuator *J. Eur. Ceram. Soc.* **21** 2011-4
[6] Cao J L, Wang X H, Gui Z L, Li L T 2008 Structural investigation s of the co-fired interface of Pb-based relaxor ferroelectrics and Ag-Pd electrode *Ceram. Intern.* **34** 2025-8
[7] Zuo R, Li L, Gui Z 2001 Interfacial development and microstructural imperfection of multilayer ceramic chips with Ag/Pd electrodes Ceram. Intern. **27** 889-93
[8] Yao K, Zhu W 1998 Improved preparation procedure and properties for a multilaye piezoelectric thick-film actuator *Sens. & Act.* A **71** 139-43

IOP Conf. Series: Materials Science and Engineering **49** (2013) 012034 doi:10.1088/1757-899X/49/1/012034

Manufacturing, structure and properties of recycled polyethylene terephthalate /liquid crystal polymer/montmorillonite clay nanocomposites

Guntis Japins[1], Rita Berzina[1], Janis Zicans[1], Remo Merijs Meri[1], Tatjana Ivanova[1], Valdis Kalkis[2], Ingars Reinholds[2]

[1]Institute of Polymer Materials, Riga Technical University, [2]Department of Chemistry, University of Latvia

e-mail of corresponding author: guntisjapins@gmail.lv

Abstract. Polyethylene terephthalate (PET)/liquid crystal polymer (LCP)/monthmorillonite clay (MMT) compositions were obtained by melt mixing. Their mechanical, structural, rheological and thermal properties were investigated.

1. Introduction

Polyethylene terephthalate (PET) is widely used thermoplastic material. As a consequence large amount of PET waste is generated every year. Biggest source of recycled PET (RPET) are soft drink and alcoholic beverage bottles. In last decades the technology of post-consumer PET recycling has increased to very high level. Now it is possible to use RPET not only for fiber or cheap product production, but also for manufacturing of new beverage bottles and high class engineering materials [1, 2].

Liquid crystal polymer (LCP) is an intermediate state between solid crystal and isotropic fluid. This group of polymers is recognizable by their rigid rod-like molecules [3]. LCP improves mechanical properties of matrix material, besides LCP makes easier processing, because of shear-thinning effect [4]. Furthermore, in extrusion or injection molding LCP easily orient in the flow direction. It gives reinforcement effect on matrix. LCP dispersed in an isotropic thermoplastic matrix, can exist in the shape of spheres, fibrils or lamellae [3].

Nano scale montmorillonite clay (MMT) is also often used for reinforcing of polymers because of its anisometric plate-like particle shape. Due to that MMT improves also thermal stability and barrier properties [3]. There are many investigations about nano clay impact on thermoplastic polymers [5].

Simultaneous effect of both of these modifiers on polymers is scarcely investigated. In this study impact of MMT and LCP on RPET structure and properties is investigated.

2. Experimental

2.1. Materials

As matrix material RPET, obtained from post-consumer beverage bottles, is used. As modifier MMT clay from Laviosa Chimica Mineraria S.P.A. (Italy) is used. Liquid crystal polymer is synthesized from p-acetoxybenzoic acid and polyethylene terephthalate.

2.2. Preparation of RPET/LCP/MMT compositions

RPET has melt mixed with MMT (1 to 5 wt. %) and LCP (5 wt. %) by using twin screw extruder, which have five heating zones (Thermo PRISM TSE 16 TC). Temperatures along the barrel length have been set in this way: 245 °C (feeding zone), 255, 260, 270 °C, 275°C (head zone). Screw rotating speed has been 80 rpm. Prepared compositions are listed in Table 1.

Table 1. Prepared RPET/LCP/MMT compositions

Composition	RPET (wt. %)	LCP (wt. %)	MMT (wt. %)
1	98	2	-
2	95	5	-
3	90	10	-
4	94	5	1
5	93	5	2
6	90	5	5

2.3. Preparation of specimens

Specimens for tensile tests have been produced by using injection moulding machine Mini Jector, Model 55. Temperatures have been set 250-290-300 °C from hopper to die.

2.4. Measurement of physical properties

Mechanical properties have been investigated by using universal material testing machine Zwick Roell BDO-FB020TN at constant speed of the upper cross-head - 50 millimeters per minute.

Rheological properties have been estimated by determination of melt flow rate by using ИИРТ – AM device at temperature of 265 °C and load of 2, 16 kg.

Calorimetric characteristics have been investigated by using Mettler Toledo DSC 1 differential scanning calorimeter. DSC curves have been recorded in three steps: first – heating from 25 to 300 °C; second – cooling from 300 to 25 °C; third – repeated heating from 25 to 300 °C. Heating or cooling rate has been 10 °C / minute. The data for analysis have been taken from the third run.

Density has been determined by using hydrostatic weighing method. Sartorius YDK 01 density measurement kit mounted on Sartorius KB BA 100 electronic balance has been used. Measurements have been made at 25°C temperature. Ethyl alcohol has been used as liquid media.

3. Results and discussion

3.1. Mechanical properties

Firstly it is necessary to find most appropriate RPET/LCP composition for modification with MMT. As base composition for modification with MMT, RPET blend with 5 wt. % of LCP has been chosen. This decision is based on the data from composition tensile test. Mechanical properties of various RPET/LCP blends are show in figure 1. Yield stress and stress at break (figure 1a) of the investigated RPET blends show maximum values at LCP content of 5 wt. %. Increase of Young's modulus at this LCP weight content also is relatively larger than at higher modifier concentrations (figure 1b). It is interesting to note, that up to 5 wt. % of LCP, strain at break values of the blends also increases.

The effect of MMT on the tensile properties of the RPET/5%LCP based nanocomposites is shown in Figure 1c. As shown, by rising MMT content considerable increase of Young's modulus is observed. The reason for the modulus increase most probably is reinforcement of the composite because of the orientation of the platelike clay particles in the direction of the applied load. Contrary to Young's modulus, strain values of RPET/5%LCP/MMT composites considerably decrease along with raising MMT content. It is, however, well know that adding inorganic filler (in our case MMT clay) system strain values decreases [5].

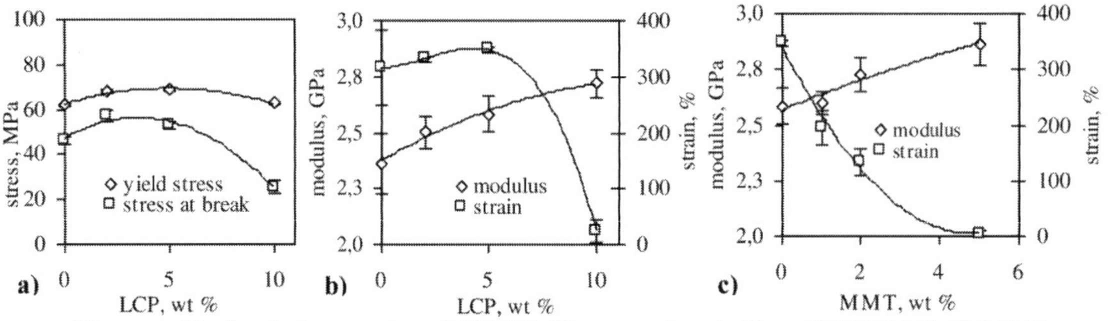

Figure 1. Mechanical properties of RPET/LCP composites (a, b) and RPET/5%LCP/MMT nanocomposites (c): yield stress and stress at break (a), Young's modulus and strain at break (b, c) as functions of modifier weight content.

3.2. Structural properties

The influence of LCP and MMT on the RPET density is illustrated in figure 2a for RPET/LCP and figure 2b for RPET/5%LCP/MMT composites.

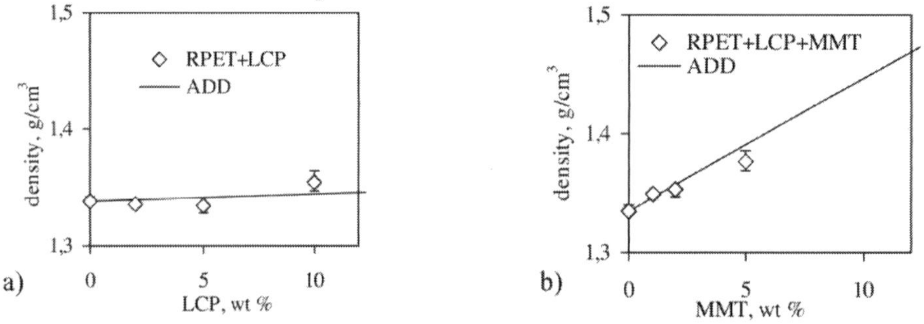

Figure 2. Density of RPET/LCP (a) and RPET/LCP/MMT (b) composites as function of modifier content. With continuous line additive density values, which are calculated by using equation 3.1., are shown

Experimental density values are determined by means of hydrostatic weighting. Theoretical density values are calculated by using equation 3.1.

$$\rho_{add} = \sum_{i=1}^{n} \rho_i \varphi_i \tag{3.1}$$

were ρ_{add} – additive (theoretical) density, g/cm^3, ρ_i–densities of the components, g/cm^3, φ_i volumetric contents of the components in the composite.

It has been found, that the experimental density values are almost equal to the theoretical ones, especially in the case of RPET/LCP blends, for which deviations are below 1%. Indirectly this can testify about certain compatibility of the blend components, most probably due to the presence of ethylene terephthalate moieties in the LCP macromolecule.

The effect of LCP and MMT on the crystallinity degree (χ) and glass transition temperature (T_g) of RPET have been investigated by means of differential scanning calorimetry (table 2). It is observed that χ increase by ca 50 % along with addition of LCP to RPET. Addition of MMT to RPET/5%LCP blend also raises crystallinity of the system by approximately the same level. Increase of χ probably is the main reason why real density of the RPET/10%LCP blend is above the theoretically calculated. In the case of MMT containing RPET/5%LCP based nanocomposites this effect is hindered because of greater incompatibility of the polymeric constituents of the blend and MMT. Increased RPET χ values testify that both modifiers promote crystallization of RPET matrix, being another factor, which is responsible for increased stiffness and strength characteristics of the investigated compositions. There is also some increase of T_g of RPET along with raising the content of either LCP or MMT. Rising of

T_g approves, that MMT, and most probably also LCP, reduce chain flexibility of RPET, thus stiffening the material. This is another reason why stiffness of the investigated compositions rises.

Table 2. Crystallinity degree and glass transition temperature of the investigated compositions.

Composition	Crystallinity degree χ (%)	Glass transition temperature T_g (°C)
RPET	21	74
RPET+2%LCP	24	76
RPET+5%LCP	24	77
RPET+10%LCP	31	81
RPET+5%LCP+1%MMT	30	78
RPET+5%LCP+2%MMT	31	79
RPET+5%LCP+5%MMT	33	80

3.3. Rheological properties

Rheological properties of the investigated compositions have been estimated by means of melt flow rate (MFR) measurements. It is well known that LCP in thermoplastic polymer compositions acts as a processing aid by decreasing viscosity and respectively increasing melt flow rate of the matrix polymer [3]. In figure 3a RPET melt flow rate dependence on LCP content is illustrated. As we can see only at LCP concentration 10 wt. % there is significant increase in MFR. In composites filled with MMT situation is quite unusual (figure 3b). Contrary to the widely accepted assumption that addition of inorganic filler increase viscosity (decrease MFR), in this case an opposite effect is observed. Similar situation is also referred by other authors [5]. It is thought that under shear stress MMT layers orient in the flow direction, thereby decreasing viscosity.

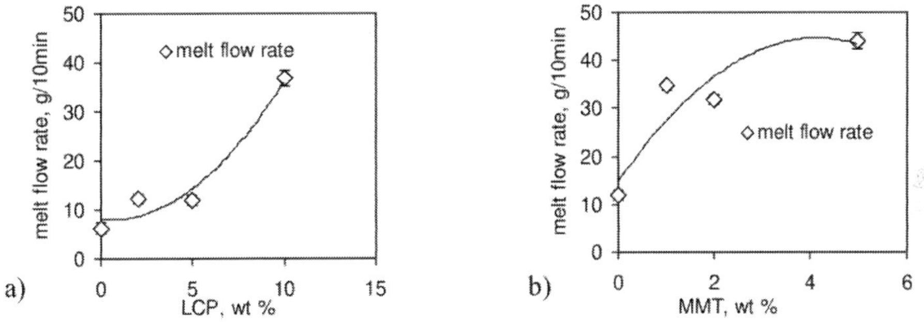

Figure 3. Melt flow rate of RPET/LCP blends (a) and RPET/5%LCP/MMT nanocomposites (b) as functions of modifier weight contents.

4. Conclusions

Increase of mechanical properties of RPET by modification with rigid-rod LCP and plate-like nanolevel MMT has been achieved. Positive effect of LCP and MMT nanoparticles on the crystallinity degree and glass transition temperature of RPET matrix is another reason for stiffness and strength increase. Addition of aforementioned modifiers eases also flowability of RPET, which is important for manufacturing of thin-wall items.

5. Acknowledgements

The research is carried out within the framework of the ERDF project Nr. 2010/0209/2DP/2.1.1.1.0/10APIA/VIAA/028

References

[1] Sheirs J 1998 *Polymer Recycling: Science, Technology & Applications.* John Wiley & Sons, 121-182
[2] Hannawi K, Kamali-Bernard S and Prince W 2010 *Waste Manage.* **30** 2312-2320
[3] Lee M W, Hu X, Li L, Yue C Y, Tam K C and Cheong L 2003 *Comos. Sci. Technol.* **63** 1921-1929
[4] Cidale M T, Fernandez M, Filipe S and Santamaria A 2012 *Polym. Test.* **31** 290-296
[5] Pavlidou S, Papaspyrides C D. 2008 *Prog. Polym. Sci.* **33** 1119-1198.

IOP Conf. Series: Materials Science and Engineering **49** (2013) 012035 doi:10.1088/1757-899X/49/1/012035

Influence of some system parameters on silica surface patterns by sol-gel phase separation method

T Kangur[1,2], **L Nurmis**[1,2], **M Järvekülg**[1,2]

[1] Institute of Physics, University of Tartu, Riia 142, 51014Tartu, Estonia
[2] Estonian Nanotechnology Competence Centre, Riia 142, 51014 Tartu, Estonia

E-mail: kangur@ut.ee

Abstract. We have studied the effect of different coating methods and precursor compositions on the size, surface density and shape of round surface structures prepared by phase separation-based surface patterning method with potential application in preparing superhydrophobic optically functional structural coatings. Increase in solvent polarity resulted in larger micro- and nanosize surface features. Variation in precursor concentration and extent of initial polymerization were shown to result in different surface densities and geometries of formed features. The effect of different surface patterns on wettability was evaluated by WCA measurements.

1. Introduction

Sol-gel method offers a variety of well established, facile and versatile technologies for flexible design of materials for a wide range of applications, including optical devices [1], sensors [2], biomaterials [3], coatings [4]. In addition to the possibility of preparing different pure [5], doped [6, 7], mixed oxides [8] and hybrids [8, 9] at relatively low temperatures, sol-gel also enables in-situ control over the nano- and microstructure of prepared materials during transition from sol to gel [10]. Micro- and nanostructure of surfaces has a critical role in the design of superhydrophobic coatings [11] and sol-gel has been applied in phase separation based strategies [12, 13] to achieve suitable micro- and nanoroughness. In the case of self-cleaning windows, optical properties of obtained hydrophobic surfaces are also a concern [14]. Sol-gel-based methods that are simple and can be applied on industrial scale have been used to prepare films that, when coupled with low-surface energy, can have water contact angles as high as 167° [15]. Nevertheless, such film coatings can compromise the transparency of the glass due to numerous reflections and refractive surfaces. We have reported a strategy based on phase separation in silicon alkoxide solutions that results in a single layer of round surface features [16]. This method also holds considerable potential in designing optically functional coatings, as each surface feature can be designed to act as a micro-lens. We hereby present our results on the variation of several system parameters such as extent of precursor concentration, pre-polymerization and solvent polarity on the formation of round silica surface features. The results demonstrate the potential of the method for the development of single-step coating technology for structural coatings with simultaneous optical and superhydrophobic functionality.

2. Experimental

Patterned structures were prepared from solutions of partially hydrolyzed tetraethylorthosilicate (TEOS) in different solvents. All chemicals were purchased from Sigma-Aldrich. Conventional acid-catalyzed hydrolysis and polymerization of TEOS was carried to prepare sols. TEOS was mixed with solvent on a magnetic stirrer, followed by the addition of water containing 3.5 % nitric acid to initiate hydrolysis. Obtained solution was stirred for 1 h before coating on soda lime glass slides (2.5 x 2.5 cm) by spin coating or spraying. Glass slides were pre-cleaned with acetone in ultrasonic bath, rinsed with methanol and dried with gaseous nitrogen. For investigating the impact of solvent type, sol concentration and water-alkoxide molar ratio on silica structures, different process parameters were applied. Solvent ratios (solvent:TEOS) 1, 5, 9, 13 and 17 (S) as well as water-alkoxide molar ratios 0.4, 0.8, 1.2, 1.6, 2 and 4 (R) were tested with solvents propanol, ethanol and methanol. During synthesis relative humidity of surrounding environment was fixed at 15 % to 20 % at temperatures 22-23 °C. All samples were subsequently heated at 200°C for 20 h to transform gel material into oxide.

Scanning electron microscope (SEM) graphs of silica domes were acquired with Tescan VEGA II SBU with accelerating voltage of 10 kV. The samples were previously sputter coated with a 5 nm layer of gold using a SC7640 Auto/Manual High Resolution Sputter Coater. To estimate surface hydrophobicity water contact angle (WCA) measurements were performed. A water droplet of 4 µl volume was placed at the surface and measured at three different spots using Canon EO5 650D with objective Canon MP65 and GIMP for measuring WCA from images.

3. Results and discussion

In this study we examined the influence of reaction parameters such as solvent type, sol concentration and water-alkoxide molar ratio on sol-gel phase separation process and subsequent formation of surface features. Mostly, we obtained symmetric round surface structures with diameter ranging from 200 nm to 20 µm. Sol-gel phase separation is known to start with the nucleation of isolated domains of separated phase, droplets of either solvent-rich or solvent-poor phase inside continuous solution phase domain [12]. The process then proceeds to formation of bi-continuous two-phase system with a large phase interface. In most cases where porous oxides are prepared, this kind of structure is also preserved by optimized speed of solvent evaporation. In our case the volatile solvent is optimized to evaporate before the evolution of continuous organosilica-rich phase (explained in figure 1). Nucleation is followed by growth of the droplets and gradual diminishing of solvent-rich phase due to solvent evaporation. In most cases we observed the formation of isolated round surface features formed from the droplets on the surface after partial flattening of not fully gelled silica-forming precursor.

As expected, different coating methods, such as spin-coating and spraying, lead to qualitatively different results. Spin-coating is more sensitive to variations in precursor compositions, results in sharper change in the size of surface features. During the spin-coating most of the material, which is applied on the surface, spins off the surface, and the thickness of the precursor layer in which the phase separation occurs, depends on the viscosity. The latter is influenced by both solvent concentration and extent of precursor polymerization. With spraying, the domes tend to be larger in diameter and height due to the larger thickness of precursor left on the surface. The amount of precursor that stays on the substrate is also less influenced by the viscosity of the precursor. Additionally, as observed, some of the surface features formed in case of spray coating were less flattened, in some cases not even to half-sphere geometry. That is the result of

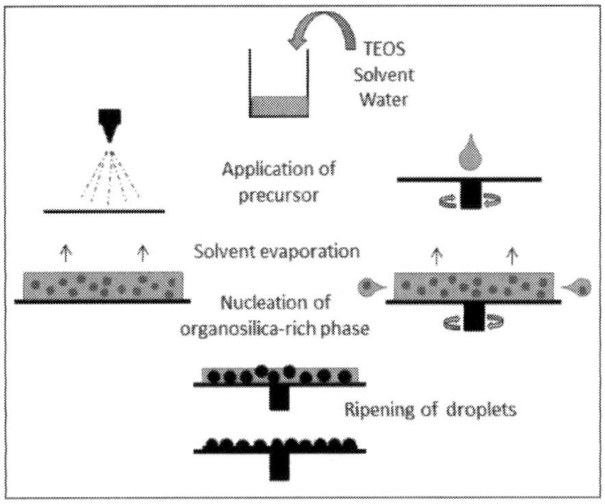

Figure 1. Schematic illustration of used coating processes and formation of surface patterns.

solvent evaporation from the coated layer being slow enough compared to gelling, so that the viscosity of the silica-forming phase preserves the droplet shape once the solvent has fully evaporated.

Varying parameters did not lead to clear linear changes in size and shape. Nevertheless, increase in solvent polarity (propanol < ethanol < methanol) and volatility led to increase in the diameter of the surface features, which varied from about 200 nm to 20 μm. At the same time, the used solvents acted differently in response to varying sol concentration and water-alkoxide molar ratio.

Varying amount of added propanol resulted in decrease in the surface density of structures from 200 to 5 domes per 100 μm^2 as propanol concentration increased due to the lower material amount, which has stayed on the substrate during spin-coating (figure 2a-d). However, dome diameter was not significantly affected and remained about 200 nm. By spraying, there was no such variation in density. As propanol has comparatively low polarity among used solvents, phase separation has not occurred completely, resulting in formed surface featured being covered by thin film layer. As R was varied, there was not noticeable change in dome size or density.

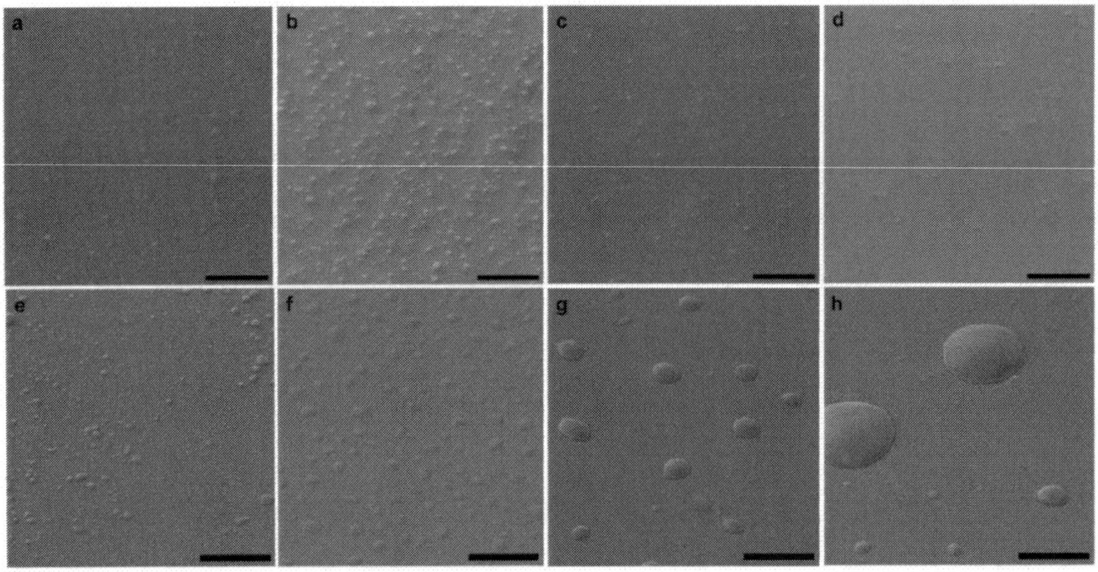

Figure 2. SEM images of series with variation of concentration in propanol (a – d) and methanol (e – h) solutions. Scale bars correspond to 5 μm (a – d) and 20 μm (e – h).

When using ethanol as a solvent, the difference in coating method was clearly seen. Silica structures via spraying were larger in diameter and height compared to the spun films. For example, TEOS:solvent ratio 1:9 with R = 1.2 led to size change of about 10 times (figure 3a,b). Increasing R led to slight increase in the surface density of the structures. With R = 1.6, the most spherical shape features were observed (figure 3c).

Methanol behaved differently as a solvent compared to propanol and ethanol that were relatively similar tendencies. Methanol is the most polar of the three, leading to fast complete phase separation but is also the most volatile so TEOS component has not polymerized significantly by the time the

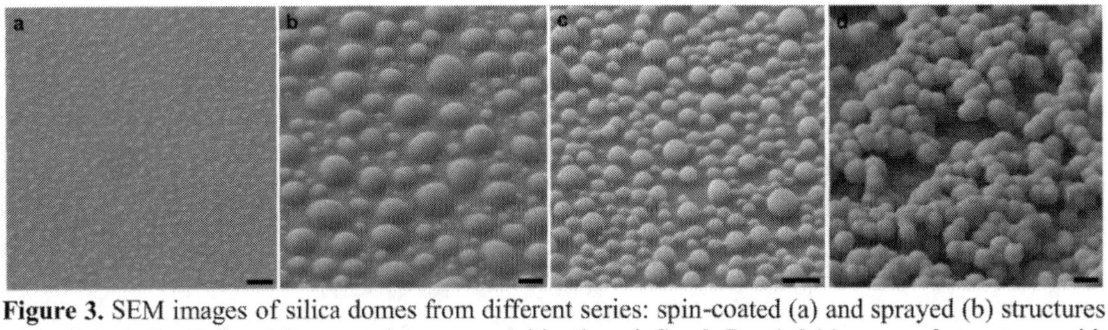

Figure 3. SEM images of silica domes from different series: spin-coated (a) and sprayed (b) structures with ethanol, S = 9, R = 1.2; sprayed structures with ethanol, S = 9, R = 1.6 (c); sprayed structures with ethanol, S = 1, R = 1.2 (d). Scale bars correspond to 2 μm.

solvent has fully evaporated and the still liquid droplets can merge on the substrate. For that reason relatively large surface features or even films were formed by spraying. As TEOS:methanol ratio was varied, spin-coating led to increasing domes as solvent concentration increased (figure 2e-h). Solvent ratio S = 17 also resulted in continuous film.

In most cases WCAs of samples prepared by spraying were higher compared to the spin-coated surfaces. Sprayed coatings with S = 1 and R = 1.2 in ethanol were estimated to have the highest hydrophobicity, with WCA about 131° (figure 3d). However, these cases may be classified as porous films. In case of surfaces patterned with round surface features, 88° WCA was achieved in case of S=9, R=1.6 in ethanol. It is important to point out that the achieved shift towards hydrophobicity is the effect of surface structure only, the surface energy resulting from surface chemistry of prepared substrates is relatively high. To demonstrate that, a film was prepared in a similar fashion using aged TEOS solution in which no phase separation occurred and a smooth film with no detectable roughness was obtained. The WCA was estimated to be 25° in case of that reference sample.

4. Conclusions

Using different coating methods and suitable precursor compositions in sol-gel phase separation surface patterning can be used for preparing round silica surface features with different size, surface density and geometry. Different solvents, concentrations and initial extent of precursor polymerization were used in spray- and spin-coating. Obtained results provide first proof of controllability and potential of phase separation-based surface patterning method in the development of optically functional superhydrophobic structural surface coatings.

Acknowledgments

This research was supported by ESA grant IUT2-25; ESF grants ETF8377, ETF8699, ETF9283, SF0180058s07; Estonian Nanotechnology Competence Center; Graduate School on Functional Materials and Technologies (GSFMT), University of Tartu and Tallinn University of Technology, EU Social Funds project 1.2.0401.09-0079; European Union through the European Regional Development Fund (Centre of Excellence "Mesosystems: Theory and Applications", TK114).

References

[1] Sanchez C, Lebeau B, Chaput F and Boilot JP 2003 *Advanced Materials* **15** 1969.
[2] Jeronimo PCA, Araujo AN and Montenegro MCBSM 2007 *Talanta* **72** 13.
[3] Avnir D, Coradin T, Lev O and Livage J 2006 *Journal of Materials Chemistry* **16** 1013.
[4] Purcar V, Stamatin I, Cinteza O, Petcu C, Raditoiu V, Ghiurea M, Miclaus T and Andronie A 2012 *Surface & Coatings Technology* **206** 4449.
[5] Francis LF 1997 *Materials and Manufacturing Processes* **12** 963.
[6] Paerna R, Joost U, Nommiste E, Kaeaembre T, Kikas A, Kuusik I, Kink I, Hirsimaki M and Kisand V 2012 *Physica Status Solidi a-Applications and Materials Science* **209** 953.
[7] Kiisk V, Kangur T, Paalo M, Taette T, Lange S, Pikker S and Sildos I 2011 *Materials Chemistry and Physics* **130** 293.
[8] Timusk M, Jaervekuelg M, Salundi A, Lohmus R, Leinberg S, Kink I and Saal K 2012 *Journal of Materials Research* **27** 1257.
[9] Paalo M, Taette T, Hussainov M, Hanschmidt K, Lobjakas M, Lohmus A, Maeeorg U, Kink I and Iop 2012. Preparation and Characterization of CNT/TiO2 Based Transparent Fiber Electrodes. *International Conference on Functional Materials and Nanotechnologies.*
[10] Saal K, Taette T, Jaervekuelg M, Reedo V, Lohmus A and Kink I 2011 *International Journal of Materials & Product Technology* **40** 2.
[11] Yan YY, Gao N and Barthlott W 2011 *Advances in Colloid and Interface Science* **169** 80.
[12] Nakanishi K and Tanaka N 2007 *Accounts of Chemical Research* **40** 863.
[13] Kajihara K, Kuwatani S, Maehana R and Kanamura K 2009 *Bulletin of the Chemical Society of Japan* **82** 1470.

[14] Ganbavle VV, Bangi UKH, Latthe SS, Mahadik SA and Rao AV 2011 *Surface & Coatings Technology* **205** 5338.
[15] Mahadik SA, Mahadik DB, Kavale MS, Parale VG, Wagh PB, Barshilia HC, Gupta SC, Hegde ND and Rao AV 2012 *Journal of Sol-Gel Science and Technology* **63** 580.
[16] Reemann P, Kangur T, Pook M, Paalo M, Nurmis L, Kink I, Porosaar O, Kingo K, Vasar E, Koks S, Jaks V and Jarvekulg M 2013 *Journal of materials science. Materials in medicine* **24** 783.

IOP Conf. Series: Materials Science and Engineering **49** (2013) 012036 doi:10.1088/1757-899X/49/1/012036

Photoinduced mass transport in azo compounds

K Klismeta, J Teteris and J Aleksejeva

Institute of Solid State Physics, University of Latvia, 8 Kengaraga Str., LV 1063 Riga, Latvia

E-mail: k.klismeta@gmail.com

Abstract. The photoinduced changes of optical properties in azobenzene containing compound thin films were studied under influence of polarized and non-polarized 532 nm laser light. Under influence of light azo compounds experience *trans-cis* isomerisation process, that can be observed in the absorbance spectrum of the sample. If the light is linearly polarized, molecules align perpendicularly to the electric field vector and as a result photoinduced dichroism and birefringence is obtained. If a known lateral polarization modulation of the light beam is present, mass transport of the azobenzene containing compound occurs. By measuring the surface relief with a profilometer the direction of mass transport can be determined. The studies of this work show that direct holographic recording of surface relief gratings can be used in optoelectronics, telecommunications and data storage.

1. Introduction

Upon irradiation of UV or visible light, azo compounds experience *trans-cis* photoisomerisation process: the stable *trans* configuration may undergo a transformation to the metastable *cis* configuration, the change can be observed in the absorption spectrum of the azo compound incorporated in polymers [1]. Light electrical field align azo moieties perpendicular to the direction of light polarization – that leads to birefringence and dichroism [2]. It was shown by P.Rochon *et al.* in 1995 [3] that stable, highly efficient surface relief gratings can be optically induced on surface of azopolymer films. One of the used methods to induce mass motion is holographic recording when a periodical intensity pattern can be obtained at the submicron scale. It was observed by X.L. Jiang *et al.* [4] that the formation of SRG on the azobenzene containing polymer films strongly depends on the polarization state of the two recording beams. Under the recording condition of two orthogonally linearly polarized recording beams +45°:-45, the maximum surface modulation was obtained. In this case polarization modulation is present on the surface of sample, but there is no intensity modulation.

In this work we present experimental investigation of photoinduced changes of optical properties, including mass motion, in Poly(Disperse Red 1-methacrylate) thin film under influence of polarized and non-polarized 532 nm laser light.

The abovementioned properties of azobenzene containing compounds suggest that these materials could have potential application in various photoactive devices, optical information storage and telecommunication, nonlinear optics and production of diffractive optical elements. [5]

2. Experimental

Poly(DR1-MA) was purchased at Sigma-Aldrich. The molecular structure of the compound is shown in figure 1.

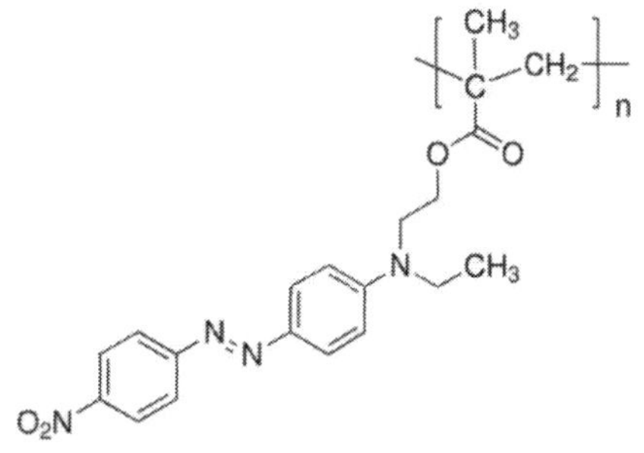

Figure 1. Molecular structure of Poly(Disperse Red 1- methacrylate).

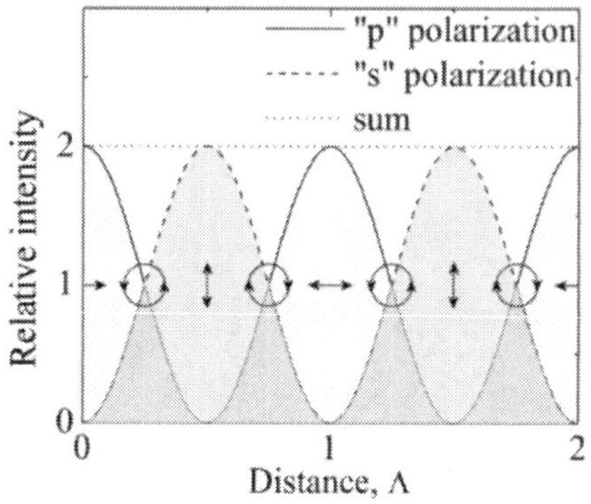

Figure 2. Polarization and intensity modulation of p and s components on the surface of sample.

Azo-dye (DR1) is chemically bound to the main polymeric chain. The powder was dissolved in trichloroethylene at room temperature. The solution was filtered with PTFE filters with pore size of 0.45 μm. The films were produced from the solution using BYK Gardner GmbH applicator, the thickness of the liquid layer was 120 μm. Glass plates were used as a substrate. The thickness of dry films was determined using Veeco Dectak 150 surface profilometer.

The experimental set-ups used in this work are shown on figure 3 and figure 4.

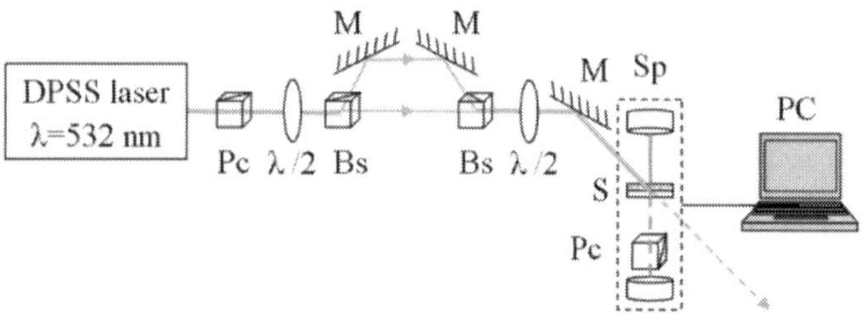

Figure 3. Experimental set-up for absorption spectra measurements.

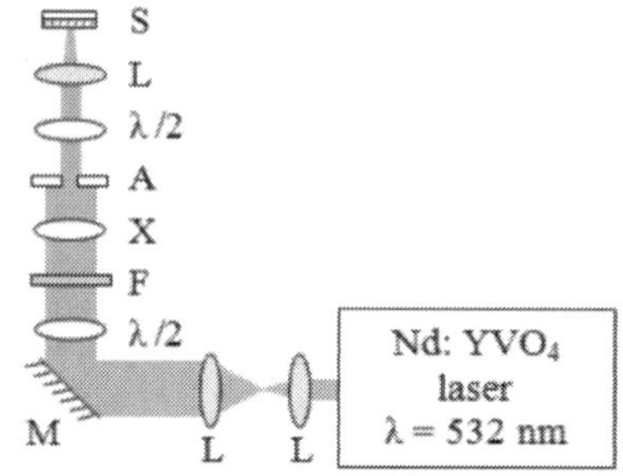

Figure 4. Experimental set-up for surface relief recording.

A – aperture, Bs – beam splitter, F – neutral filter, L – lens, λ/2 – half wave plate, M – mirror, Pc – polarisation cube, PC – computer, S – sample, Sp – spectrometer, X – element for creating polarisation modulation.

Experimental set-up for studying the absorbance spectrum while irradiating the sample with polarized and non-polarized 532 nm laser light is shown in figure 3. The polarization of laser beam was improved by Glan-Taylor polarization cube. Half-wave plate is used to control the polarization state. Due to the fact that the used laser possesses short coherence length, a non-polarized light can be obtained by balancing the vertical and horizontal polarization components. The linear polarization, orthogonal or parallel to the spectrometer probe light's polarization, can be obtained by using the half-wave plate. The spectra were measured using Ocean Optic HR4000CG spectrometer.

For inspecting the changes in the sample induced by periodically modulated polarized light an experimental set-up was created, shown in figure 4. Lateral polarization modulation of the light beam was created by using a new method - as a result during the experiment there was a uniform irradiance on the surface of the sample but polarization modulation with a set period was present (figure 2). Nd:YVO_4 laser beam is firstly expanded by two lenses according to telescope principle. For rotating the polarization, two half wave plates are needed and between them an element for creating the periodical polarization modulation. Usage of aperture allowed cutting out required number of periods. For mass transport direction investigation 2 periods were cut out. Lastly the modulated beam is focused on the sample. The created polarization pattern on the surface sample is analogical to holographic recording with +45°:-45 polarization state, the period in this experiment was 25 μm.

3. Results

Figure 5 shows the absorption spectrum of non-irradiated sample and spectra taken while irradiating the sample with polarized and non-polarized 532 nm laser light.

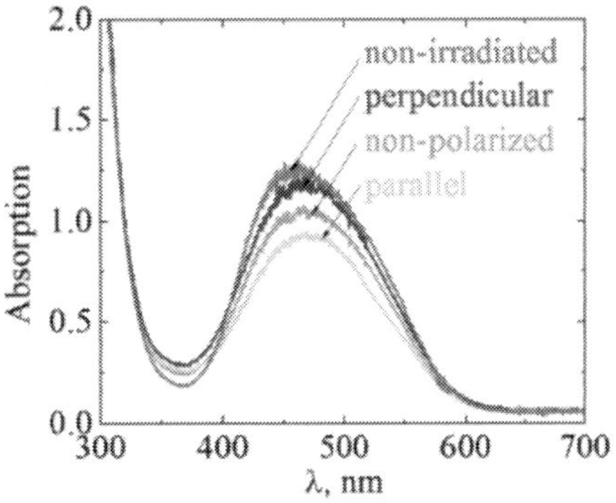

Figure 5. The absorption spectrum of non-irradiated sample and while irradiating with λ=532 nm, I=0.2 W/cm^2.

The thickness of the sample was 0.16 μm. The spectra were taken when sample was already irradiated for 10 seconds. The laser intensity was 0.2 W/cm^2. If the sample is irradiated with non-polarized light, the only effect observed is photoisomerisation – a part of the stable *trans* isomers transform into *cis* isomer. As a result the intense absorption of *trans* isomer at 420 - 500 nm decreases, while the absorption of *cis* isomer at 340 - 400 nm increases. The moieties are orientated randomly, since there is no strict direction of the electric intensity vector. If the pump beam is polarized, moieties photoisomerize and orientate. It was observed that the absorption of *trans* isomer of azobenzene molecules is greater in the direction perpendicular to the polarization of the pump beam. If the polarizations of pump and probe beams are parallel, the absorption is respectively smaller because of the dichroism phenomenon and also because in this case greater number of moieties experience photoisomerisation, thus concentration (and absorption) of *trans* isomer decreases. [6]

Figure 6 and figure 7 show the created structures on the surface of the thin film while irradiating it with laser light which has polarization modulation.

The thickness of the sample was 0.4 μm. The sample was irradiated for 2 hours with intensity of 8 W/cm^2 to obtain surface relief which can be measured with profilometer. Two configurations of the set-up were used - in first case, the polarization pattern on the sample was s-p-s-p-s and in the second case it was p-s-p-s-p. Figure 6 shows that if the "s" polarization is in the middle of the irradiated area, a "hill" is induced in the middle of the formation. If the configuration of the experimental set-up is so that "p" polarization is in the middle, a "valley" in the middle appears (figure 7).

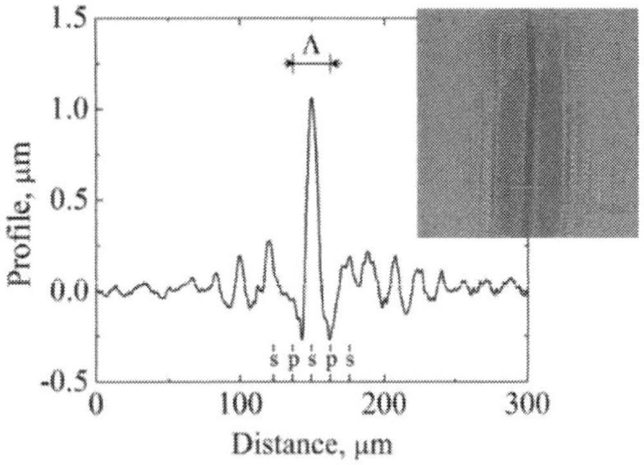

Figure 7. Surface relief created with configuration s-p-s-p-s and Λ=25 μm – measured profile and picture from optical microscope (scale is 50 μm).

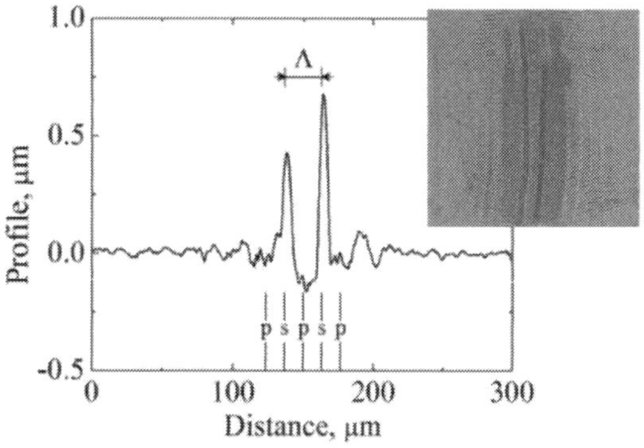

Figure 7. Surface relief created with configuration p-s-p-s-p and Λ=25 μm – measured profile and picture from optical microscope (scale is 50 μm).

The results obtained in this experiment correspond to experiments with optical slit [7]. Slit experiment is analogical to holographic recording with s:s or p:p polarization configuration. But as it was mentioned, it is known that the maximum surface modulation is reached by recording condition of +45°:-45° polarization state, which was achieved with the set-up used in this experiment.

4. Conclusions

Under influence of 532 nm laser light, moieties of poly(Disperse Red 1- methacrylate) photoisomerise and orientate so that the absorption is greater in the direction which is orthogonal to the electric intensity vector of the pump beam and smaller if the vector is parallel. If a polarization modulation of the pump beam is present on the surface of the sample, moieties move away from the region where the electric intensity vector is orthogonal to the grating and move to the region where the vector is parallel.

5. Acknowledgement

This work has been supported by the European Regional Development Fund within the project No. 2010/0204/2DP/2.1.1.2.0/10/APIAA/VIAA/010.

References

[1] Sudesh Kumar G and Neckers D C 1989 *Chem. Rev.* **89** 1915-25
[2] Todorov T, Nikolova L and Tomova N 1984 *Applied Optics* **Vol. 23, No. 23** 4309-4312
[3] Rochon P, Batalla E and Natansohn A 1995 *Appl. Phys. Lett.* **66 (2)** 136-8
[4] Jiang X L, Li L, Kumar J, Kim D Y, Shivshankar V and Tripathy S K 1996 *Appl. Phys. Lett.* **68 (19)** 2618-20
[5] Natansohn A and Rochon P 2002 *Chemical Reviews* **Vol. 102, No. 11** 4139-75
[6] Ojanen J and Rantala T T 2009 *The Open Chemical Physics Journal* **2** 37-46
[7] Teteris J and Gertners U 2012 *IOP Conf. Series: Materials Science and Engineering* **38** 1-6

IOP Conf. Series: Materials Science and Engineering **49** (2013) 012037 doi:10.1088/1757-899X/49/1/012037

LASER CONOSCOPY OF LARGE-SIZED OPTICAL CRYSTALS

A.I. Kolesnikov[1], R.M. Grechishkin[1], S.A. Tretiakov[1], V.Ya. Molchanov[2], A.I. Ivanova[1], E.I. Kaplunova[1], E.Yu. Vorontsova[1]

[1]Tver State University, Tver, Russia
[2]Acousto-Optic Research Centre, MISIS, Moscow

e-mail: rmgrech@yandex.ru

Abstract. Conoscopic interferometry provides a simple method of non-destructive control of the quality of a number of ferro-piezoelectric and optical crystals. Standard optical microscopes, including some commercial instruments, are easily adapted for implementation of conoscopic studies, though limited to small handy samples with a thickness of the order of 0.5 mm. In the present work we show that the usage of wide convergent or divergent conical laser beams in a simple benchtop configuration makes it possible to examine large-sized optical crystals by the method of conoscopy, including samples elongated along the optical axis direction. As distinct from traditional optical microscopy the conoscopic figures obtained with the aid of the laser installment may contain tens and hundreds of isochrome fringes thus increasing the informative capabilities of the method. Large-sized crystals of $LiNbO_3$ (∅57×95 mm), TeO_2 prisms (44×41×14mm) were examined experimentally at different angles between the optical axis and normal to the crystal surface. The experimental studies of different optical anomalies are confirmed by calculations based on the theoretical analysis given in a previous work of the authors.

1. Introduction

The investigation of optical properties of the birefringent crystals by the conoscopic method has been extensively exploited since long ago [1-2]. In most cases standard polarizing optical microscopes are finding their use for conoscopic examinations. Though quite satisfactory for the study of relatively thin samples with a thickness of the order of 0.5 mm this classical method might not be effective when analyzing technically important crystalline boules of large dimensions frequently exceeding 5…10 cm in length and diameter [3]. Making use of thin slices cut from a large boule does not resolve the problem, because in this case the method becomes destructive. Moreover, successful testing of a small part of a boule by no means guarantees the quality of the remaining part.

The current trends in the development of optoelectronics demand better characterization and modeling of various birefringent devices, especially acoustooptical ones, for which large crystal elements of high optical quality are required. Both theoretical and experimental studies of the conoscopic method including the rarely considered general case of arbitrary orientation of the optical axis (axes) tilt angle are needed.

In the present work we make theoretical calculations and perform direct experimental observations of the conoscopic patterns of a number of large-sized crystal boules in demand for birefringent devices.

IOP Conf. Series: Materials Science and Engineering **49** (2013) 012037 doi:10.1088/1757-899X/49/1/012037

2. Results and discussion

The equations of the isochromes based on the solution of the equation for the angle of refraction of the extraordinary ray derived without commonly adopted simplifications in [4] were used in this work. In the general case of an arbitrary orientation of the optical axis with respect to the surface the isochromes are described by curves of 4th order and higher. There is no room for a further consideration of the general case here which will be presented elsewhere. A practically important for the present particular case of the calculations of the effects of sample thickness d, angular aperture A, and refractive indices n_o, n_e on the conoscopic figures is given in figure 1.The modeling was limited by the case of 25 isochromes to restrict the computation time.

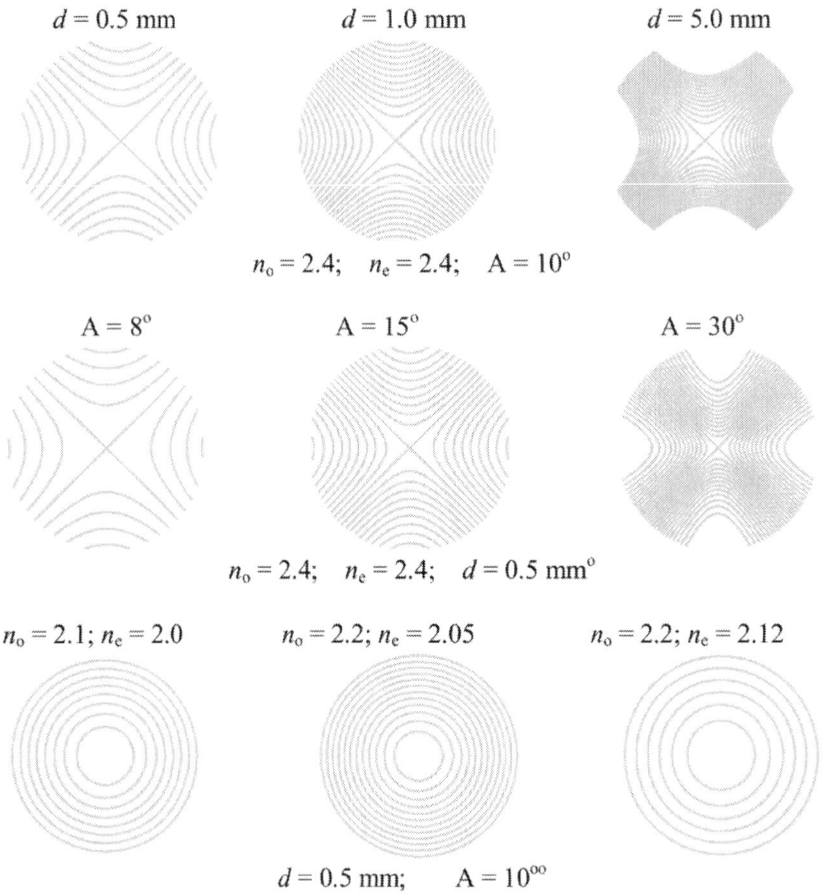

Figure 1. Effect of sample thickness d, angular aperture A, and refractive indices n_o, n_e variations on the conoscopic figures (calculation). Wavelength $\lambda = 500$ nm, focus length of lens 350 mm

A simple computer-based conoscopic benchtop tester was used in the experiment (figure 2). It consists of a source of monochromatic linearly polarized light (YAG:Na^{3+} with frequency doubling $\lambda = 533$ nm). Frosted glass filter served to expand the beam divergence and generate a smooth speckle pattern. The fringe pattern is collected on a projecting screen. It was also possible to get a converged beam with the aid of a lens of high numeric aperture for reducing the screen size and applying a CCD camera.

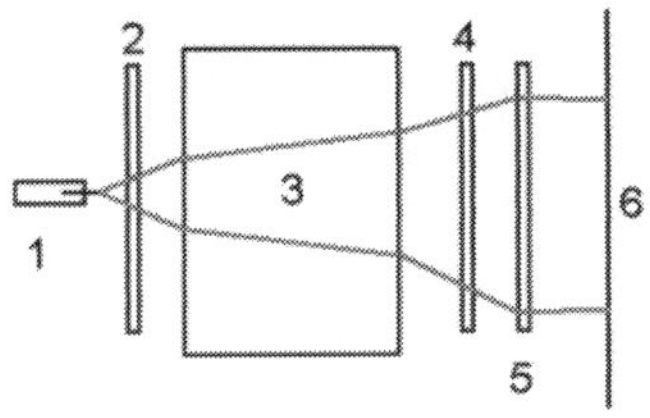

Figure 2. Basic outline of the conoscopic tester. 1 – laser source, 2 – polarizer, 3 – object (crystal), 4 – analyzer, 5 – lens, 6 – screen

Examples of the conoscopic figures obtained after quick check of several crystal bars are shown in figure 3-6. End faces of the samples were hand prepolished to enable the light transmission. It should be mentioned that with conoscopic studies the requirement for the nonparallelity of end faces is much softer compared with that of classical interferometers because in conoscopes the two (ordinary and extraordinary) beams travel along close directions in the crystal and pass similar optical paths [3].

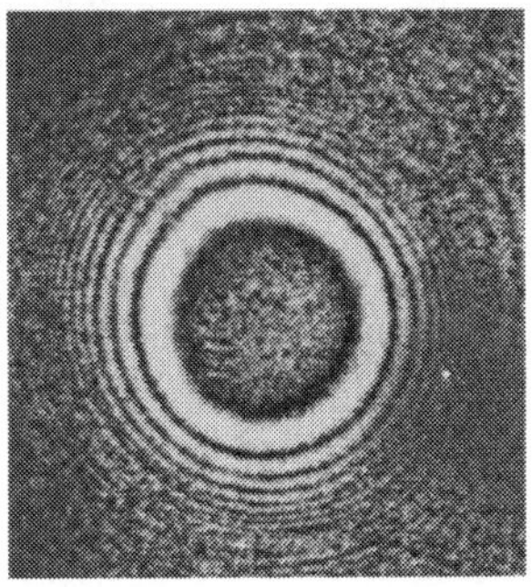

Figure 3. Conoscopic pattern produced by a polar cut of TeO_2 crystal in normal orientation

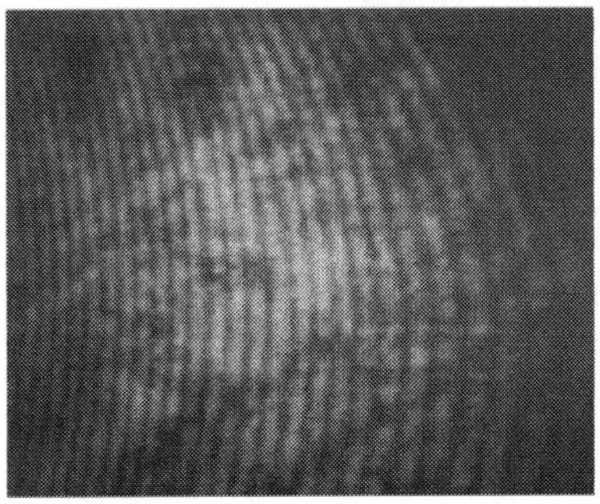

Figure 4. Conoscopic pattern produced by a TeO_2 crystal for an angle of 3° between the optical axis and light cone axis

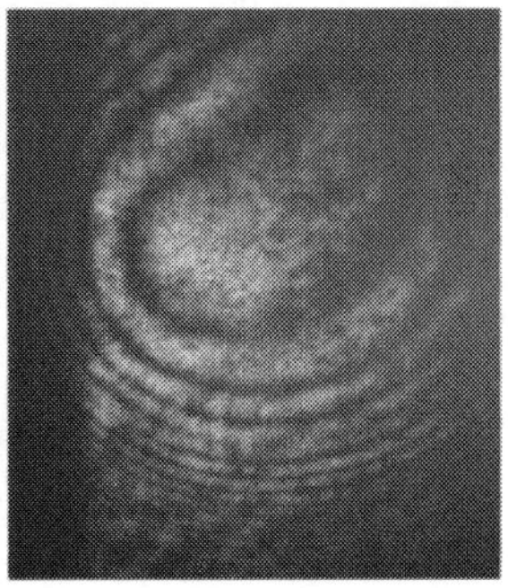

Figure 5. Conoscopic pattern distortion due to mechanical stresses at the edge of the sample

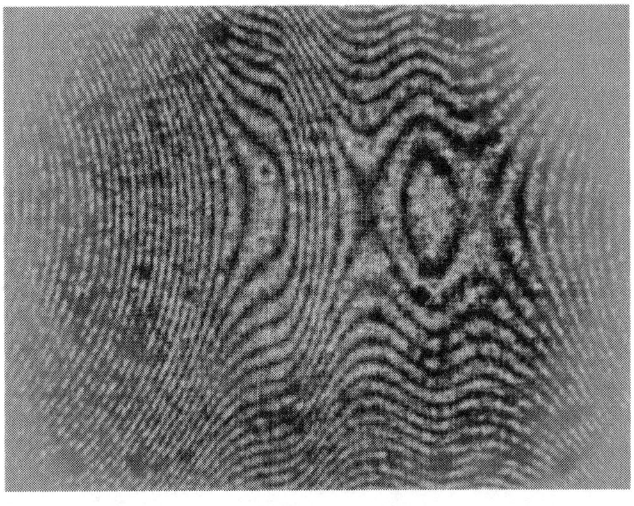

Figure 6. Effect of striations on the distortion of isochromate shapes in paratellurite crystal

Figures 3 and 4 demonstrate normal interference figures formed by a crystal region free of noticable defects. Among the least desirable defects of various types deteriorating the performance of birefringent devices are those affecting the values of refractive indices and/or their distribution in the volume of the material. They may be induced during the crystal growth process and post-growth heat treatments or by mechanical stresses arising in the course of cutting, polishing and assembling the optical components (figure 5). Local distortions of the isochromate shapes may be used for the estimation of $\Delta n = n_o - n_e$ variations, followed by the estimation of mechanical stresses δ_{ij} through the use of piezooptic coefficients $\pi_{\lambda\mu}$ and constants of elastic stiffness C_{ijkl}.

3. Conclusion

The analytical results of the earlier performed first-principle calculations of isochromate patterns in birefringent media were applied to the conoscopic examination of large-sized crystal boules unusable for examination in conventional polarizing optical microscopes with Bertrand lens. In the present work we have shown that the usage of convergent or divergent laser beams in simple benchtop configuration makes it possible to explore big samples by the method of conoscopy including crystals elongated in the optical axis direction. As distinct from traditional optical microscopy the conoscopic figures obtained with the aid of the laser installment may contain tens and hundreds of isochromate fringes, thus increasing the informative capabilities of the method. Large-sized crystals of $LiNbO_3$ (∅65×95 mm), $LiTaO_3$ (∅70×40 mm), TeO_2 prisms (50×50×30 mm) were examined experimentally at different angles between the optical axis and normal to the crystal surface. The capabilities of conoscopic observations as a fast contactless method for characterizing the quality of large boules for their potential use in optical instrumentation were demonstrated.

Among the advantages of using lasers to this end are their high intensity of illumination, monochromaticity and small angular aperture. However, a separate study of the ways of improving the laser beam uniformity and stability (pi-shapers [5], etc.) would increase the functional possibilities of the method. This would also be helpful for the development of computer-controlled scanning conoscope imaging system [3, 6].

References

[1] Born M and Wolf E 1969 *Principles of Optics*, 3rd ed. (London: Pergamon)
[2] Shuvalov L A 1980 *Modern Crystallography IV. Physical Properties of Crystals* (Berlin: Springer)
[3] Bajor A L, Salbut L, Szwedowski A, and Piatkowski T 1998 Imaging conoscope for investigation of optical inhomogeneity in large boules of uniaxial crystals, *Rev. Sci. Instrum.* **69** 1476-87
[4] Kolesnikov A I, Kaplunov I A, Ivanova A I, Tretiakov S A, Talyzin I V, Malyshkin Yu A, Grechishkin R M, Vorontsova E Yu 2012 Isochrome shapes in the conoscopic patterns of uniaxial crystals *Ferroelectrics* **441** 75-83
[5] Laskin A, Shcherbakov A S; Molchanov V, Laskin V, Makarov O 2011 Developing the refractive light beam shapers as lossless apodization systems suppressing the side-lobes in Fourier transform optical systems *22nd Congress of the International Commission for Optics: Light for the Development of the World Proc. SPIE* **8011** 80110L
[6] Bitri N, Gharbi A and Marcerou J P 2008 Scanning conoscopy measurement of the optical properties of chiral smectic liquid crystals *Physica B: Condensed Matter* **403** 3921-7

IOP Conf. Series: Materials Science and Engineering **49** (2013) 012038 doi:10.1088/1757-899X/49/1/012038

Influence of Compressive Stress on Dielectric and Ferroelectric Properties of the$(Na_{0.5}Bi_{0.5})_{0.7}Sr_{0.3}TiO_3$ Ceramics

J Suchanicz[1], G Klimkowski[1], M Karpierz[1], U Lewczuk[1], I Faszczowy[1], A Pękala[1], K Konieczny[1], M Antonova[2]andA Sternberg[2]
[1]Institute of Physics, Pedagogical University, 30-084 Krakow, Poland
[2]Institute of Solid State Physics, University of Latvia, LV-1063 Riga, Latvia
E-mail: sfsuchan@up.krakow.pl

Abstract: Good quality lead-free ceramics of $(Na_{0.5}Bi_{0.5})_{0.7}Sr_{0.3}TiO_3$(NBTS30) have been produced by a solid phase sintering process. The dependence of dielectric and ferroelectric properties on the unaxial pressure (0-1200 bar) were investigated. A shift and decrease of maximum value of ε, decrease of the thermal hysteresis and coercive field and increase of polarization with increasing pressure were observed. The results were discussed in terms of an elastic changes in inter-ionic distances in a crystal structure and switching nanoregions under the action of pressure. The NBTS30 ceramic is expected to be a new promising candidate for lead-free electronic material.

1. Introduction

At present, the most widely-used piezoelectric materials are PZT-based ceramics due their high performance and excellent piezoelectric properties [1-2]. However, environmental issues call for the use of nonhazardous substances for device fabrication. Therefore, a considerable effort has been devoted to the development of lead-free piezoelectric ceramics with properties comparable to PZT. Research activities have mainly been focused on $Na_{0.5}Bi_{0.5}TiO_3$ (NBT) and NBT-based materials.The phase of NBT at room temperature is rhombohedral (R3c) [3]. A phase transition to tetragonal phase (P4bm) takes place at about 260°C and to cubic ($Pm\bar{3}m$) phase at about 520°C. Besides, there exist two other characteristic temperatures called the depolarization temperature (T_d≈190°C) and the maximum-permittivity temperature (T_m≈320°C). At T_d, the formation of ferroelectric state occurs (on cooling), which corresponds to a small diffuse bump on electric permittivity profile. $SrTiO_3$ (ST) is a so-called quantum paraelectric, it has a perovskite structure and a cubic symmetry (Pm3m) at room temperature. Below -168°C a transition to a tetragonal phase (14/mcm) is reported [4]. NBT-ST solid solutions show relaxorbehaviour and have been the starting material for the development of many other lead-free materials [5]. However, the properties of these materials are not yet well understood. There are only a few reports on their temperature-dependent dielectric and ferroelectric properties [6-7]. We notice here that in many applications, piezoelectric materials are often subjected to combined mechanical and electrical loading at different temperatures. It is therefore important to determine the dielectric and ferroelectric properties of these materials as function of applied stress and temperature. In this work we concentrate to investigate the influence of uniaxial pressure (applied parallel to ac field) and temperature on the dielectric and ferroelectric properties of $(Na_{0.5}Bi_{0.5})_{0.7}Sr_{0.3}TiO_3$ (NBTS30) ceramics. Such type of investigations have been performed on NBTS30 for the first time.

2. Experimental procedure

Powders of $(Na_{0.5}Bi_{0.5})_{0.7}Sr_{0.3}TiO_3$ were synthesized by solid phase reaction from high purity grade oxides and carbonates: Na_2CO_3, $SrCO_3$, Bi_2O_3 and TiO_2. The mixture of raw materials was homogenized and milled in an agate ball mill in ethanol for 24 hours, then dried and calcined. Due to

multicomponent composition the two step calcination was chosen: first at – 4 hours at 850°C and then at – 2 hours at 950°C. The calcined powder were reground, cold pressed (100 MPa) and sintered by conventional ceramic technology for 3 hours at the temperature of 1190°C.The obtained ceramics were cream-coloured and translucent. They exhibited very good mechanical properties (see below).
Dielectric studies were carried out for silver electrodedsamples using BM LCR meter in the temperature range of 30-500°C. The compressive stress in the range of 0-1200 bar was applied parallel to the measuring electric field with the use of a lever and a weight.

3. Results and discussion

X-ray powder studies proved that NBTS30ceramics has single phase of perovskite type with cubic structure (figure 1a). They did not show any evidence of macroscopic distortions in XRD patterns.

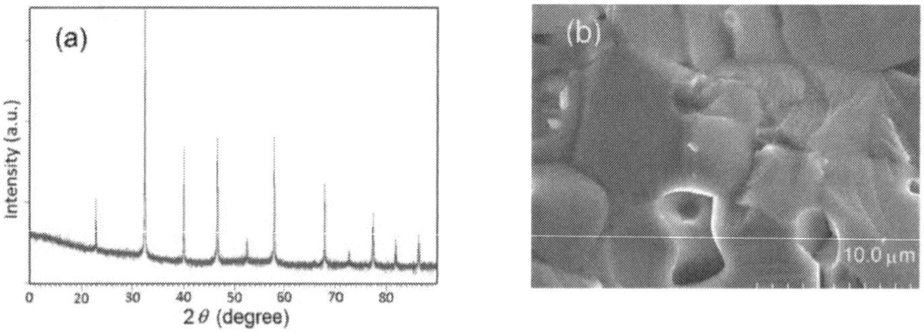

Figure 1. XRD pattern (a) and SEM micrographs (b) of the NBTS30 ceramic (magn. 5000x).

Figure 1(b) shows the SEM micrographs of the NBTS30 ceramics. The ceramic are well sintered. The polyhedral grains and grain boundaries accompanied with the occurrence of small pores are observed. The average grain size of the grains is ca. 5 μm. The EDS analysis confirmed the high purity and the expected quantitative composition of the sample.The samples were characterized by the bulk density which exceeds 95% of the theoretical density, unit cell parameter a=3.904±0.02Å, a spontaneous polarization P_s=21μC/cm^2, a coercive field E_c=30kV/cm; piezoelectric coefficient d_{33}= 446 pm/V; elastic constants: Young modulus E=148.41±3 GPa, shear modulus G=58.77±0.1 GPa and Poisson`s ratio υ=0.263±0.006 (all parameters are at room temperature).

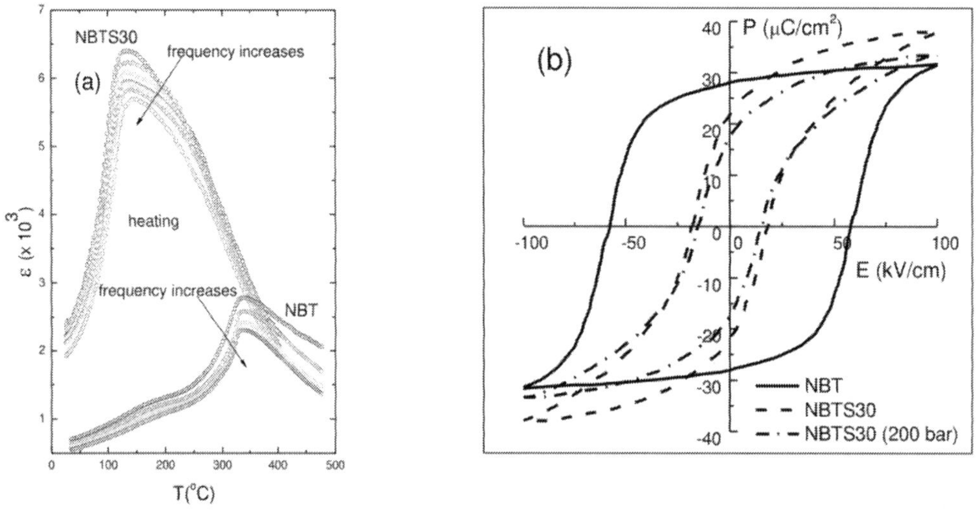

Figure 2. Temperature/frequency dependence of ε (a) and hysteresis loops (b) of NBT and NBTS30 (b) ceramics. Hysteresis loop of NBTS30 at p=200 bar is also show.

In general, after Sr doping to NBT (figure 2) the dielectric and ferroelectric properties are enhanced: electric permittivity (ε) increases, their maximum (ε_m) is more diffuse and is shifted to lower temperature;ε_m decreases and is shifted to higher temperature as the frequency increases, similar to relaxor ferroelectrics; polarization and coercive field increases and decreases, respectively.Figure 3(a) shows Raman spectra for NBT, ST and NBTS30 ceramics. No modes are Raman active in the cubic phase and seven ones are predicted for tetragonal phase of ST [8]. ST displays two broad bands at about 300 and 650 cm^{-1} along with weak bands at about 156 and 185 cm^{-1} exist. These broad bands can be assigned to second-order Raman scattering [8]. On the other hand, the modes of NBT (R3c phase) are mainly at about 100-200, 200-400, 450-650, and 750-900 cm^{-1}. The low-frequency one is associated with Na-O vibrations, the second one-with Ti-O vibrations, whereas the higher frequency bands can be assigned to vibrations of the TiO_6octahedra [9]. The high-frequency range above 700 cm^{-1} can be caused by the vibrations resulting from the shift of oxygen [10] and also correlated to present of oxygen vacancies. Raman spectra of NBTS30 show shifting and broadening for most of the bands. Moreover, a new mode is visible shown as a shoulder at the right side of the low-frequency mode (~138 cm^{-1}, indicated by arrow) and is evident of a structural change associated with A-site symmetry change. The mode in second region (~256 cm^{-1}) appear as two overlapping modes. Another additional mode appeared as a shoulder at the lower side of 450-650 cm^{-1} mode (indicated by arrow). In spite of the cubic symmetry (figure 1a), the existence of first-order Raman scattering indicates that the phase has a considerable local departure from the macroscopic cubic structure. Indeed, the appearance of additional modes in Raman spectra indicate that the rhombohedral phase of NBT changesto a cubic phase through an intermediate noncubic phase or both phases may be coexist (e.g. small rhombohedral islands in cubic matrix-polar nanoregions in nonpolar matrix).

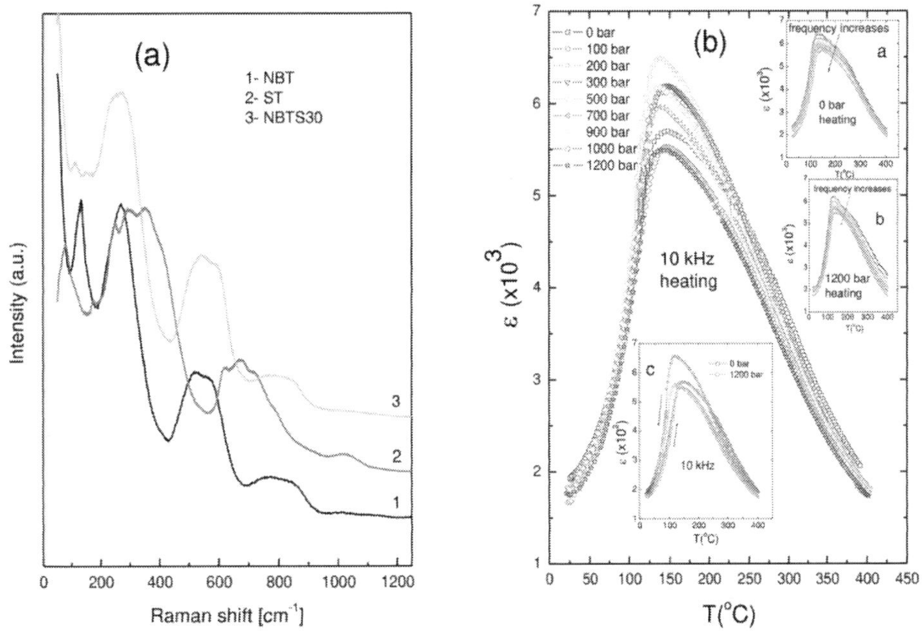

Figure 3. Raman spectra of NBT, ST and NBTS30 ceramics (a) and temperature/pressure dependence of ε of NBTS30 ceramic(b). The insert *a* and *b*in figure 3(b)show temperature/frequency dependence of ε of NBTS30 ceramic at p=0 and 1200 bar, respectively. Insert *c* show temperature dependence of ε of NBTS30 ceramic on heating/cooling at p=0 and 1200 bar.

The temperature/pressure dependence of ε shown in figure 3(b) reveal five main features with increasing pressure: (1) the ε(T) maximum (ε_m) seems to be gradually less diffuse for pressure up to 900 bar and then becomes more diffuse,(2) the temperature of $\varepsilon_m$first shifts towards lower temperatures of approximately -70°C/kbar (up to pressure about 200 bar), and then increases of approximately +13°C/kbar, (3) the dielectric dispersion slightly increases above T_m(insert*a* and *b* in figure 3(b)),(4) $\varepsilon_m$first increases (up to pressure about 200 bar, $\partial\varepsilon_m/\partial p$=+3920±20/kbar), and then decreases ($\partial\varepsilon_m/\partial p$=-988±20/kbar),(5)the thermal hysteresis decreases (inserts *c* in figure 3(b)). Thus, dielectric measurements of NBTS30 ceramic under uniaxial pressure show the occurrence of the threshold pressure p_{thresh}≈200 bar. It has been found that character of ε changes below p_{thresh} is different to that above this pressure. This effect can be attributed to a manifestation of the clamping pressure, which can be overcome by external pressure higher than p_{thresh}. This clamping pressure may appear at the interphase boundaries or local imperfections like dislocations, where point defects hoard during ageing and consequently accumulate mechanical stress.

In general, the changes in the properties of materials under compressive stress in the present experiments can arise from (1) an elastic change of distances between ions in the crystal structure, (2) a change in nanoregions structure, and from (3) creating/anihilating of defects [11]. The first effect can lead to the mainly decrease of dipole moments, causing the decrease of their contribution to dielectric response. In NBTS30 electric and elastic nanoregions coincide, they can be reoriented both by electric field and uniaxial pressure. The mechanical load (0-1200 bars) was large enough to induce nanoregions orientation. In present experiments, the uniaxial pressure reduces the number of nanoregions oriented in the direction of applied and thus leads to a decrease in the nanoregions ordering. The hysteresis loops area is found to decrease with increasing the stress applied parallel to the electric field direction (decrease of polarization is observed, figure 2(b)), confirming this suggestion. The hysteresis loop area represents the unit-volume polarization dissipation energy of a ferroelectric material subjected to one full cycle of electric field loading. Polarization dissipation energy is related directly to the amount of domain participating in the switching process during an electric field loading cycle. A decrease in the polarization dissipation energy with increasing compressive stress indicates that less and less nanoregions are remained(they are oriented by the stress in an opposite direction) and cannot be reoriented by the electric field, i.e. they cannot participate in the polarization reversal. As a consequence, the polarization become lower with increasing stress. The coercive field seems to be decreases under pressure (figure 2(b)). The decrease in permittivity is likely from the constraining of the interphase walls under stress, resulting in a decrease of interphase walls mobility.

4.Conclusions

Good quality lead-free ceramics of $(Na_{0.5}Bi_{0.5})_{0.7}Sr_{0.3}TiO_3$ (NBTS30) have been produced by a solid phase sintering process. The present investigations show that Sr-doping to NBT enhanced their dielectric and ferroelectric properties. The uniaxial pressure strongly affects their dielectric and ferroelectric properties shown by change of electric permittivity maximum, a shift in T_m,a decrease of thermal hysteresis, coercive field and of polarization. The obtained data were discussed basing on the strength of the elastic change in distances between ions in the crystal structure and nanoregions ordering under the influence of pressure. It is obvious that NBTS30 ceramics are promising materials for high-frequency transducer and actuator applications.

References

[1] Fu H and Cohen R E 2000*Nature* (London)**403** 281
[2] Ćwikiel K and Nogas-Ćwikiel E 2007 *Phase Transitions***80** 141
[3] Suchanicz J1997 *Ferroelectrics***200** 1
[4] Jauch W and Palmer A1999 *Phys.Rev.* B**3** 2961
[5] Mercurio J P, Marchet P, Saïd S, Gomah-Petry J R and Suchanicz J 2003 *CeramicTranslations***136** 373
[6] Hiruma Y, Imai Y, Watanabe Y, Nagata H and Takenaka T 2008*Appl.Phys.Lett.***92**262904
[7] Park S E and Hong K S 1997*J.Mater.Res.***12** 2152
[8] Nilsen W G and Skinner J G 1968*J.Chem.Phys.***48** 2240
[9] Petzelt J, Kamba S, Fabry J, Noujni D, Porokhonskyy V, Pashkin A, Franke I,Roleder K, Suchanicz J, Klein R and Kugel G E 2004*J.Phys.: Cond.Matter***16** 2719
[10] Kreisel J, Glazer A M, Jones G O, Thomas P A, Abello P A and Lucazeau G 2000*J.Phys.Condens.Matter***12**, 3267
[11] Suchanicz J, Kim-Ngan N-T, Konieczny K, Jankowska-Sumara I and Balogh A G2011*J.Appl.Phys.* **109** 104105

Acidic Ionic Liquids as Composite Forming Additives for Ion-conducting Materials

L Lasmane[1], E Ausekle[1], G Vaivars [1,2], A Priksane[1],
[1] University of Latvia, Department of Chemistry, Latvia
[2] University of Latvia, Institute of Solid State Physics, Latvia

E-mail: Guntars.Vaivars@cfi.lu.lv

Abstract. This paper represents the material conductivity investigation of several AILs (Acidic Ionic Liquids) containing an alkane sulfonic acid group covalently bonded to pyridine and N–alkylimidazole cations. Three different anions (HSO_4^-, $H_2PO_4^-$ and TsO^-) have been introduced in the structure of ionic liquids to evaluate the impact of this factor on material conductivity. Ion conductivity values in all studied ionic liquids obtained from impedance measurements in temperature range from 20 to 120 °C. Values of electroconductivity depending on ionic liquid's structure vary in rage from $1.34 \cdot 10^{-5}$ to $1.05 \cdot 10^{-2}$ S/cm at 25 °C.

1. Introduction

Rising interest in the field of ionic liquids during the last decades has resulted in a great deal of scientific information concerning characteristics of these compounds - covering research about their physical, as well as chemical properties. Ongoing physicochemical studies of ionic liquids keep providing more and more comprehensive and detailed information that leads to a deeper insight into properties and application possibilities of these compounds not only for laboratory purposes, but also for potential industrial use [1,2].

Former evaluation of thermophysical properties of ionic liquids as crucial factor for industrial purposes, has shown application of ionic liquids, although with certain drawbacks, as possible alternative to currently used fluids [3]. One of the perspective fields of research using ionic liquids is proton conducting polymer electrolytes such as fuel cells [4-6] due to high temperature stability and low partial vapor pressure.

Because of the considerable number of possible cation and anion combinations, and structure modifications in ionic liquids as "designer compounds" [7] the aim of this study is to give an insight in ionic conductivity of alkane sulfonic acid group containing acidic ionic liquids (AILs) (Figure 1) to describe this class of ionic liquids. Acidic ionic liquids synthesized and examined in this paper include 3-(1-pyridinio)-1-propanesulfonate [PyPS] hydrogen sulfate (**1a**), dihydrogen phosphate (**1b**) and *p*-toluenesulfonate (**1c**), 3-(1-methyl-3-imidazolio)-1-propanesulfonate [MeImPS] hydrogen sulfate (**2a**), dihydrogen phosphate (**2b**) and *p*-toluenesulfonate (**2c**) as well as 3-(1-butyl-3-imidazolio)-1-propanesulfonate [BuImPS] hydrogen sulfate (**3a**), dihydrogen phosphate (**3b**) and *p*-toluenesulfonate (**3c**).

2. Experimental

1a: X=HSO_4
1b: X=H_2PO_4
1c: X=TsO

2a: R=CH_3, X=HSO_4
2b: R=CH_3, X=H_2PO_4
2c: R=CH_3, X=TsO

3a: R=C_4H_9, X=HSO_4
3b: R=C_4H_9, X=H_2PO_4
3c: R=C_4H_9, X=TsO

Figure 1. Structures of AILs investigated in this study.

2.1. Chemicals and instruments

Reagents for synthesis of ionic liquids were used as received without further purification: 3-(1-pyridinio)-1-propanesulfonate (98% purity), 1,3-propanesultone (98% purity), 1-methylimidazole (99% purity), 1-butylimidazole (98% purity) purchased from Sigma Aldrich, *p*-toluenesulfonic acid monohydrate (98% purity) was supplied from Fluka, sulfuric acid (95%) and phosphoric acid (85%) were purchased from Stanlab.

Toluene and diethyl ether were dried over P_2O_5 and redistilled before use.

^{1}H NMR spectra of ionic liquids were recorded on Varian 200 MR in D_2O. TG/DTA was done on DTG-60/60H Shimadzu. Impedance measurements were performed on Autolab PGSTA-30.

2.2. General procedure for synthesis of AILs

The acidic ionic liquids (AILs) used in this paper (Figure 1) except **1a-c**, were synthesized in two step synthesis similar to the methods described in literature [8, 9].

Ionic liquid **2a** was prepared as follows: 1-methylimidazole (0.05 mol) was dissolved in 30 mL of dry toluene, 1,3-propanesultone (0.05 mol) was added gradually. The mixture was stirred at 60 °C for 24 h. The resulting white zwitterion salt was filtered, washed repeatedly with hot ethanol and diethyl ether to remove non-ionic residues and dried in vacuum (2 Torr, 100 °C, 2h). Then, a stoichiometric amount of sulfuric acid was added and the mixture stirred for 12 h at 80 °C. The resulting viscous ionic liquid was repeatedly washed with dry toluene and after removal of the solvent dried in vacuum (2 Torr, 100 °C, 2h).

^{1}H NMR of **2a** (200 MHz, D_2O, TMS): δ 2.01-2.16 (m, 2H), 2.69 (t, 2H), 3.66 (s, 3H), 4.13 (t, 2H); 7.22 (s, 1H), 7.29 (s, 1H), 8.53 (s, 1H).

Ionic liquids **2b-c** and **3a-c** were prepared using the method described above. In case of ionic liquids **1a-c** the commercially purchased zwitterion salt PyPS was used. All ionic liquids were analyzed by ^{1}H spectroscopy, and the spectral data agreed with their structures and data given in literature [8, 9].

2.3. Electrochemical measurements

Autolab PGSTA-30 was used to do the impedance and cyclic voltammetry measurements. The measuring cell consisted of two stainless steel electrodes and platinum wire as a reference electrode. It was heated up to 120 °C and kept for 30 min under argon flow prior the measurements. The impedance spectra were measured in a two-electrode mode in a frequency range from 0.1 Hz to 1000 Hz with AC potential amplitude of 0.01 V in a temperature range from 120 °C to 20 °C (during the cooling process). The cell resistance was obtained by extrapolating to higher frequencies. Additional cyclic voltammetry measurements (including reference electrode, 10 mV s^{-1}, in potential range +/-2V) were also carried out and no water electrolysis was observed showing no influence of possible presence of free acid or water in the studied compounds.

3. Results and discussion

To examine ionic conductivity of propane sulfonic acid group containing acidic ionic liquids, in total nine compositionally different ionic liquids were examined.

As it can be seen from the data given in Table 1 change of anion gives considerable impact on the melting point of ionic liquid. [$H_2PO_4^-$] anion containing [PyPS] and [MeImPS] show higher melting point temperatures, whereas [HSO_4^-] anion in the structure produces liquid state ionic liquids in the room temperature.

Since ionic liquids are likely to absorb water from the environment, the conductivity values were obtained during cooling cycle after electrochemical cell was pre-dried at 120°C. The temperature dependence of conductivity in a studied temperature range is close to linear in Arrhenius coordinates and it was used to obtain activation energies (E_A).

Table 1. Physical properties of AILs depending on their structure.

	Melting point (°C)[a]	**Melting point energy (J/g)**[a]	**Maximum stability point (°C)**[a]	**E_A (eV)**	**σ (S/cm), at 25 °C**
1a	105	688	240	0.86	$6.57 \cdot 10^{-4}$
1b	123	291	262	1.03	$1.34 \cdot 10^{-5}$
1c	106	394	270	1.04	$1.35 \cdot 10^{-4}$
2a[b]	-	-	250	0.70	$1.05 \cdot 10^{-2}$
2b	123	547	279	0.92	$1.14 \cdot 10^{-4}$
2c	80	408	276	1.21	$2.80 \cdot 10^{-5}$
3a[b]	-	-	223	0.78	$1.57 \cdot 10^{-3}$
3b[b]	-	-	232	1.02	$2.04 \cdot 10^{-5}$
3c	68	355	238	1.05	$2.72 \cdot 10^{-5}$

[a] Determined by TG/DTA analysis;
[b] AILs are in liquid state at room temperature;

Data given in Figure 2a, b, c, which represent conductivity measurements depending on the anion, show no significant drop of conductivity in higher temperature regions in case of all examined ionic liquids.

If [MeImPS] and [BuImPS] cations are compared then a decrease of conductivity can be observed in case of the ionic liquid with longer alkyl chain in the cation structure. The impact of the anion on conductivity as represented in Figure 2d, show that ionic liquids with [HSO_4^-] anion, in case of [MeImPS] as cation, give noticeably higher values comparing to [$H_2PO_4^-$] and [TsO^-] anions.

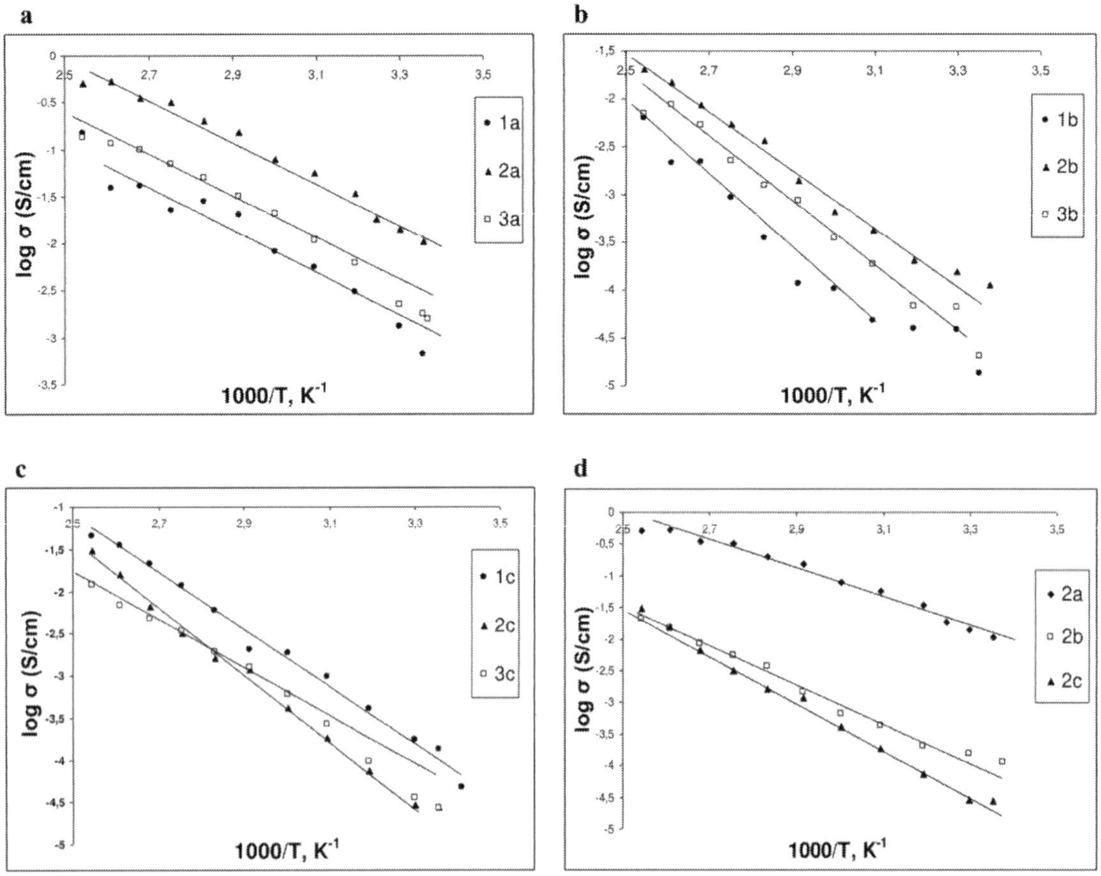

Figure 2 a, b, c. Electroconductivity change in various temperatures depending on the cation in ionic liquid's structure, **d** Electroconductivity change in various temperatures depending on the anion in [MeImPS] cation containing ionic liquids.

4. Conclusion

The hydrogen sulfate anion increased conductivity by one order to much higher values as compared to phosphate or tosylate anion. Prolonging of alkyl chain decreased electroconductivity.

References

[1] Hallet J P and Welton T 2011 *Chem. Rev.* **111** 3508-76
[2] Greaves L T and Drummond C J 2008 *Chem. Rev.* **108** 206-37
[3] Franca J M P, Nieto de Castro C A, Lopes M M and Nunes V M B 2009 *J. Chem. Eng. Data*, **54**, 2569–75
[4] Deligoz H and Yilmazoglu M 2011 *Journal of Power Sources*, **196** 3496-3502
[5] Chandan A, Hattenberger M, El- kharouf A, Du S, Dhir A and Self V 2013 *Journal of Power Sources* **231** 264-78
[6] Garaev V, Kleperis J, Pavlovica S and Vaivars G 2012 IOP Conf Ser: Mater Sci Eng **38**
[7] Rogers D R and Seddon R K 2003 *Science* **302** 792-3
[8] Cole C A, Jensen L J, Ntai I, Weaver K J, Weaver J K, Forbes C D and Davis H J 2002 *J. Am. Chem. Soc.* **124** 5962-3
[9] Xing H Wang T Zhou Z and Da Y 2005 *Ind. Eng. Chem. Res.* **44** 4147-50

Experimental setup for analysis of sorption and desorption of tritium in liquid lithium under different external conditions

A Lescinskis[1,2], G Kizane[1,3], A Vitins[1], E Platacis[2], O Lielausis[2], A Romanchuks[2] and K Kravalis[2]
[1] Institute of Chemical Physics, University of Latvia, 4 Kronvalda Blvd, Riga, LV – 1586, Latvia
[2] Institute of Physics, Agency of University of Latvia, 32 Miera Street, LV – 2169, Latvia

E-mail: gunta.kizane@lu.lv

Abstract. An original complex device has been designed and created specially for this research. Some of tritium sorption and thermal desorption in liquid lithium experiments has been successfully done already. First experiment series were carried out under reduced pressure (vacuum) to rule out other operating gases (like Ar) that may squeeze out tritium from lithium. Experiments showed good tritium desorption from lithium which highly depends on the temperature. A proportional gas counter tritium monitor was used to collect data from tritium thermal desorption experiments.

1. Introduction

The concept of using flowing liquid metals for all plasma facing components has been proposed as a potential and final solution to the critical plasma – material interaction issues [1]. Liquid lithium is a perspective material and it's intended to be used in the DT fusion devices (Jet, Tokamak) as a cooling agent, which protects plasma facing components from complete thermal destruction. For example, temperature at the surface of divertor (inside the lower part of the plasma chamber) during fusion reaction can reach 800 ^{0}C [2, 3] and has to be constantly cooled down. Besides the divertor, plasma facing limiters are important as well. The most popular material in use for today's Tokamak first wall components is graphite [4]. In liquid limiter case, limiter represents a non stop flowing liquid metal wall between fusion plasma and first solid wall of the plasma chamber. And one of the main candidates for this solution is lithium [4-6]. Lithium also is one of the most advantageous blanket materials not only because of its high heat transfer property to compose a self – cooled fusion reactor system but it also has a high tritium breeding ratio [7]. Around 7 grams of tritium per year as a by-product is generated in the Li loop [8]. Therefore, the concept of liquid lithium is one of the most suitable and perspective materials as than others. The main idea of the liquid metal cooling concept is that liquid lithium is flowing over the plasma facing components and cools it down. During this process lithium gets polluted (adsorbing) with plasma chamber gases (including fuel) and with chemically eroded particles of divertor surface as well. In order to not disrupt complex process of nuclear plasma, concentration of tritium in lithium shall not exceed certain acceptable standards (less than 1 appm) [9]. If too much fuel is absorbed, replenishment the plasma at a fast enough rate may be

[3] To whom any correspondence should be addressed.

impossible [10]. Lithium keeps a very strong ionic bonding with hydrogen isotopes [7]. That fact makes it difficult to recover or extract tritium back from lithium waste. Development of tritium recovery method is one of the most important issues for the design of self – cooled Li blanket loop [7]. There are several published methods for the tritium separation from liquid lithium, like permeation window method [7-9, 11, 12], liquid eutectic method [11], molten salt recovery method [9, 11], cold trapping method [9, 11] and various distillation methods [9]. Some authors also list another one method as an individual method called gettering method [9, 13]. But technically, permeation window method in itself includes not only tritium separation but also collection which usually is done by the adsorption. This means, that permeable window in itself is or it has integrated tritium adsorbers, commonly referred as getters. That's why listing a gettering method as a separate method is not quite necessary, basically it's a part of the same permeation window method. Each of these tritium separation methods has its own drawbacks which mostly associated with insufficiently low concentration of tritium recovery. For example, the cold trap process can only reduce tritium concentration to about 400 appm, which is far above the ITER designed goal of reducing tritium concentration in lithium to ~1 appm [9]. Nevertheless, experimentally proved that using yttrium permeation window method it's possible to reduce tritium concentration even from 0.0007 to 0.04 ppm [7], which is more than enough for ITER designed goal. But this method is quite slow even with few grams of lithium and must be done in laboratory conditions with great care. This means it would not be so easy to deal with hundreds of tons of liquid lithium which need to be separated from tritium as soon as possible. That's why these all listed methods are promising methods for tritium separation and none of these individual methods hasn't been denied or accepted as the best yet.
The particular research aims is to study tritium sorption and thermal desorption processes in molten lithium under different external conditions (temperature ramp, magnetic field, atmosphere and others). Investigate lithium - tritium chemically bonded forms which also includes a number of polymorphic forms (Li_nT) [14]. Develop and optimize conditions for effective separation of tritium from lithium.

2. Experimental

0.20 g of solid pure lithium of quoted purity 99.9 % was placed in a 15.7 cm^3 stainless steel reactor tube and hermetically closed. All procedures were done in a hermetic glove box with an argon atmosphere. A vacuum of 0.0l mbar was chosen as the working atmosphere for the reactor tube. Reactor tube with lithium was placed in the programmable electric furnace and started a ramp heating to 220 ^0C at 10 ^0C / min. After reaching temperature of 220 ^0C, a 10 MBq of tritium was injected into reactor. After 2 hour of constant heating at 220 ^0C reactor was slowly cooled down to the initial room's (20 ^0C) temperature at the 13 ^0C / h. At this point thermal desorption process can be started. Lithium reactor was slowly re-heated to 750 ^0C at 240 ^0C / h and this temperature was held constant for 2 hours. During this process liquid lithium reactor was continuously purged thru by the purge gas (He + 0.1% H_2). After 2 hour of constant heating and purging at 750 ^0C, reactor was slowly cooled down (40 ^0C / h) to initial room's temperature without interrupting the purging. Thermally desorbed tritium was detected by the tritium monitor (Mab Solutions TEM-2100A) and all temperature data was collected by the data logger (Agilent 34970A). P10 (Ar + 10 % CH_4) was used as a counting gas for tritium monitor. A liquid nitrogen cold trap was used to prevent possible entering of lithium vapour in detector during purging process. To avoid possible moisture of the purge gas, a hot zinc granule furnace (400 ^0C) was placed right after the cold trap. An empty Drechsel bottle (2.3 L) was used to dilute the initial high concentrations of tritium. Figure 1 shows schematic diagram of the experimental apparatus and setup. Figure 2 shows the temperature program for sorption of tritium in liquid lithium. Whole sorption experimental process including ramp heating and cooling to 20 ^0C ran for 18 hours.

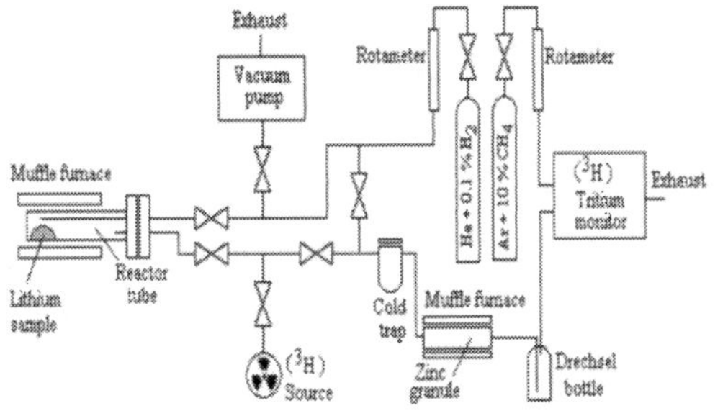

Figure 1. A schematic diagram of the experimental apparatus and setup.

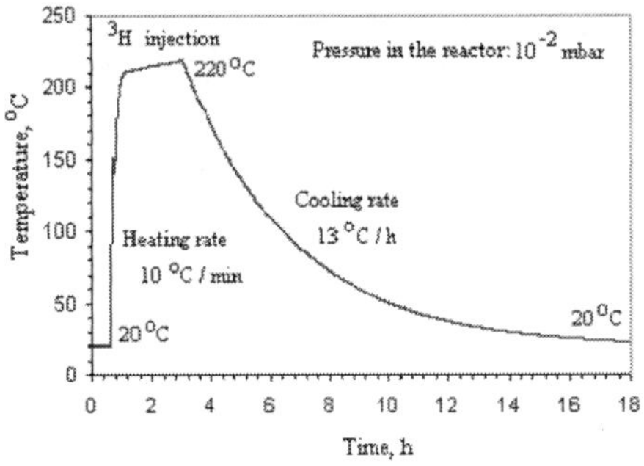

Figure 2. Temperature program for tritium sorption in liquid lithium.

3. Results and discussion

Purging tritium from the cold reactor (Figure 3) shows that large amount of tritium (3.52 MBq) is not adsorbed or chemically bonded with lithium even after two hour of heating at 220 ^{0}C in which lithium is in liquid state.

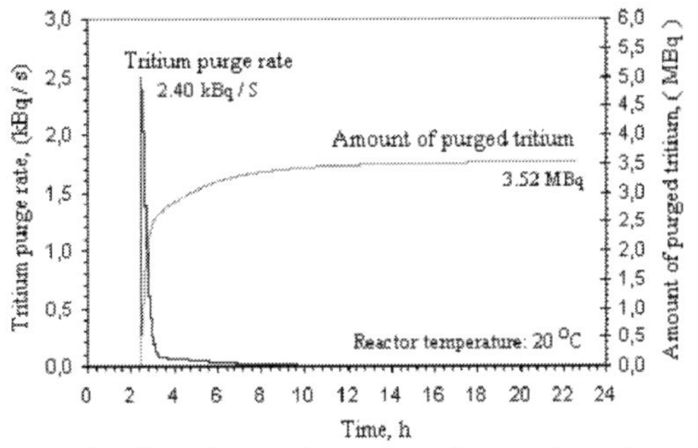

Figure 3. Purging of none adsorbed and none chemically bonded tritium from the cold reactor.

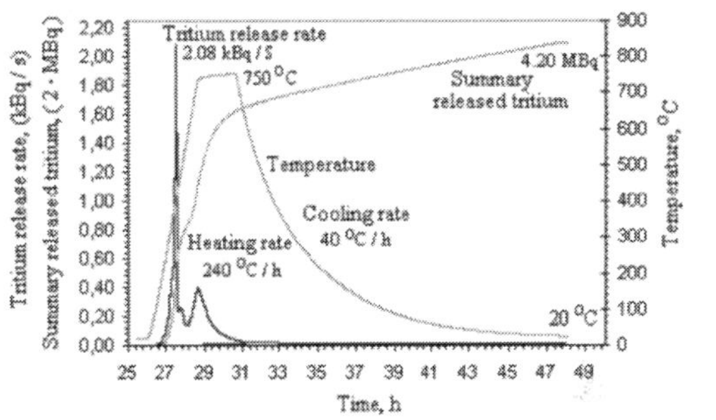

Figure 4. Tritium release from the lithium during thermal desorption process.

Purging tritium during thermal desorption (Figure 4) shows that tritium release rate rapidly increases by increasing the temperature of the reactor tube. That fact demonstrates good tritium release tendency from liquid lithium in effect of temperature. Maximum tritium release rate reaches 2.08 kBq / s and it's approximately the same (2.40 kBq / s) as purging the cold reactor from none adsorbed tritium (Figure 3). Total amount of released tritium during thermal desorption is 4.20 MBq (Figure 4).

Total amount of purged tritium including non adsorbed and thermally desorbed tritium is 7.72 MBq. Initial amount of injected tritium was 10 MBq, which means, that quite large amount of tritium (2.28 MBq) still stays in lithium. The fact shows that single thermal desorption process does not provide so simple and straightforward separation of tritium. And this type of thermal distillation has to be used combined with other methods or other external conditions, like magnetic field and various atmosphere gases in the reactor that may help to squeeze out tritium from lithium. A minimal amount of lithium was found in the cold trap, which means that lithium vapours may be transferred thru the system during thermal purging despite the fact that lithium has quite low vapour pressure. No lithium vapour was observed in the system tubes located after the cold trap, indicating that lithium vapour easily condenses on the cooler surfaces. Further investigations under different external conditions are in progress.

4. Conclusions

Experiments showed that tritium release from liquid lithium during thermal desorption is highly dependent on the temperature. Single thermal desorption process does not provide so simple and straightforward separation of tritium.

Acknowledgments

This work was supported by the European Regional Development Fund, project No. 2010/0262/2DP/2.1.1.0/10/APIA/VIAA/176 "Advanced lithium-ion technology for plasma treatment equipment (divertor) for the protection of the active surface".

References

[1] M.J. Baldwin, R.P.D., S.C. Luckhardt, R. Seraydarian, D.G. Whyte, R.W. Conn, *Plasma intereaction with liquid lithium: MEasurments of retention and erosion.* Fusion Engineering and Design, 2002, **61 – 62,** 231 – 236

[2] J. Roth, A. Kirschner, W. Bohmeyer, S. Brezinsek, et al., *Flux dependence of carbon erosion and implication for ITER,* Journal of Nuclear Materials, 2005, 970 – 974

[3] J.N. Brooks, J.P. Allain, R.P. Doerner, A. Hassanein, R. Nygren, T.D. Rognlienl, D.G. Whyte, *Plasma-surface interaction issues of an all – metal ITER,* Nuclear Fusion, 2009, **49,** 6 p

[4] G. Martin, M.L., *Li3: fist step studies of a aliquid lithium limiter.* Fusion Engineering and Design, 2002, **61 – 62**, 237 – 243

[5] S.V. Mirnov, V.B.L., S.M. Sotnikov, V.A. Evtikhin, I.E. Lyublinski, A.V. Vertkov, *Li - CPS limiter in tokamak T - 11M.* Fusion Engineering and Design, 2003, **65**, 455 – 465

[6] I. Lyublinski, A.V., Experience and technical issues of liquid lithium application as plasma facing material in tokamaks. Fusion Engineering and Design, 2010, **85**, 924 – 929

[7] S. Fukada, Y. Maeda, M. Kinoshita, T. Muroga, *Verification to recover tritium in neutron-irradiated Li by Y plate*. Fusion Engineering and Design, 2007, **82**, 2152 – 2157

[8] M. Kinoshita, S. Fukada, N. Yamashita, T. Muroga, M. Nishikawa, *Experimental study of tritium recovery from liquid lithium by yttrium.* Fusion Engineering and Design, 2006, **81**, 576 – 571

[9] D. K. Sze, R.F. Mattas, J. Anderson, R. Haange, H, Yoshia, O. Kventon, *Tritium Recovery from Lithium Based On Cond Trap.* Third International Symposium on Fusion Nuclear Technology, University of California, Los Angeles, June 27 - July 1, 1994, 19 p

[10] R. Stubbers, W. Olczak, M. Nieto, D.N. Ruzic, *Measurment of hydrogen absorption in flowing liquid lithium in the flowing lithium retention experiment (FLIRE).* Journal of Nuclear Materials, 2005, **337 – 339**, 1033 – 1037

[11] R.M. Alire, *Extraction of Tritium from Liquid Lithium by Permeation.* Lawrence Livermore laboratory, University of California. Report number: 52455, 1978

[12] Y. Edao, S. Fukada, S. Yamaguchi, Y. Wu, H. Nakamura, *Tritium removal by Y hot trap for purification of IFMIF Li target*, Fusion Engineering and Design, 2010, **85**, 53 – 57

[13] T. Takeda, A. Ying, M. A. Abdou, *Analysis of tritium extraction from liquid lithium by permeation window and solid gettering process*, Fusion Engineering and Design, 1995, **28**, 278 – 285

[14] C.H. Wu, R.O. Jones, *Stability and structure of Li_nH molecules (n=3–6): Experimental and density functional study.* Journal of Chemical Physics, 2004, **120**, 5128 – 5132

IOP Conf. Series: Materials Science and Engineering **49** (2013) 012041 doi:10.1088/1757-899X/49/1/012041

Multispectral analysis and cone signal modelling of pseudoisochromatic test plates

K Luse[1], M Ozolinsh[1,2], S Fomins[2] and A Gutmane[1]
[1] University of Latvia, Optometry and Vision Science Dept., Ķengaraga street 8, Riga LV-1063, Latvia
[2] Institute of Solid State Physics, Ķengaraga street 8, Riga LV-1063, Latvia

E-mail: kaiva.luse@gmail.com

Abstract. The aim of the study is to determine the consistency of the desired colour reproduction of the stimuli using calibrated printing technology available to anyone (*EpsonStylus Pro 7800* printer was). 24 colour vision assessment plates created in the University of Latvia were analysed right after their fabrication on august 2012 and after intense use for 7 months (colour vision screening on 700 people). Multispectral imagery results indicate that the alignment of the samples after seven months of use has maintained on the *CIExy* confusion lines of deutan deficiency type, but the shift towards achromatic area in the diagram indicate decrease in the total colour difference (ΔE^*_{ab}) of test background (achromatic) areas and stimuli (chromatic) areas, thus affecting the testing outcome and deficiency severity level classification ability of the plates.

1. Introduction

Altered colour vision affects approximately 8% of male and 0.4% of female population and congenital red-green colour vision deficiency (M cone dysfunction - protan deficiency and L cone dysfunction - deutan deficiency) makes up most part of above mentioned percentage [1,2]. Regardless of the design used for the creation of pseudoisochromatic (PIC) tests, they are based on evoking different grouping responses in people with normal trichromatic and abnormal colour vision. Colours for such plates must be chosen in specific alignment to *CIExy* colour diagram confusion lines according to each deficiency type [1,3]. By careful choice of coloured symbol chromatic values and its corresponding background colour, tests characterized by high test validity can be obtained [4,5]. The amount of total colour difference ΔE^*_{ab} of the test plate determines the level of difficulty of the specific plate. The perception of colour in case of colour vision deficiencies is still studied and tests for quantitative measurement of deficiency severity are developed [6,7]. Previous studies have shown that high desired chromaticity value accuracy for PIC test creation can be obtained using calibrated inkjet printing [8]. The analysis of durability of the prints and chromatic value validity for colour vision testing purposes is the aim of this study. Previous research shows that colours from the same test plates but different editions have significant chromatic variations that result in clinical misevaluation [9].

2. Methods and experiment setup

2.1. Development of the colour vision test plates

The PIC plates were designed using the principles described in the book of J.Birch [1]. Neutral colours (colours confused with gray) were chosen for the study. The design of the experiment for acquiring corresponding pairs of achromatic and chromatic colour matches is described earlier [8]. A set of 24 PIC plates (10 for protan deficiency (5 reddish, 5 greenish) and 14 for deutan deficiency (9 reddish and 5 greenish)) was created (see Figure 1 for examples).

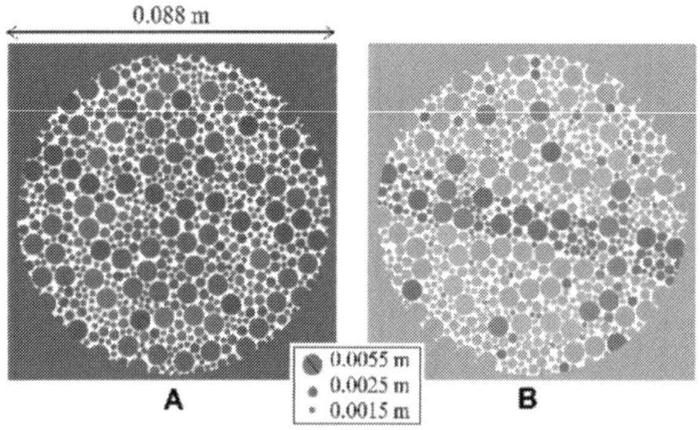

Figure 1. Two plates of twenty four plate set are shown. A – plate for medium deutan colour vision deficiency assessment ($\Delta E^*_{ab} = 27$), B - plate for severe deutan colour vision deficiency assessment ($\Delta E^*_{ab} = 37$).

1.1. Multispectral image capture

For multispectral imaging *CRI Nuance Vis 07* multispectral camera with *Nikon AF Micro-Nikon 60mm f2.8D* objective was used. Multispectral image capture was performed in dark room, the test plates were inserted in Qualitest CT-100W1 light booth (attached to the wall at equal height at all times, distance from the camera 55±5 cm, and illuminated from above via standard D65 light source (T=6500K)). 1290 x 920 pixel spatial images in range from 420 to 720 nm (step 10 nm) were taken. Images were transformed using cone sensitivities into cone excitation images (see [10] for details). Selected image areas of 89 pixels were analysed and *CIExy* diagram (x,y,Y) values acquired for each pixel. Image capture was performed right after the production of the plates on August 2012 and after seven months of use on March 2013. The chromaticity coordinates x,y of 3 deutan deficiency testing plates are discussed in this paper (with total colour differences 9 (reddish), 27 (reddish) and 37 (greenish) units).

3. Results and discussion of plate multispectral analysis

3.1. Fading of the pigments and sample relations to the confusion lines.

To analyse the change of the chromatic values of the plates after seven months of use, the distance between pixel sets in *CIExy* diagram was described using ellipses (radius of the ellipse can be defined as SD of data in x and y directions). Ellipse shift differences are plotted in Figures 2 and 3. Meaningful shifts of the chromatic areas towards the achromatic area of the diagram were obtained for all lightness levels of chromatic stimuli and all three saturation levels of the plates. Picture 3 illustrates a significant shift of the coloured areas (1) and (2) toward the achromatic area (3) and (4), the correspondence to confusion lines is not altered, similar fading was observed in case of reddish symbol plate of colour difference 27 units (data not shown). The analysis from the perspective of sample relation to confusion lines was performed as follows. Five areas in each plate were analysed, each area was measured twice. Additional interplate analysis was not performed for the consistency of chromatic values was shown earlier [8]. Figure 2 shows alignment of light chromatic (1), light achromatic (2) and white paper (3) areas in the least saturated plate (dark chromatic and achromatic areas are not shown to avoid overcrowding of the data). Figure 3 shows similar condition in case of the most saturated plate. The measured pixels of the image of the plates are in good correspondence to a single deutan colour confusion line both prior and post use.

3.2. Colour coordinate dispersion analysis

For dispersion description Gaussian curve standard deviation (SD) of image pixel colour coordinate mean values was calculated. For dispersion amplitude comparison statistics tool *independent sample t-test* was used with confidence level 95%. The pixel value deviations from mean values before and

IOP Conf. Series: Materials Science and Engineering **49** (2013) 012041 doi:10.1088/1757-899X/49/1/012041

after test use was compared for each plate's 4 areas (light chromatic, light achromatic, dark chromatic and dark achromatic) along both chromaticity coordinates (x,y). Statistically in 11 out of 12 cases the colour dispersion is significantly larger (the ellipse is wider) in x axis direction and in 9 cases (out of 12) colour dispersion was significantly larger (the ellipse is wider) in y axis direction if compared to newly printed pigment area colour dispersions along both axis in *CIExy* diagram. The sizes of the ellipses have changed from (x=0.002, y=0.0026) to (x=0.0032, y=0.0036) in case of light samples and from (x=0.0028, y=0.003) to (x=0.0038, y=0.0044) in case of dark samples.

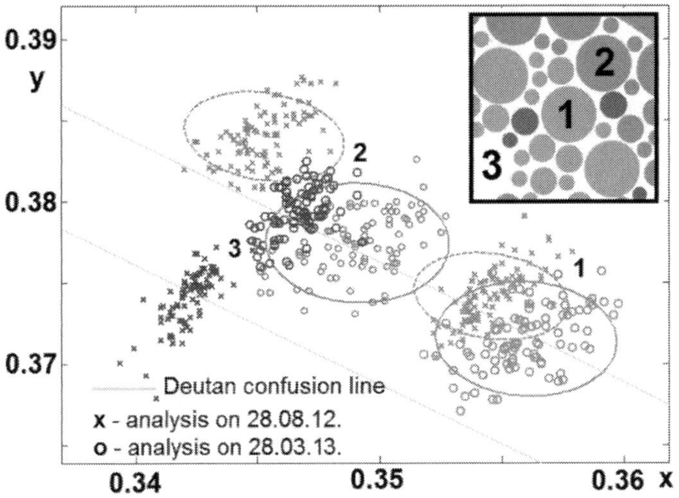

Figure 2. Multispectral analysis of deutan deficiency plate of least saturation (ΔE^*_{ab} = 9) before (x) and after (o) seven months of use. Pixel values are shown and areas marked with ellipses which centres represent mean pixel data set values and each ellipse size indicates SD's for each pixel set.

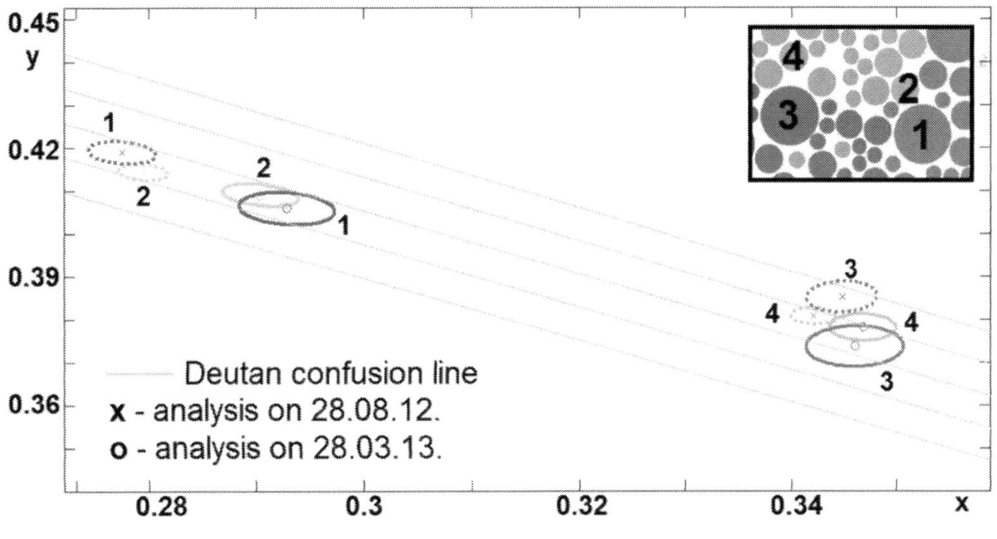

Figure 3. Multispectral analysis of deutan deficiency plate of most saturation (ΔE^*_{ab} = 37) before (dotted line) and after (continuous line) seven months of use.

3.3. Cone signal modelling

Multispectral data provide a possibility to test the developed test plate's validity by mathematical modelling. To perform M and L cone response simulation in case of protan and deutan deficiency, spectral images obtained were transformed to cone signals using cone sensitivity functions (method in detail described in [10]). Figure 4 shows the results in case of plate designed for deutan deficiency testing (total colour difference is 27 units). Orange line notes the resolution limit.

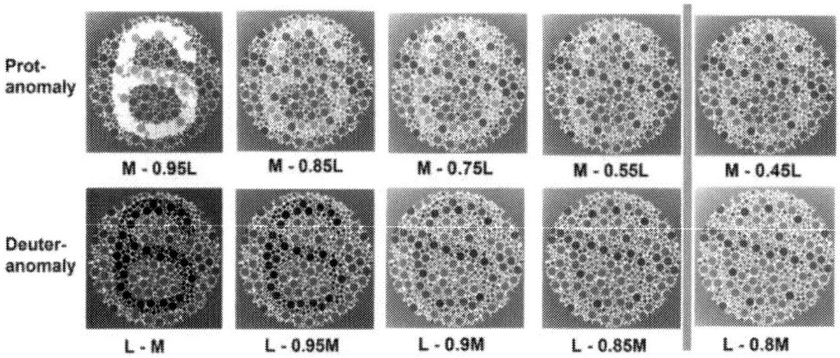

Figure 4. Cone signal modelling for both red-green deficiency types and various deficiency severity levels (indicated by the decrease of the multiplier value). Simulation shows that the plate is indeed more suitable for deutan deficiency testing as the symbol vanishes with smaller decrease in M cone response.

Conclusions

Multispectral analysis has shown significant fading of the chromatic colour test plate areas after seven months of use. Fading has affected plates of both saturations and both directions across *CIExy* chromaticity diagram (reddish and greenish plate types respectively). The alignment of the samples after seven months of use has maintained on good correspondence to the confusion line of deutan deficiency type, but the shift towards achromatic area in the diagram indicate significant changes in the total colour difference (ΔE^*_{ab}) of background samples and stimuli samples thus affecting the testing outcome and deficiency severity level classification ability of the plates. The dispersion of the plate chromatic and achromatic stimuli along *CIExy* x and y axis was larger after seven months of use.

Acknowledgments

This work was supported by projects «Support for Doctoral Studies at University of Latvia» and ERAF project No.2010/0259/2DP/2.1.1.1.0/10/APIA/VIAA/137.

References

[1] Birch J, 1993 *Diagnosis of Defective Colour Vision* (London) Butterworth Heinemann, 184 p., 42,55.
[2] Neitz J, Neitz M, 2011 The genetics of normal and defective color vision, *Vision Research*, doi:10.1016/j.visres.2010.12.002
[3] Smith VC, Pokorny J,1975 Spectral Sensitivity of the Foveal Cone Photopigments between 400 and 500 nm, *Vision Research*, **15**, pp. 161 – 171.
[4] Cole BL, Lian KY & Lakkis C, 2006 The new Richmond HRR pseudoisochromatic test for colour vision is better than the Ishihara test. *Clin Exp Optom*; 89(2): 73-80.
[5] Cole BL, 2007 Assessment of inherited colour vision defects in clinical practice. *Clin Exp Optom*, 90: 3: 157–175.
[6] Barbur JL, Rodrigues-Carmona M, Harlow JA, Mancuso K, Neitz J and Neitz M, 2008 A study of unusual Rayleigh matches in deutan deficiency, *Visual Neuroscience*, 25: 507-516.
[7] Rabin J, Gooch J and Ivan D, 2011 Rapid quantification of color vision: the cone contrast test, *Invest Ophthalmol Vis Sci.* Feb 9: 52 (2) : 816-20. doi: 10.1167/iovs.10-6283.
[8] Luse K, Fomins S, Ozolinsh M, 2012 Pseudoisochromatic test plate colour representation dependence on printing technology, *IOP Conf. Ser.: Mater. Sci. Eng.* 38 012024
[9] Lee DY, Honson M, 2003 Chromatic Variation of Ishihara Diagnistic Plates, *Color research and Application*, **28**(4), pp. 267-276.
[10] Fomins S, Ozolinsh M, 2011 Multispectral Analysis of Color Vision Deficiency Tests, *Materials Science (Medžiagotyra)*, **17**(1), pp. 104-108.

The effect of aging on the superconducting transition temperature and resistivity of Y-Ba-Cu-O ceramics after high temperature treatment

S.K.Nikoghosyan[1,2], V.V.Harutyunyan[2], V.S.Baghdasaryan[2], E.A.Mughnetsyan[1], E.G.Zargaryan[1] and A.G.Sarkisyan[1]
[1]International Scientific-Educational Center of NAS RA, Armenia,
[2]A.I. Alikhanyan National Science Laboratory (Yerevan Physics Institute) Foundation, Armenia

E-mail: nick@mail.yerphi.am

Absract. Some changes in superconducting transition temperature T_c and normal resistivity r of $YBa_2Cu_30_x$ ceramics are investigated as a result of the aging stimulated by high temperature treatment and slow cooling to room temperature followed by the storage under environmental conditions during 255 hours. It is shown that the changes of these parameters fall on three specific intervals of storage time t_a. In addition, different types of aging processes: fast and slow ones are observed for the heat-treated samples. It is found that the fast processes are accompanied by a slight increase of r and T_c within the first interval of storage time t_a ($t_a \leq 162$ hrs.), while the slow processes correlate with a sharp decrease of T_c and simultaneous increase of r within the second aging interval (162 hrs. $\leq t_a \leq$ 206 hrs.). It is revealed that the fast and slow aging processes result from the redistribution of atoms in intergranular weak and strong links, correspondingly. The probable reason of the observed results is the formation and coexistence of various structural phases including the semiconducing one.

1. Introduction

It is known that in the oxygen-deficient high-temperature superconducting (HTS) $YBa_2Cu_30_x$ samples obtained by quenching in liquid nitrogen after high temperature treatment and subsequent storage at room temperature, annealing of their characteristic parameters occurs [1-4]. Here, with increasing storage time, a significant increase in the temperature of superconducting (SC) transition T_c is observed due to processes of oxygen ordering in Cu-O chains of the sample unit cell [1,2]. This effect is known as "aging phenomenon", and a relationship between SC properties and structural parameters is revealed [1,2]. It is determined that the appearance of this phenomenon depends on the oxygen stoichiometry, as well as the sample treatment temperature and the conditions of its future storage (temperature, ambient atmosphere) [1,2,4]. Later, it was found that the phenomenon of aging takes place in HTS samples with intermediate shielding capacity, which in turn depends on temperature and the applied magnetic field [5]. Such a complicated dependence on various factors makes the results on aging effect of characteristic parameters (T_c and the corresponding values for resistivity r) of HTS rather ambiguous, and sometimes even controversial [1-4]. The above-mentioned data as well as the recent results of thermal annealing [6] show the importance of investigating the aging effects induced in heat-treated HTS materials not only after fast quenching, but also after slow cooling from 400°C to room temperature under environmental conditions and their subsequent aging in the same conditions. The paper is aimed at this problem.

2. Experimental methods

Two nearly identical rectangular $YBa_2Cu_30_x$ samples measured 2 mm×1 mm× 8 mm were used in our experiments. They were cut from the ceramic tablets synthesized in the air by the ordinary powder technology. Study of the aging phenomenon was carried out by registration of voltage-current characteristics using standard four- probe technique in the presence of the Earth's magnetic field [7]. Contacts on the samples were made using conductive silver paste. One of the two samples was heated in a tubular furnace at 400°C in the air for 30 minutes, cooled together with the furnace to room temperature at a rate of 2.5°C/min and then kept under environmental conditions during 255 hrs. (approximately 11 days). The other sample was kept in the same conditions and used as the control. During the whole storage time t_a, specific resistance r versus dc transport current was periodically recorded for both samples at 78 K. T_c was determined from the middle point of the sharp decrease on the r (I) curves within the temperature range from 77 to 300K. Note that the control sample parameters did not change during the whole storage period.

3. Results and discussion

Figure 1 presents a family of r (I) curves measured at 78 K after the storage of heat-treated samples during various t_a. Figure 2 shows r vs. t_a curves for fixed I measured at 78 K, and Figure 3-T_c and

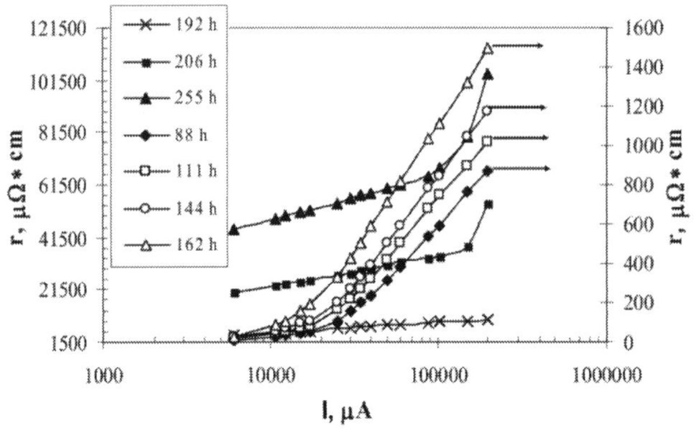

Figure 1. r (I) curves at 78 K for different storage times t_a.

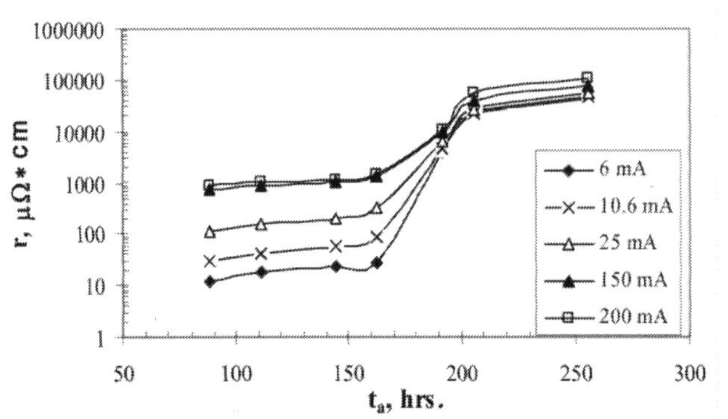

Figure 2. r (t_a) dependences of heat-treated samples for different transport currents I.

r(290K) dependences on t_a. For T_c and r (T, I) parameters, it is possible to distinguish three characteristic intervals of t_a. It is seen that for both measurement temperatures, r (t_a) curves for $t_a \leq 200$ hrs. are qualitatively similar. Besides, with increasing t_a, r in the first interval (88 hrs. $\leq t_a \leq$ 162 hrs.) grows much more slowly than in the second interval (162 hrs. $\leq t_a \leq$ 206 hrs.). Note that with increasing t_a, r (290K) in this interval increases reaching its maximum at t_a = 206 hrs. and then decreases with increase of t_a in the third interval ($t_a >$ 206 hrs.). While in all intervals r(78K) increases with increasing t_a, in the third interval this increase occurs at much lower rates than in the second interval (see Figures 2 and 3). Moreover, at 78 K for the lower I and t_a the observed increase in r is smaller due to the fact that in this case the concentration of undestroyed weak links is small [7]. However, for the later aging stages (higher t_a) the concentration of destroyed weak links increases and, consequently, r of the sample increases. Note that in the first aging interval, when I changes within the range from 6 to 200 mA, the increase in r (78 K) makes 50 to 75 times, whereas for $t_a \geq$ 192 hrs. the increase is only makes a factor of 2-3 (Figure 1). What this means is in the heat treated samples, rapid

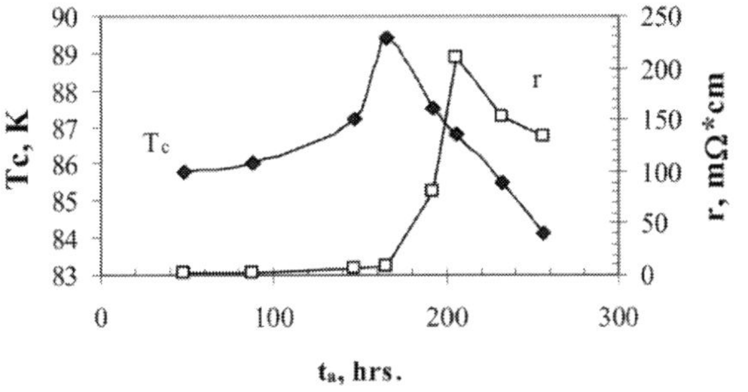

Figure 3. T_c and r (290K) dependences of heat-treated samples on the storage time t_a.

and slow aging processes are stimulated. For fast processes, the redistributions of oxygen atoms in weak linked copper-oxygen chains of the sample unit are responsible [1,2], and for slow processes - the redistributions of oxygen atoms in strongly bound copper-oxygen planes. Figure 3 also shows that t_a dependences of both T_c and r (290K) parameters have maxima at 162 hrs. and 205 hrs., correspondingly. Here, the increase in T_c with increasing t_a correlated with the simultaneous weak increase of r, and the sharp decrease in T_c after reaching its maximum is accompanied by rather rapid increase in r. The increase in T_c suggests that during the aging, first the (fast) ordering processes in the oxygen sublattice of the unit cell occur due to the redistribution of weakly bound oxygen atoms in the boundaries of superconducting grains, which is consistent with the result of [3]. However, if for yttrium based ceramics heat-treated at 78 K the observed temporary growth of T_c^i (responsible for intergranular weak links) correlates with the corresponding decrease in r [3], in our case it correlates, on the contrary, with the weak growth in r. It is seen that within the same interval of aging, at fixed t_a r(I) curves reveal a certain threshold I = I_s, below which with increasing I a relatively weak increase in r(78K) occurs, and for I> I_s the increase becomes significantly steeper (Figure 1). With increasing t_a, I_s (that creates the sample self-magnetic field), shifts to lower values due to the fact that the concentration of destroyed weak links increases, and the change of r(I) growth rate already occurs at lower I_s [4,7]. Qualitatively similar pattern is also observed for r(I) curves in the second aging interval (slow processes). However, the initial growth rate for slow processes is less than that for fast processes almost by an order of magnitude, and the corresponding I_s is more by a factor of 20. It speaks in favor of the assumption that in these aging processes, sharp increase in r with increasing I in I> I_s region is caused by the destruction of strong links in grains (or in superconducting planes). This behavior of r(I) (see Figure 1) is probably due to the depinning of self-magnetic field vortex lines from amorphous interface regions inside the grains and their movement into the grains [6]. Note that if with increasing t_a in the first interval I_s decreases from 30 to 10 mA, in the second interval I_s decreases from 150 to 100 mA. The probable reasons for the sharp increase of r in the first interval can also be the co-existence of two structural phases and changes in their volume fraction as well as in their degree of order during aging. This hypothesis is supported by our result on the availability of semiconducting phase and increase of their concentration (not included in this paper), and by the results of other investigations [6,8]. In particular, the observed increase in T_c can be explained by the transfer of O1 oxygen atoms from O5 position to their original sites, and the sharp growth of r in the second interval of t_a by the transition of atoms from O4 site into vacant O5 sites [2,8]. As for the formation of amorphous defect regions causing a sharp increase of r (I) for I> I_s in the third t_a interval, OH^- groups, which penetrate into the sample from ambient air moisture and then occupy O2 or O3 positions in Cu-O planes of the unit cell, could be responsible [6]. However, to get final clarification on the causes of the obtained results, further research is required. Such research is important in view of both the practical application and obtaining/expanding our understanding of the high-temperature superconductivity mechanism.

4. Conclusions

Study of the phenomenon of aging induced in $YBa_2Cu_30_x$ samples after heat treatment at 400°C and slow cooling to room temperature followed by the storage under environmental conditions, leads to the following conclusions. The changes of characteristic parameters during the progress of aging fall on three specific intervals of storage time t_a. In all intervals, r(78 K) increases with t_a. In the first and second intervals of storage time t_a, the behaviors of r (78K) and r (290K) curves are qualitatively similar: their growth in the first interval with increasing t_a is much slower than in the second interval. However, in the second interval this increase for r(78 K) is more than two orders of magnitude faster than for r(290K). In the third interval, r(78 K) continues to increase with increasing t_a, but much more slower than in the second interval. While r(290K) decreases after reaching its maximum at t_a = 206 hrs., T_c also increases with increasing t_a in the first interval, passes through a maximum at t_a = 162 hrs. and decreases rapidly in the third interval. Moreover, in this range the increase of T_c correlates with weak increase in r(78 K) and r(290K). It is possible to assert that ordering processes are fast processes, and degradation processes are slow ones. Besides, the ordering is due to the redistribution of intergranular weak links, and slow processes are due to disordering in both weak and strong links (or inside the grains) of the HTS sample unit cell.

5. References

[1] Veal B, Paulikas A, You H, Shi H, Y. Fang, and Downey J 1990 *Phys. Rev.* B **42** 6305
[2] Jorgensen J, Pei S, Lightfoot P, Shi H, Paulikas A and Veal B 1990 *Physica* C **167** 571
[3] Sinchenko A 1989 Abstracs of 2nd All-Union Conf. High Temperature Superconductivity, September 25-29, Kiev, 1989, vol 3, p 103
[4] Svoboda P, Vasek P, Smrckova O, Sykorova D, Plechacek V and Nevriva M 1990 *Sol. St. Commun* **75** 331
[5] Li M, Norblad P and Kawamura H 2001 *Phys Rev Lett* **86** 1339
[6] Bobylev I, Zyuzeva N *Fizika Tverdogo Tela* 2013 **55** 855
[7] Nikoghosyan S, Harutunyan V, Baghdasaryan V, Mughnetsyan E, Zargaryan E and Sarkisyan A 2013 *Solid State Phenomena* **200** 267
[8] Ceder G, McCormack R, and de Fontaine D 1991 *Phys. Rev.* B **44** 2377

Effect of holographic grating period on its relaxation in a molecular glassy film

A Ozols[1], P Augustovs, V Kokars, K Traskovskis, and D Saharov

Faculty of Material Science and Applied Chemistry, Riga Technical University, Azenes iela 14/24, LV-1007, Riga, Latvia

E-mail: aozols@latnet.lv

Abstract. Holographic grating (HG) relaxation has been experimentally studied in 5,5,5-triphenylpentyl 4-((4-(bis(5,5,5-triphenylpentyl)amino) phenyl) diazenyl) benzoate molecular glassy film for HG periods (Λ) of 0.50, 2.0 and 8.6 μm. A strong effect of HG period on its relaxation is found manifesting itself differently in the volume and on the surface. The volume part of HG is fairly stable during 40 days if $\Lambda > 0.50$μm whereas the surface part of HG (most probably, surface relief grating) exhibits relaxational self-enhancement which is maximal at $\Lambda = 8.6$μm. It is proposed that thermostimulated directional mass transfer in the process of relaxation can be responsible for this relaxational self-enhancement. Weak HG recording and relatively fast HG decay takes place at Λ=0.50 μm. Therefore, effective chromophore photoorientation domain of about 0.2 μm is supposed.

1. Introduction

Azobenzene-containing materials have found numerous applications in optoelectronic devices, including those based on hologram recording [1-4]. However, the relaxation processes after the hologram recording has been studied much less than the recording itself [5].The studies of relaxation processes are not only practically important. They are also of large physical interest because rheological processes, especially in disordered media , are far from clear. Relaxational (dark) self-enhancement (RSE) of holograms is an unexpected manifestation of rheological processes after the hologram recording. It was for the first time reported by R.G.Brandes et al in 1970 in bromide–doped As-S films [6].Further RSE was reported and studied in additively-coloured KBr crystals [7] and a-As_2S_3 films [8]. Detailed studies of RSE in undoped a-As_2S_3 films have been made by A.Ozols et al [7]. Record diifraction efficiency increase up to 18 times in a relaxation process was found there. RSE of holographic gratings was later reported and studied also in $Bi_{12}Ti_{0.76}V_{0.24}O_{20}$ crystals [10], in dichromated gelatin and polyvinyl alcohol films [11], and in photopolymers [12].

In this paper, we have experimentally studied holographic grating (HG) relaxation in an azobenzene molecular glassy thin film choosing different HG periods. The obtained results show that relaxation processes strongly depends on HG period and that relaxation processes are different in the volume and on the surface. Noticeable RSE was observed only on the surface.

[1] To whom any correspondence should be addressed.

IOP Conf. Series: Materials Science and Engineering **49** (2013) 012043 doi:10.1088/1757-899X/49/1/012043

2. Samples and experiments

Two 2 μm thick samples of 5,5,5-triphenylpentyl 4-((4-(bis(5,5,5-triphenyl-pentyl)amino) phenyl) diazenyl) benzoate molecular glassy film (briefly **K-RJ-9** film) were identically prepared by spin-coating and used for experiments. Their glass transition temperature was T_g=66^{0}C. Film consisted of molecules bound together by weak van der Waals forces .Chemical structure of molecules is shown in Fig.1. In Fig.2 absorption spectrum is presented. **K-RJ-9** molecule is a push-pull molecule of donor-acceptor type consisting of central -N=N- azochromophore groups surrounded by donor (=N-) and acceptor (esther O=C-O-) groups. This molecular structure enables intramolecular charge transfer facilitating *trans-cis* photoisomerization.

We have used the following experimental procedure. First, full HG recording was made by two equal power simmetrically incident 532 nm laser beams monitoring both transmission self-diffraction efficiency (SDE_t) and reflection self-diffraction efficiency (SDE_r) to evaluate the efficiency of recording and its time scale (Fig.3). Self-diffraction efficiencies were defined in the following way: $SDE_t = P_{dt} / P_i$, $SDE_r = P_{dr}/P_i$ where P_{dt} and P_{dr} are the beam powers diffracted in the plus first transmission and reflection orders, respectively, for the upper recording beam, P_i is the incident lower recording beam power.

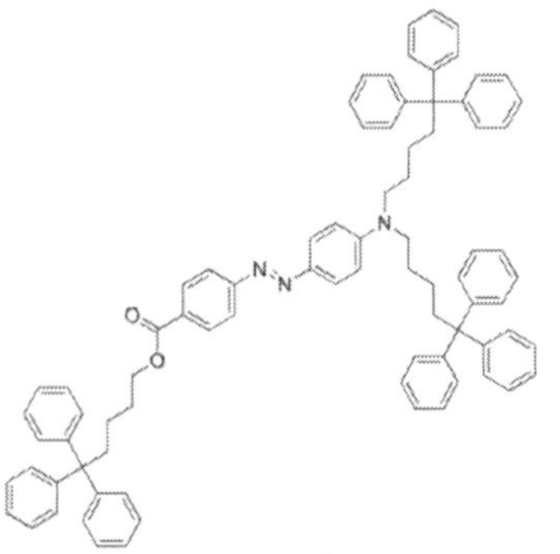

Fig. 1.Chemical structure of a **K-RJ-9** molecule.

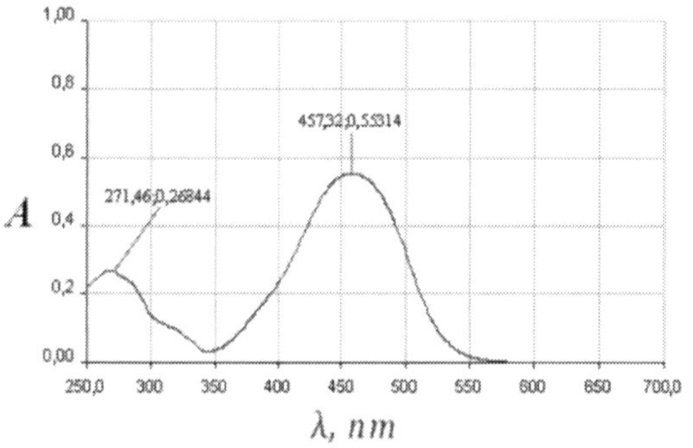

Fig. 2. Absorption spectrum of **K-RJ-9** in methylene chloride (CH_2Cl_2) solution.

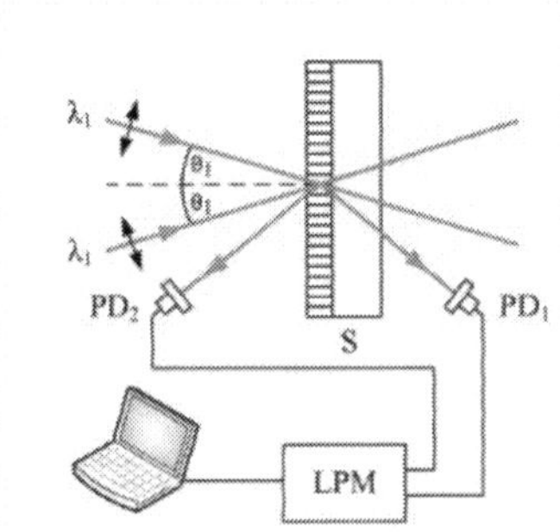

Fig. 3. The first experimental setup .Notations: S-sample, λ_1-recording beam wavelength, PD_1-photodiode measuring transmission self-diffraction efficiency SDE_t, PD_2-photodiode measuring reflection self-diffraction efficiency SDEr, LPM-Ophir Laserstar laser power meter connected to PC, θ_1-recording beam incidence angle. Recording beams were p-polarized.

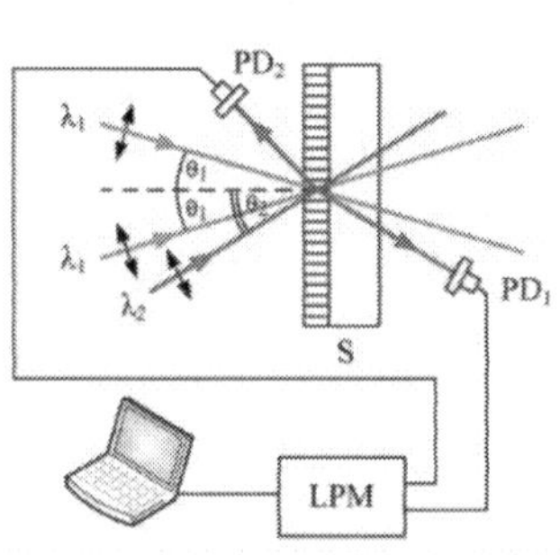

Fig. 4. The second experimental setup. Notations: S-sample, λ_1-recording beam wavelength, λ_2-readout beam wavelength, PD_1-photodiode measuring DE_t, PD_2-photodiode measuring DE_r, LPM-Ophir Laserstar laser powermeter connected to PC, θ_1-recording beam incidence angle, θ_2-readout beam incidence angle. All beams were p-polarized.

Second, transmission HG with the periods (Λ) of 0.50, 2.0 and 8.6 μm were recorded with two equal symmetrically incident 532 nm laser beams and read out simultaneously by a 632.8nm ≈ 633 nm beam until the maximum readout light transmission diffraction efficiency (DE) defined as $DE_t = P_{dt}/P_i$ (P_{dt} being the light power diffracted in the minus first transmision order, P_i being the incident light intensity) was reached (Fig.4). Simultaneously, reflection DE ($DE_r=P_{dr}/P_i$)was measured, here P_{dr} being the light power diffracted in the plus first reflection order. After the recording, DE_t and DE_r were continously monitored during about 30 minutes. Further, DE_t and DE_r measurements of recorded HG were repeated during 40 days.

Klastech DENICAFC 532-300 diode pumped solid state laser (λ_1 = 532 nm) with a coherence length exceeding 100 m was used for HG recording. Recording light intensity was 0.40-0.46 W/cm^2. Melles Griot 25LH928-230 He-Ne gas laser (λ_2 = 632.8 nm) was used for HG readout. Readout beam power varied in the 0.27-1.8 mW, mainly it was 0.50 mW. It was proved not to influence HG. All the diffracted light powers was measured by Ophir Laserstar laser power meter and stored in the PC memory. Diffraction efficiency time dependences in characteristic cases are presented in the next section. All the measurements started from zero at 30 seconds. The estimated measurement errors did not exceed 5%.

3. Results and discussion

The main experimental results are presented in Figs.5-11.

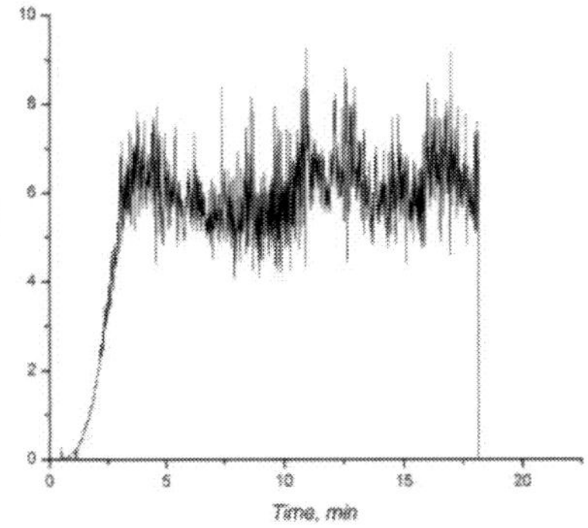

Fig. 5. Transmission SDE exposure time dependence. $\lambda_1 = \lambda_2$=532 nm, Λ=2.0 μm.

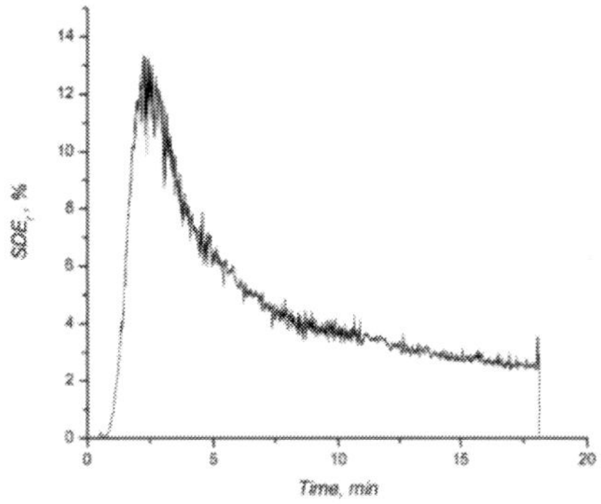

Fig. 6. Reflection SDE exposure time dependence. $\lambda_1 = \lambda_2$=532 nm, Λ=2.0 μm.

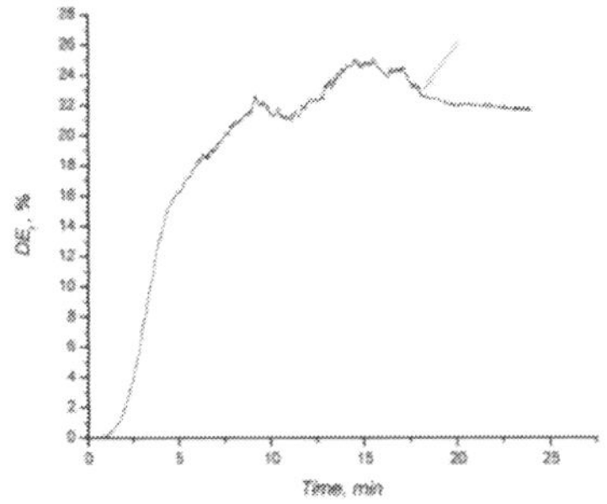

Fig. 7. Transmission DE exposure (until the green arrow) and relaxation (after the arrow) time dependences. λ_1=532 nm, λ_2= 633 nm, Λ= 2.0 μm. This quasiconstant relaxation was observed also further 24 minutes (not shown).

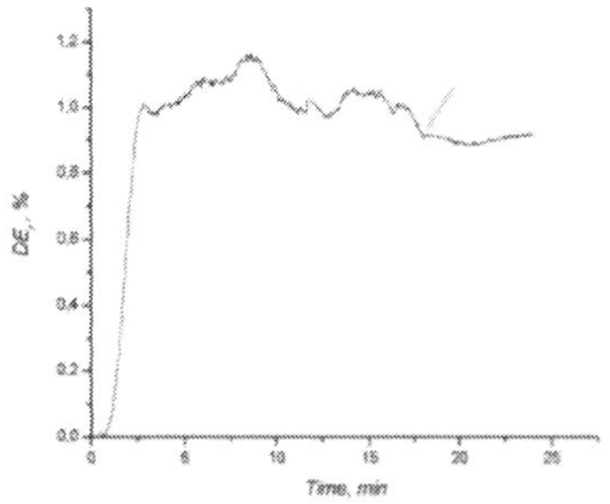

Fig. 8. Reflection DE exposure (until the green arrow) and relaxation (after the arrow) time dependences. λ_1=532 nm, λ_2= 633 nm, Λ= 2.0 μm. This quasiconstant relaxation was observed also further 24 minutes (not shown).

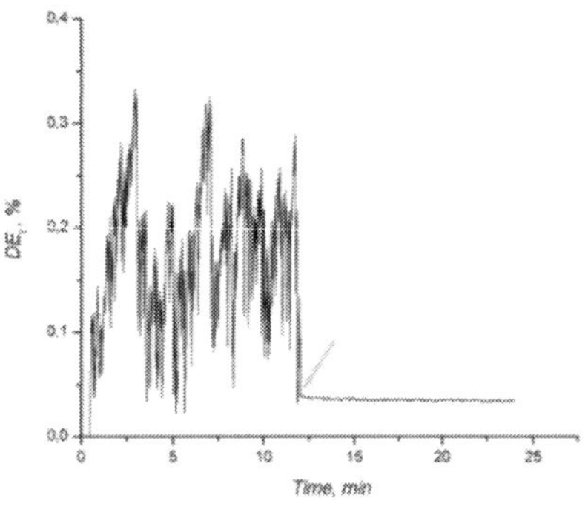

Fig. 9. Transmission DE exposure (until the green arrow) and relaxation (after the arrow) time dependences. λ_1=532 nm, λ_2= 633 nm, Λ= 0.50 µm. This quasiconstant relaxation was observed also further 24 minutes (not shown). Note the weak oscillatory recording behaviour! It was impossible to measure DE$_r$(t) for such a small grating period because the +1st and higher diffraction orders do not exist

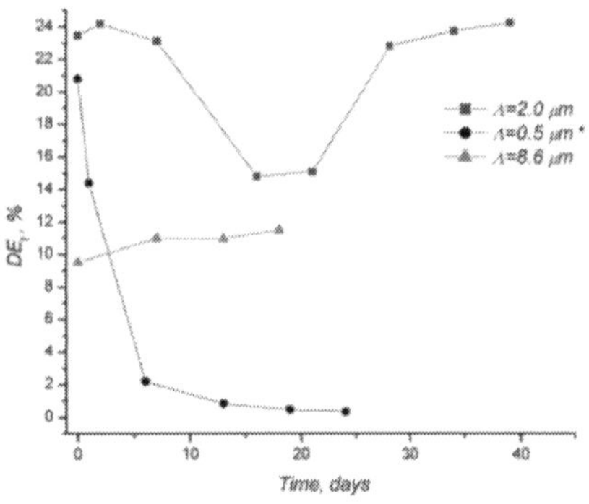

Fig. 10. Transmission DE relaxation time dependences for the gratings periods of 2.0µm (blue rectangles), 8.6µm (red triangles) and 0.5µm (black circles). In the case of 0.5 µm period 100 times larger DE$_t$ values are shown. Rectangles, triangles and circles are experimental points. Lines are to guide the eye.

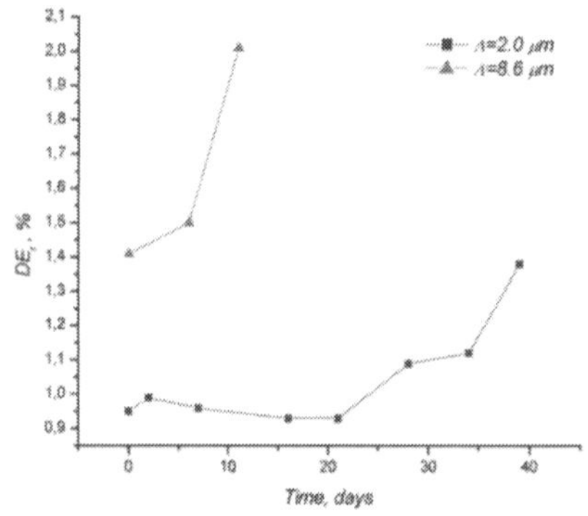

Fig. 11. Reflection DE relaxation time dependences for the gratings periods of 2.0µm (blue rectangles) and 8.6µm (red triangles). Rectangles, triangles and circles are experimental points. Lines are to guide the eye.

The DE$_t$(t) and DE$_r$(t) dependences at Λ = 8.6µm were qualitatively the same as for Λ = 2.0µm. Therefore, they are not shown to save the restricted space.

The following observations can be made from Figs. 5-11 supported also by other curves not shown.

1. SDE$_t$(t) and SDE(t)$_r$ kinetic dependences are markedly different .
2. HG recording at Λ=0.50µm is much weaker and with oscillatory DE$_t$(t) dependence when compared to the cases with Λ=2.0µm and Λ=8.6µm (Figs. 7-9).
3. All HG are practically stable immediately after the recording during 30-40 minutes, as shown by kinetic DE$_t$(t) and DE$_r$(t) curves Figs.5-9).
4. HG relaxation as judged by DE$_t$(t) dependences strongly depend on HG period Λ (Fig.10):

1) fast (in 10 days) decay in the case of Λ = 0.50μm;
2) nonmonotonical HG relaxation in the case of Λ = 2.0μm;
3) slight increase in the case of Λ = 8.6μm.

5. HG relaxation as judged by $DE_r(t)$ dependences also depend on Λ but in different fashion when compared to $DE_t(t)$ case. RSE takes place and it is much more pronounced for Λ = 8.6μm than for Λ = 2.0μm (Fig.11) . In 10 days DE_r (t) increases more than 1.4 times Λ = 8.0μm whereas it is approximately constant at Λ = 2.0μm.
6. Diffraction efficiency dependences on HG period are different for transmission and reflection cases: $DE_t(\Lambda$ = 2.0μm)>$DE_t(\Lambda$ = 8.6μm) whereas $DE_r(\Lambda$ = 2.0μm)<$DE_r(\Lambda$ = 8.6μm) (Figs. 10 and 11).

We know from literature [2,4,13]and our previous experiments (e.g., [14]) that HG recording in azocompound films at 532 nm is based on *trans-cis* photoisomerization including chromophore photoorientation perpendicularly to light electric field vector and mass transfer. Repeated *trans-cis-trans* photoisomerization cycles soften the matrix, especially near the surface [13].

From observations 2 and 4 we conclude that spatial resolution limit of our **K-RJ-9** film is close to 0.5 μm. Therefore, we suppose that effective photoreorientation domain is about 0.2 μm (a little bit less than $\Lambda/2$ if we suppose that there should be at least two photoreorientation domains per period to have recording) and it is much larger than the size of molecule (about 30 Å) due to the cooperation motion of molecule's environment.

Observations 1, 4,5,6 show that photoinduced HG recording processes and HG relaxation processes are different at the surface (determining mainly SDE_r and DE_r) and in the volume (determining mainly SDE_t and DE_t). Thus a HG consists of two components-surface HG, most probably surface-relief grating (SRG), and volume grating. SRG are recorded on the surface, and volume phase gratings in the volume. This situation was observed also in our previous experiments with other azocompound films [15]. It was reported in amorphous chalcogenide As-S-Se films, too [16].

Thermostimulated *cis-trans* isomerization and the motion of molecules due to the dipolar interactions and under influence of mechanical stresses caused by HG recording (*cis*-isomer is bulkier than *trans*-isomer[4]) are , most probably, the main relaxation processes of HG. Motion of molecules in our **K-RJ-9** film is facilitated by fact that this is a molecular glassy film in which molecules are bound together only by weak van der Waals forces. As the result, directional mass transfer is possible enhancing SRG. The possibility of such process is considered also by O.N. Oliveira, J.Kumar, L.Li and S.K. Tripathy in 2002 [4]. However, it should be confirmed by AFM measurements.

Higher DE_r at Λ=8.6μm when compared to Λ=2.0 μm (observation 6) can be explained by the specific backward diffraction properties of SRG [17].

RSE properties of our **K-RJ-9** film are quite different from those of a-As_2S_3 film [9] .RSE in a-As_2S_3 is much stronger. DE_t increase up to 18 times was observed at Λ = 1.0μm. In **K-RJ-9** film RSE takes place only at reflection, and DE_r increase only up to 1.5 times was observed at much larger period of Λ = 8.6μm. RSE in a-As_2S_3 is initiated by glassy layer network crosslinking during the HG recording by photoinduced structural changes, and it is explained by periodically distributed stress relaxation model . Compressive stress produced in HG maxima pushes unfilled sites towards minima thus causing density modulation proportional to $1/\Lambda$ [9]. No such dependence is found in **K-RJ-9** film, thus the role of a mechanical stress is different.

4. Conclusions

HG period strongly influences its relaxation. This effect is different at the surface (corresponding to the SRG) and in the bulk (corresponding to the volume grating).

RSE of SRG is found whereas volume gratings are fairly stable during 40 days if the HG period is not less than 2.0μm. It is proposed that directional mass transfer in the process of relaxation can be responsible for RSE of SRG. However, this proposal should be confirmed by AFM measurements.

Fast HG decay and weak HG recording takes place at 0.5 μm HG period. Therefore, we suppose that effective photoreorientation domain is about 0.2 μm.

References

[1] 2008 *Introduction to Organic Electronic and Optoelectronic Materials and Devices* ed S Shajing and L R Dalton (Boca Raton: CRC Press)
[2] 2009 *Smart Light-Responsive Materials Azobenzene-Containing Polymers and Liquid Crystals* ed Y Zhao and T Ikeda (Hoboken , New Jersey: John Wiley & Sons, Inc)
[3] Tsutsumi N, Kinashi K, Sakai W, Nishide J, Kawabe Y and Sasabe K 2012 *Optical Materials Express* **2** 1003
[4] Sekkat Z and Knoll W 2002 *Photoreactive Organic Thin Films* (California: Academic Press)
[5] Oliveira O N ,Jr, Kumar J, Li L and Tripathy S K in [4], pp. 429-483
[6] Brandes R G, Laming F P and Pearson A P *Appl. Optics* **9** 1712
[7] Ozols A 1979 *Latv.J. Phys.Techn. Sci.* 5 45 (in Russian)
[8] Reinfelde M, Salminen O and Ketolainen P1992 *Opt. Lett.***17** 1085
[9] Ozols A, Salminen O and Reinfelde M 1994 *J.Appl.Phys.* **75** 3326
[10] Rickermann F, Riehemann S, Buse K, Dirksen D and von Bally G 1996 *J.Opt.Soc.Am.*B**13** 2299
[11] Grzymala R and Keinonen T 1998 *Appl. Opt.* **37** 6623
[12] Lougnot D J and Turck C 1992 *Pure Appl. Opt.* **1** 269
[13] Yager K G, Barrott C J 2006 Light-Induced Nanostructure Formation Using Azobenzene Polymers *Polymeric Nanostructures and Their Applications* , ed H S Nalwa **0** (American Scientific Publishers) pp. 1-38
[14] Ozols A, Kokars V, Augustovs P, Uiska I, Traskovskis K, Saharov D 2011 *Centr. Eur. J.Phys.***9** 547
[15] Ozols A , Reinfelde M , Saharovs D , Kundzins K, Kampars V, Kokars V. 2008 *Thin Solid Films* **516** 8887
[16] Reinfelde M and Teteris J 2011 *J.Optoelectronics Adv. Mat.* **13** 1531
[17] Moharam M G, Pommet D A and Grann F B 1995 *J.Opt.Soc.Am* **12** 1077

Investigation of carbonized layer on surface of NaAlSi glass fibers

E Pentjuss, A Lusis, G Bajars and J Gabrusenoks

Institute of Solid State Physics, University of Latvia, 8 Kengaraga Street, Riga, LV-1063, Latvia

Email: evalds.pentjuss@cfi.lu.lv

Abstract. There are presented and discussed experimental results about carbonate shell on the sodium rich alumosilicate (NaAlSi) glass fibers and carbonization in wet air atmosphere and water uptake kinetic of such fiber fabrics. The analyzes of water uptake kinetic by regression technique, leaching and heating of carbonized glass fabrics helped to separate stages of fast and slow processes between fiber and carbonate shell and air atmosphere. The shell contains mixture of trona and hydrated sodium carbonate. Heating converts both substances to sodium carbonate. The weight uptake after heating encounters two fast exponential processes associated with water absorption on the surface of carbonated shell and its diffusion into volume. The slow process associates with CO_2 and H_2O absorption from air, hydration and sodium carbonate conversion to trona.

1. Introduction

Glass fiber fabrics are used in dry and wet atmosphere and in water at elevated and changeable temperatures. The changes in environment can initiate processes in the glass fabric that lead to a new its equilibrium state and changed physical properties. Some of them may be irreversible. It is accepted [1,2] that interaction of alkali silicate glasses with water or mineral acids proceeds by ionic exchange by diffusion of Na^+ ions to glass surface and H^+ or H_3O^+ from water into the bulk of glass to fill the vacancies of Na^+ ions. It looks that Na^+ ions and H_2O and CO_2 from surrounding atmosphere form the shell of $Na_2CO_3{\cdot}H_2O$ (hydrated sodium carbonate) and $Na_3H(CO_3)_2{\cdot}2H_2O$ (trona) mixture on the surface of elementary glass fibers. Such shell can be dissolved in water and acids (leached) [3]. The heating can leads to weight loss by dehydration of $Na_2CO_3{\cdot}H_2O$ over 100 ^{0}C, by decomposition reaction of

$$2(Na_3H\,(CO_3)_2{\cdot}2H_2O) = 2(Na_2CO_3{\cdot}NaHCO_3{\cdot}2H_2O) \rightarrow 3Na_2CO_3 + CO_2\uparrow + 5H_2O\uparrow \quad (1)$$

at temperature over 55-57 ^{0}C [4], and free water evaporation. There are studied the weight loss of stored at about 20 ^{0}C and 20% - 35% RH (room conditions in autumn and winter), stored and leached by acid Na-Al-Si glass fabrics, and leached powder (LP) of carbonates after its heating at different temperatures and followed long term uptake of weight in room atmosphere. The values of experimental weight uptake-time points were analyzed by regression technique.

2. Samples

The unleached and leached by acid samples of K-glass fabric with initial composition of [(18-22)Na_2O(3-5)Al_2O_3(73-79)SiO_2] from JSC "Valmiera Glass Fiber" were studied. The threads of fabric consist of 600 elementary glass fibers with diameter of 6 μm. The samples leached by acid have thickness about 0.34 mm. The unleached samples had been stored during the years at room conditions. In a result of interaction with surrounding air, the Na^+ ions leached from glass fibers to its surface and formed here the sodium carbonate shell. The long term limit of thickness of shell can exceed 2 μm. The thickness of carbonized fabric was in range 0.53 – 0.58 mm. The leached powder (LP) was prepared by dissolving the shells in water and water evaporation.

3. Results and Discussion

XRD measurements showed that LP after receiving consists of trona and hydrated sodium carbonate. Heating over 110 ^{0}C (2 h) leads to its converting to sodium carbonate. Storing in air at room conditions causes inversed reaction (1). After 2000 h long storing, LP was converted back to trona (60%) and hydrated carbonate. This slow process needs to absorb water and CO_2 from air before mentioned reaction. An artificial increase of moisture content of unheated fabric samples over its equilibrium values with surrounding air showed that time, necessary for returning to equilibrium weight are some hours only. There is no weight loss and stabilization of weight has to be associated with free water evaporation to equilibrium with environment. It indicates that beginning (about 0, 25 h) of weight increase has to be associated mainly with water absorption. There are observed the fast uptake of weight ΔM(t) during the first tens of minutes after heating (Fig.1) and much slower (hundreds of hours) uptake up to equilibrium weight later (Fig.2) for not leached (a1n-a3n), leached samples S1 and S2 of fabric and LP.

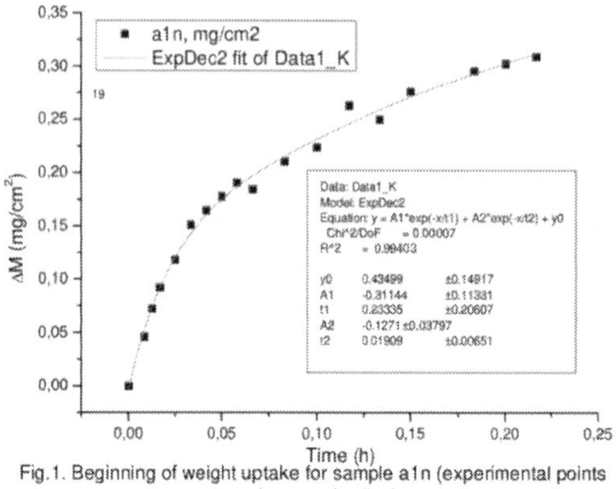

Fig.1. Beginning of weight uptake for sample a1n (experimental points and regression line)

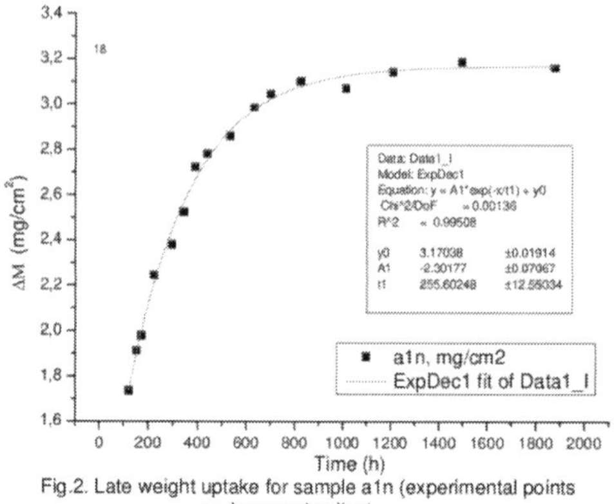

Fig.2. Late weight uptake for sample a1n (experimental points and regression line)

The experimental weight uptake–time ΔM (t) points for the beginning period are well-fitted ($R^2 > 0.99$) to regression of

$$\Delta M\,(t) = A_0 - A_1 \exp(-t/t_1) - A_2 \exp(-t/t_2), \qquad (2)$$

where A_0, A_1, A_2 are weight constants and t_1, t_2 are time constants ($t_1 \le t_2$), lim $\Delta M(t)=A_0$ when $t\to\infty$ and lim $\Delta M_1(t)=A_1\exp(-t/t_1)=A_1$, and lim $\Delta M_2(t)=A_2$, when $t\to 0$. In (2) $A_0=A_1+A_2$. Velocity of weight uptake can be owned from (2):

$$v(t) = \frac{A_1 \exp(-t/t_1)}{t_1} + \frac{A_2 \exp(-t/t_2)}{t_2}. \qquad (3)$$

The main fitting parameters for beginning of weight uptake are presented in Table 1 and for the late uptake in Table 2. The fitting results indicate to simultaneous water absorption on surface of shell with rate of 3.9 – 6.6 mg/h·cm^2 ($t_1 \approx 0.017 - 0.04$ h) and absorption associated with water that leaves the surface to volume (possible by diffusion) with rate of 1.0 – 2.2 mg/h·cm^2 ($t_2 \ge 0.17$ h) - for unleached samples (Table 1). Heating temperature increase leads to increase of irreversible weight loss (Table2 and Fig.4) or equilibrium weight decrease.

Table 1. Parameters of weight uptake beginning.

Sample	^{0}C/h	A_0 (mg/cm^2)	A_1(mg/cm^2)	t_1(h)	A_2(mg/cm^2)	t_2(h)	$v_1(0)$, (mg/h cm^2)	$v_2(0)$, (mg/h cm^2)
a1n	90/2	0.435	0.1271	0.0191	0.311	0.233	6.65	1.34
a2n	130/0.167	0.539	0.1581	0.0173	0.383	0.173	9.14	2.2
a3n	130/2	1.09	0.155	0.039	0.932	8.1	3.94	1.15
S1	110/2	1.3	0.26	0.0074	0.769	0.064	35	11.8
S2	170/2	1.37	1.36	0.096	-	-	14.2	-
LP	110/2	0.07	0.07	0.012	-	-	5.9	-

Table 2. Parameters of late weight uptake.

Sample	^{0}C/h	A_0 (mg/cm^2)	A_1 (mg/cm^2)	t_1(h)	A_2 (mg/cm^2)	t_2 (h)	Weight loss (mg/cm^2)
a1n	90/2	3.17	-	-	2.3	256	0.21
a2n	130/0.167	3.54	-	-	3.13	171	0.548
a3n	130/2	3.0	-	-	1.91	125	2.03
S1	110/2	1.15	-	-	0.04	143	0.051
S2	170/2	1.59	0.123	22.6	0.102	195	0.232
LP	110/2	2.18	-	-	2.65	255	5%

The looks of fabrics after leaching are changed from matted to shine. The leached sample S1 heated at 110 ^{0}C (Table 1) has notably less values of t_1 and t_2 (about 0.007 h and 0.06 h) and higher values of surface absorption and desorption rates (about 35 and 12 mg/h·cm^2). The reason has to be associated with smoother surface of glass fibers and absence of notable carbonized shell compared to surface of not leached fibers ones. In this case there is only one time constant in beginning phase (Table 1). Increase of heating temperature to 170 ^{0}C causes significant increase of time constant t_1 and irreversible weight loss (sample S2, Table 2) in spite of absence of carbonated shell. It should be associated with pore volume (deepness) increase by loss of matter.

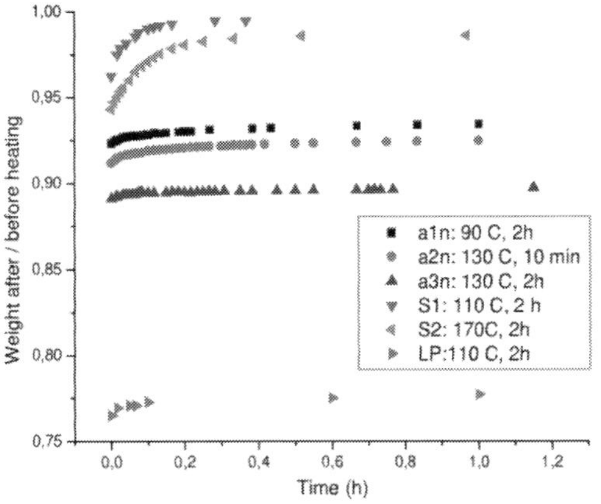

Fig.3. Ratio of weight after and before heating for beginning of weight uptake

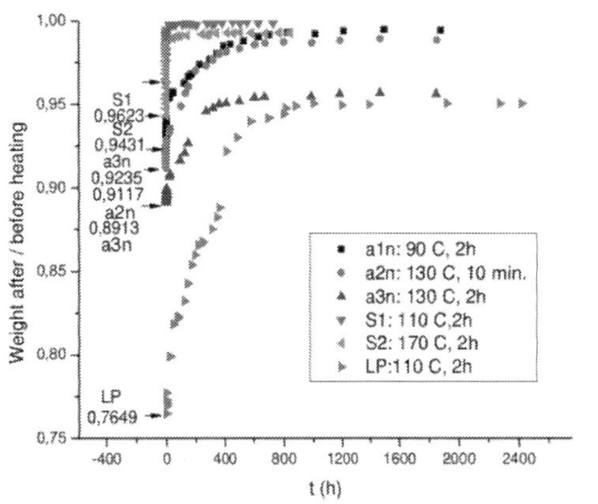

Fig.4. Ratio of weight after and before heating for late weight uptake

Following (1) the proportion of weights of heated and unheated sample (Fig.3, 4) has to be equal or more than 0.7035. In a case of LP (60% trona) the coincidence is good (0.7649 by weighing and 0.764 by XRD). In the other cases the proportions are significantly higher than mentioned value (0.7035) because in the weight are included the glass of fibers.

The late uptake of weight can be described with only one exponential function in equation (2). The values of late t_2=120–250 h have to characterize the saturation with water (free and crystallized) and conversion sodium carbonate to trona in the all volume of shells of unleached fabric. There exists some slow water insertion in surface of glass of leached fabric also. One of reasons of slow rate of conversion to trona should be associated with low concentration of CO_2 in air. Late weight uptake dominates in unleached samples and beginning uptake in leached ones (Fig.4).

4. Conclusions

Leached in water powder from unleached and carbonized NaAlSi glass fiber fabrics consist of mixture of trona and hydrated sodium carbonate. Heating of powder above 110 ^{0}C (2 h) leads to dehydration and trona conversion to sodium carbonate. Storing the heated powder at about 20 ^{0}C and 20% - 35% RH during the months causes sodium carbonate conversion back to hydrated sodium carbonate and trona.

Heating of carbonized NaAlSi glass fabric causes weight loss that partly restores during the months in air at room conditions. The beginning stage of weight uptake consists of two exponential processes that should be associated with water absorption on the surface of carbonated shell and its diffusion to volume. The first process has time constant of 0.017 - 0.04 h and second of 0.2 - 8 h and velocity of 4 - 9 mg/h·cm^2 and 1.2 - 2.2 mg/h·cm^2 accordingly. The late phase of weight uptake appears after some hundreds hours and has time constant about 200 h. This slow process should be associated with CO_2 and H_2O absorption from air and converting sodium carbonate to hydrated sodium carbonate and trona.

The leached by acid NaAlSi glass fabric has initial weight uptake process more fast (about 47mg/h·cm^2) and more expressed compared to unleached ones (5 - 11 mg/h·cm^2).

References

[1] Veal B W, Lam D J and Karim D P 1980 *Nuclear Technology* **51** 136

[2] Tsong I S T, Houser C A, White W B, Power G L and Tong S S C 1980 *Non-Cryst.Solids* **38-39** 649

[3] Pentjuss E, Lusis A, Bajars G, Gabrusenoks J and Jekabsone L 2012 *IOP Conf. Series: Materials Science and Engineering* **38** doi:10.1088/1757-899X/38/1/012021

[4] Gärtner R S and Witkamp G J 2007 *Hydrometallurgy* **88** 75

The study of correlation between microstructure of ferrites and their complex permeability spectra

J. Jankovskis[1], N. Ponomarenko[1], N. Mironova-Ulmane[2], D. Jakovlevs[1]

[1]Riga Technical University, [2]Institute of Solid State Physics, University of Latvia
e-mail: nikolaj.ponomarenko@gmail.com

Abstract. Theoretical model for complex permeability of polycrystalline ferrite (based on effects arising from realities of typical microstructure of sample) is correlated with experimental data. In the case of NiZn-ferrites there is observed a close agreement with the model and the data; for samples of MnZn-ferrites this agreement is only for small-sized cores; for bigger ones the dimensional effects (dimensional resonance) appear and the model cease to be valid.

1. Introduction

One of the leader of nowadays technologies - electronics are developing very fast (according to Moore's Law) largely because of app materials and their technologies. Nevertheless there are branches, especially magnetics-based, that are lagging behind because of insufficient progress. As such, in fact, are polycrystallyne ferrites (PF) that may be thought of as main high frequency magnetic materials.

One of main characteristic of ferrites at high frequencies is their complex magnetic permeability (CMP) $\dot{\mu}(f) = \mu'(f) - j\mu''(f)$, specified by its real $\mu'(f)$ (dispersion, DCp) and imaginary $\mu''(f)$ (absorption, ACp) components. Obviously CMP is housing large amount of information on intrinsic processes of PF. The interpretation of the data coming from these spectra has been in progress for several decades but as a rule they are only qualitative (except one specific point of the spectrum – the static permeability $\mu(0)$). Specifically, these interpretations as well as direct experiments show on close correlation between the features of spectrum and the microstructure (MS) of sample (within frequencies up to the ending of the most dominant range of dispersion – the large-amplitude broadband one at radiofrequencies [1]). But closer examination of this correlation is difficult principally since PF as materials are not well defined systems because of great variety of influencing factors [2].

The only known quantitative theoretical analysis of this correlation is the modelling [3] based on the realities of MS: account of effects of polycrystal grain sizes and their distribution that, after all, brings to closed-form relation for $\mu''(f)$ which in its turn allows for $\mu'(f)$ through the use of numerical integration of Kramers-Kronig relations (KKR) [4]. By this way gained theoretical curves for $\mu'(f)$ and $\mu''(f)$ provides the means of presentation of both relaxation and resonance types of MS thus indicating on universality, at least in principle. But problems related with the details of presentation of CMP (true values of parameters, their actual correlation with MS) still remain. Answers are a matter both of experimental verification and theoretical analysis, which in this investigation is performed for samples from both of two main groups of ferrites, i.e. NiZn and MnZn ones. Within this correlation process the correctness of every used experimental $\dot{\mu}(f)$ additionally is verified by KKR.

2. The essence of modelling.

As it was mentioned before the modelling of $\dot{\mu}(f)$ is based on realities of MS of PF (i.e., sample is viewed as aggregate of polycrystal grains of the current size D, distributed by log-normal probability density function having the median D_{med} and the standard deviation σ_D) as well as on assumption that these grains are forming magnetically independent elementary volumes (as evidenced, likely, by linear dependence of

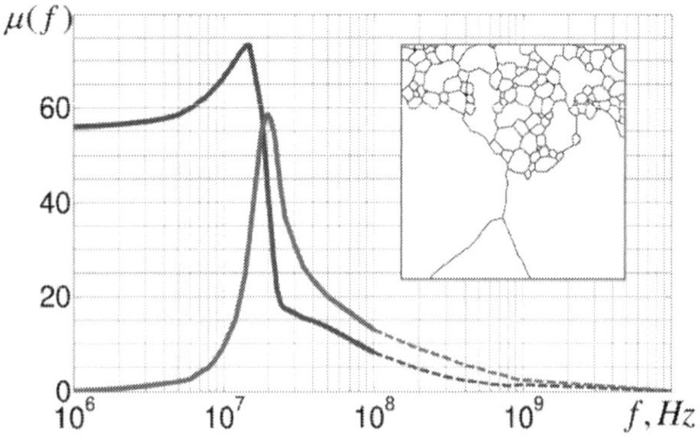

Fig.1. CMP of ferrite with unhomogenous MS

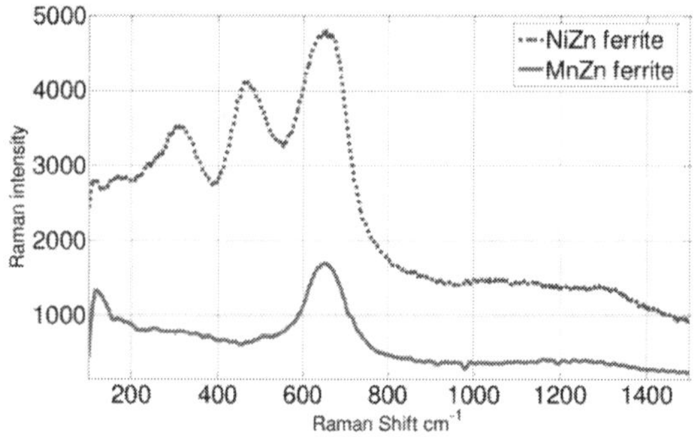

Fig.2. Raman spectra of MnZn and NiZn PF

$\mu(0)$ on average grain size D_a both by experiments [2] and theory [3]) and on hypothesis that characteristics of magnetization process within current grain D resembles the ones observed in variety of real PF samples in relation to their D_a. Then the domain wall (DW) (or equivalent DW) in grain D (without intragrain defects) acts as independent oscillator having the resonance frequency $f_0(D)$; averaging of the oscillators over the sample gives [3]:

$$\mu''(f) = \mu''_{max} exp\left[-\frac{(logf - logf_u)^2}{2\sigma^2}\right] \quad (1)$$

where f_u and μ''_{max} are characteristics of extremum of absorption, i.e. $\mu''_{max}=\mu''(f_u)$ and $\sigma = 2\sigma_D$. This three parameters (f_u, μ''_{max} and σ) Eq.1 gives rise to so called symmetrical ACp (plotted along typical *logf* axis) which would be realizing in the case of ideal MS (when there are no intragrain defects) and giving

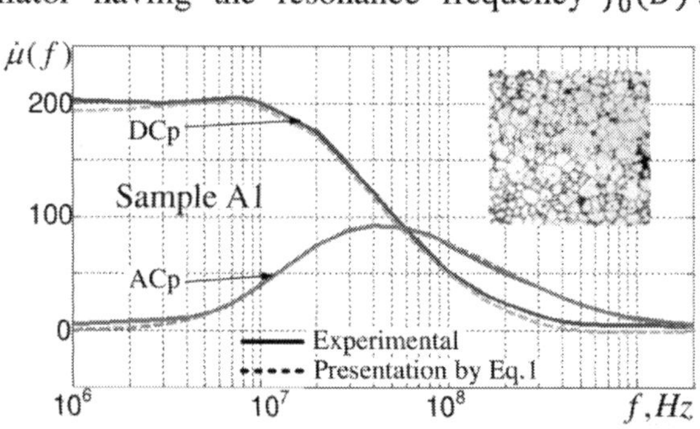

Fig.3. CMP of NiZn PF with homogenous MS

$\sigma \approx 0.4$ [8]. However, most of the experimental ACp are considerably asymmetric. It was showed in [5] that in this case (because of action of intragrain defects) there is the need in Eq.1 to operate with two values of σ: the one σ_a for $f \leq f_u$ and the other, σ_b for $f \geq f_u$. The values can be obtained from [3]:

$$\sigma_a = |log(f_u/f_a)|/\sqrt{2ln[\mu''_{max}/\mu''(f_a)]} \quad (2)$$

in which the data from experimental ACp are inserted: $f_a \leq f_u$ (often such as $\mu''(f_a) \approx 0.5\mu''_{max}$); for σ_b, f_a is simply replaced by $f_b > f_u$ (if for estimation of σ_a and σ_b exact half-level absorption is used, then Eq.2 simplifies to: $\sigma_a = 0.85|log(f_u/f_{0.5})|$, where $f_a = f_{0.5} < f_u$ and $f_b = f_{0.5} > f_u$). This line of the reasoning in derivation of Eq.1 shows on its claim to present $\dot{\mu}(f)$ as the *intrinsic* CMP of samples; nevertheless in the case of MnZn ferrites this can be obscured by dimensional resonance [6].

3. The samples and results

This research (finally aiming to either validate or disprove the proposed model) is based on experimental data from ring samples of different sizes (Table 1) and known MS (which were obtained from micrographs by the method of Saltikov [7]). Within the model there is assumed that MS is homogeneous (i.e. have the log-normal distribution) that was confirmed by all there used samples (having, Table 1, $\sigma_{lnD} \approx 0.5$, that is the criteria for normal, homogenous MS [9]). In addition the micrographs of the ferrites (Fig.3…5) show that there are numerous defects within grains, which leads to asymmetrical CMP spectra, to the need to use both σ_a and σ_b for presentation of $\dot{\mu}(f)$. Inhomogenity of MS, as observed in [2], immediately appears as a peculiarity on $\dot{\mu}(f)$ (Fig.1; experimental spectrum for $f>10^8$ Hz is extrapolated on for possibility to use KKR). Raman spectra for samples A2…A5 and B1…B5 (Fig. 2) show that MnZn-ferrites are more inverse than NiZn-ferrites.

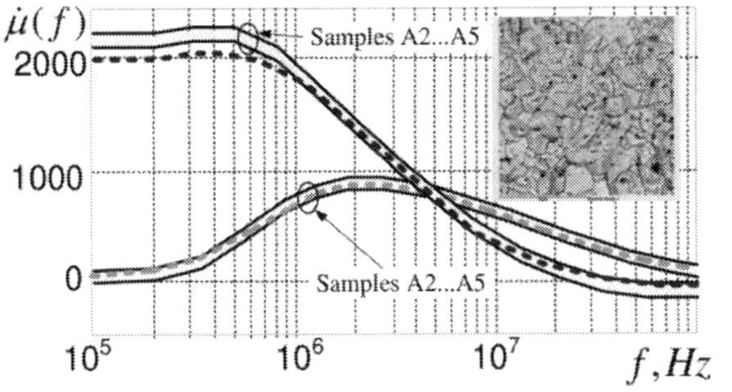

Fig.4. Quantitative model for A2 PF

Table 1. Data of NiZn and MnZn samples used in experimental proof of the modelling

Group	№	Dimensions, mm	A, mm^2	$\mu(0)$	$\mu(0)_{exp}$	D, μm	σ_{lnD}	σ_a	σ_b	σ_{av}
NiZn	A1	R16x7x6	27	200	203	7.7	0.49	0.47	0.66	0.56
	A2	R12x8x6	12	2000	2200	7.9	0.55	0.47	0.79	0.63
	A3	R20x10x6	30	2000	2100	7.9	0.55	0.44	0.80	0.62
	A4	R31.5x20.5x6	33	2000	2125	7.9	0.55	0.43	0.81	0.62
	A5	R40x23.5x6	51	2000	2140	7.9	0.55	0.42	0.84	0.63
MnZn, 6000HM1	B1	R10x4.3x10	28.5	6000	6750	12.9	0.53	0.41	0.53	0.47
	B2	R20x9.1x10	54.5	6000	6530	12.9	0.53	0.43	0.55	0.49
	B3	R25x12x10	65	6000	6650	12.9	0.53	0.44	0.54	0.49
	B4	R30x14.3x10	78.5	6000	6850	12.9	0.53	0.44	0.57	0.50
	B5	R33x10x10	115	6000	6620	12.9	0.53	-	-	-

CIP spectra was measured up to adequately high frequencies to ensure all of ACp large amplitude dispersion region be included. Measurements were taken by R&S network analyzer ZVR-E, with attached specially designed measurement fixture with known parameters.

CMP spectrum (Fig.3) of medium-permeability NiZn ferrite with Co-additives (sample A1, Table 1) and of small size, measured some time ago [10], shows the correlation with its MS similar to now measured high-permeability ones of different sizes, but of similar MS (A2…A5; Fig. 4: practically coinciding set of spectra). This similarity is achieved by the use of special process in the production of samples – by waterjet assisted mechanical cutting from single Ferroxcube 4S60 tile [11]. To relate the spectra with their MS (like as in [8]) averaged dispersion parameters $\sigma_{av} = (\sigma_a + \sigma_b)/2$ are given in Table 1; σ_{av} shows that MS of the samples is rather homogenous but with considerable amount of defects within grains. In all cases approximations of experimental $\dot{\mu}(f)$ is excellent (Fig. 2) or good (Fig. 3).

In this context it is possible to state that situation for MnZn samples (as well cutted from single tile of 6000HM-1 ferrite [12]) depends on sizes: for rather small size cores (B1…B4, Fig. 5) picture is similar with already discussed NiZn samples; for cores of higher sizes (e.g., B5, Fig. 5) dimensional resonance [6] takes place – CMP ceases to have correlation with its MS, approximations by Eq. 1 are not physically meaningful.

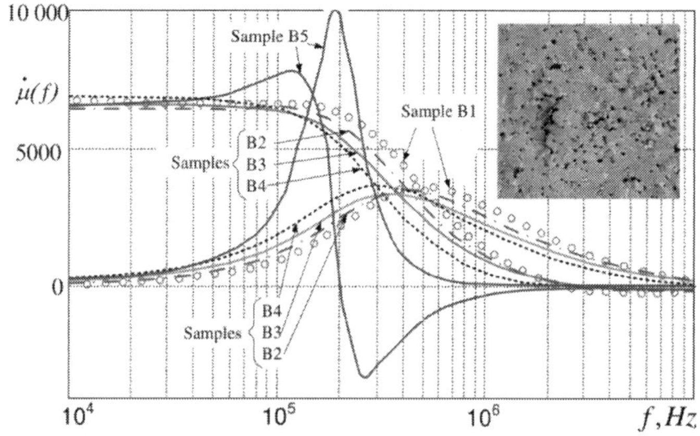

Fig.5. Experimental CMP curves for samples B1...B5

4. Conclusion.

A correlation made between the theoretical model of $\dot{\mu}(f)$ (based on account of effects coming into being from realities of actual MS of PF) and the experimental data evidenced that the mode is well suited for presentation of CMP as intrinsic characteristic of PF. The parameters needed for the presentation are clear physically and well understood in relation to typical, homogeneous MS of PF; as such the model may serve for prediction of magnetic spectra as well. In the full measure these statements corresponds to NiZn-ferrites; in the case of MnZn-ferrites the size of magnetic core exerts some effect on $\dot{\mu}(f)$: for small sizes there are no influence (the measured $\dot{\mu}(f)$ corresponds to intrinsic properties of PF); for bigger ones $\dot{\mu}(f)$ gradually turns into extrinsic characteristic which eventually identifies ferrite as resonator (exhibiting dimensional resonance).

Literature references

[1] Guyot M., Merceron T., Cagan V., Messekher A. 1988 Phys. Stat. sol. (a), v.106, No.2, 595-612
[2] Globus A., 1977 J. phys (Fr.), v.38, No4 (suppl), C1-1-C1-15
[3] Jankovskis J.. 2006 Advances in Scien. & Techn., v.45, 2560-2565
[4] Smith J., Wijin H.P.J. 1959 Ferrites. Philips' Techn. Library, Eidhowen
[5] Jankovskis J.. 2004 J. Magn. Magn. Mater., v. 272-276, e1847-e1849
[6] Brockman F.G., Dowling P.H., Steneck W.G. 1950 Phys. Rev., v.77 No.1, 85-93
[7] Saltikov S. A. 1976 *Stereometricheskaja metallografija. Izd. Metallurgija*, Moskow (in Russian)
[8] Jankovskis J., Ponomarenko N., Stepins D. 2013 Key Engineering Materials, Vol. 543, 507-510
[9] Kurtz S. K., Carpay F. M. A. 1980 J. Appl. Phys 5 5725
[10] Rankis G., Yurshevich V., Brivkalns K., 1986 Proc. Latvian Acad. Sc., No 5, 38-45
[11] Ferroxcube data book, 2008
[12] www.rusgates.ru, 6000HM-1 material (in Russian)

IOP Conf. Series: Materials Science and Engineering **49** (2013) 012046 doi:10.1088/1757-899X/49/1/012046

Photo-induced formation of surface relief in amorphous As_2S_3 films

E Potanina[1], **J Teteris**
Institute of Solid State Physics, 8 Kengaraga Str., Riga, LV-1063, Latvia

E-mail: elina.potanina@gmail.com

Abstract. In this report we study the formation of surface relief in arsenic trisulfide films under laser light radiation with modulated polarization direction. The formation of surface relief is obtained with method what is different from holographic recording - only one laser beam with modulated polarization direction was used. Samples with different thickness were made using vacuum evaporation. The photo-induced changes on the surface were initiated by DPSS laser light (λ = 532nm). Profiles of surface reliefs were observed and relief depth dependence on film thickness and exposure was studied.

1. Introduction

Structure like surface relief grating (SRG) can be used as diffractive optical element and as Bragg reflector. The new method was used to study SRG formation in chalcogenide films. Well-known chalcogenide – amorphous arsenic trisulfide was used for this purpose. As_2S_3 is known as inorganic polymer, which is often used in holography due to large photo-induced changes of its chemical and optical properties [1.2.3]. It has been studied since 1950', a lot of interesting optical phenomena [1.2.3.4.5.6] were discovered, but still there are things to discuss. Photo-induced optical anisotropy such as birefringence and dichroism has been studied before [6] and is closely connected to photo-induced mass transport. Studying SRG formation process mainly consists of finding appropriate conditions for efficient photo-induced mass transport. Surface relief can be obtained with holographic record as well as by illumination through optical slit [8.9], but this study is based on one beam record. Since there is no interference, set-up is stable to vibrations and there is no need in high coherence. Experimental set-up consisting of polarization direction modulator is shown in Figure 1.

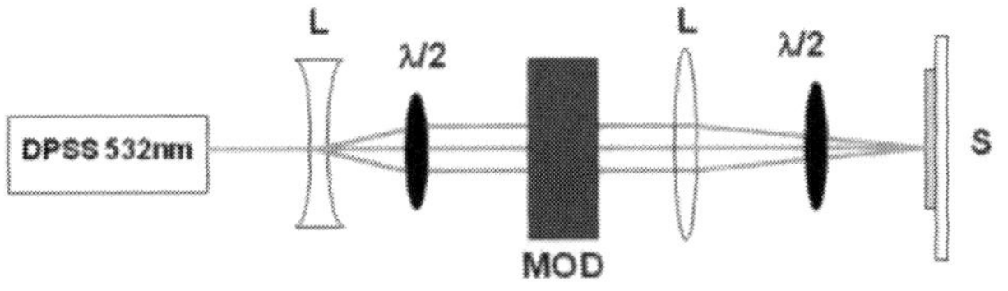

Figure 1. Experimental set-up for recording surface relief grating. **L** - lens; **λ/2** – half-wave plate; **MOD** – polarization modulating element; **S** – sample. Recording was performed with DPSS laser λ = 532nm.

[1] To whom any correspondence should be addressed.

2. Experimental

Surface relief gratings (SRG) are usually obtained with holographic method based on two beam interference [8.9]. Depending on recording beam polarization direction it can provide an interference pattern with light intensity or polarization modulation what is necessary for SRG formation. In this study we observe formation of surface relief, which is obtained with new method, different from holographic recording. The method is based on modulating light polarization direction in one laser beam. Intensity modulation pattern, what forms on samples surface, is quite similar to interference pattern. By using MOD and half-wave plate polarization is modulated according to already known configurations. For example, the S-S polarization modulation is a result of two linearly polarized wave's interference. Modulated polarization directions in one laser beam and its analogy in holography are shown in Figure 2.

	S-S	P-P	S-P	45^0; -45^0
1	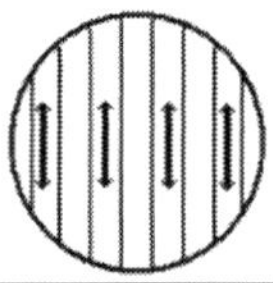	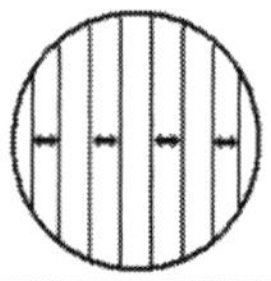	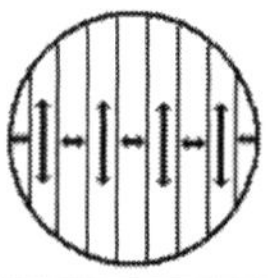	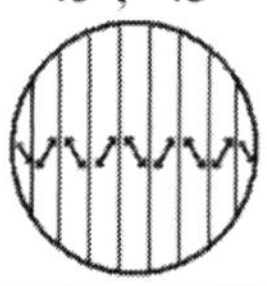
2	Interference of two S-polarized waves	Interference of two P-polarized waves	Interference of R and L (or 45^0; -45^0) polarized waves	Interference of S and P-polarized waves

Figure 2. Simplified scheme of polarization distribution; ↕ - electric field vector oscillation direction.
1 – Modulated polarization direction in one laser beam.
2 – Similar distribution with two beams (holographic analogy).

For direct SRG record amorphous arsenic trisulfide films where used as photoresist. Samples with different thickness (from 1µm to 10µm) were prepared by vacuum evaporation onto glass substrate. Surface reliefs were recorded using modulated DPSS laser beam with 532 nm wavelength (Figure 2.). Intensity of laser beam was constant during the recordings – 3,77 W/cm^2. SRGs were recorded with a constant period – 50 µm. Depth and profile of recorded SRG was measured with profilometer. For polarization modulation change the half- wave plate is rotated, but in case of S-S or P-P configuration polarization cube is inserted after the second half-wave plate.

3. Results and discussion.

First, SRG recording with different polarization modulations was studied. Recording with all four configurations gave positive results (Figure 3.), but SRG recorded with S-P modulation was the deepest.

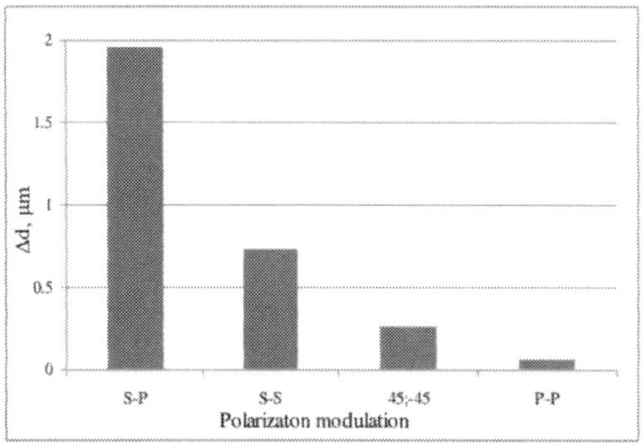

Figure 3. Δd – SRG depth in micrometers. Recording beam intensity – 3, 77 W/cm^2; Exposure time – 2h.

Intensity distribution is the same for S-P and 45^0;-45^0, but the results differ ~8 times. S-S and P-P configurations are similar, but still there is a big difference in results. It means that electric field vector oscillation direction is principal photo-induced mass transport mechanism. In SRG formation process polarization modulation plays the key role. S and P – polarizations are orthogonal, so the electric field gradient has biggest value. It is electric field gradient [8] what induces mass transport, but mass transport direction depends on polarization direction. The velocity of mass transport process is not constant and growth of SRG is not infinite process. SRG formation process, increasing exposure, is shown in Figure 4. SRG formation process consists of two parts: linear, fast growth (t = [0..12]h) and saturation process, the part of slow growth(t = [12..70]h). It supposes to reach the maximum, i.e., biggest possible SRG depth value and after that growing stops. Thinner films are most likely to reach the maximum faster, because there is same amount of received energy, but less mass to transfer. Form of graphic, process tendency, is the same for all three films with different thickness. Recording SRG in 1 μm thick film is possible at smaller exposure or intensity. At 10 h exposure with 3,77 W/cm^2 intensity SRG starts to disband. Damage of SRG can be caused by large amount of received energy, illuminated area heating or large period influence.

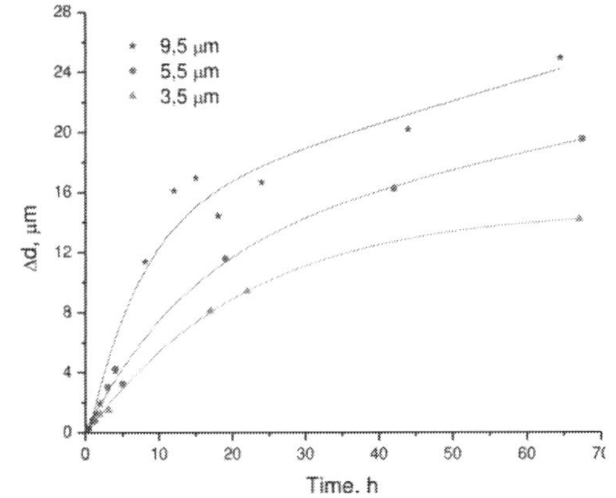

Equation	$y = A_1 e^{-\frac{x}{t_1}} + A_2 e^{-\frac{x}{t_2}} + y_0$					
	* 9,5	±Δ	● 5,5	±Δ	▲ 3,5	±Δ
y_0	14.9	3.0*10^{-16}	138.2	11.9	39.2	2.0
A_1	-7.5	7.9*10^{-9}	-123.6	11.9	-11.8	0.1
t_1	21.7	2.8*10^{-8}	800.2	81.4	13.5	0.1
A_2	-7.5	7.9*10^{-9}	-15.8	1.9*10^{-2}	-27.4	1.8
t_2	21.7	2.8*10^{-8}	7.0	1.1*10^{-2}	201.5	19.5
R^2	1		1		1	

Figure 4. SRG depth dependence on exposure. Δd – SRG depth in micrometers; samples with thicknesses 3,5 μm, 5,5 μm and 9,5 μm were used; recording beam intensity – 3, 77 W/cm^2. Second order exponential decay function is used to fit experimental data.

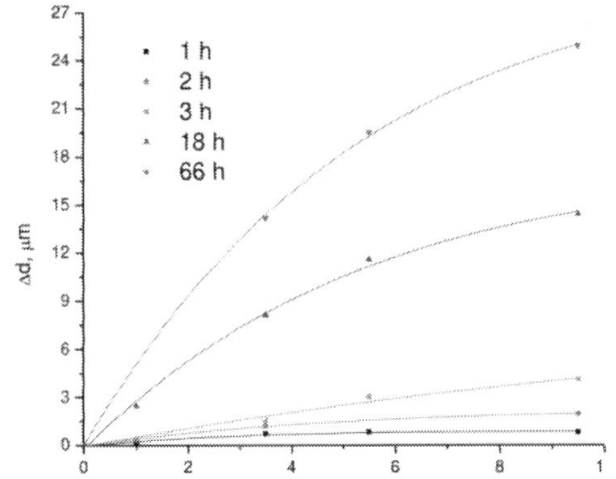

Equation	$y = Ae^{-\frac{t}{x}} + y_0$			
Coeff.	A	t	y_0	R^2
▼ 66 h	-30.23	5.40	30.20	0.99966
±Δ	1.11	0.41	1.13	
▲ 18 h	-18.01	5.49	17.78	0.9975
±Δ	1.34	0.89	1.42	
◀ 3 h	-9.55	15.71	9.41	0.98172
±Δ	7.69	17.11	7.82	
● 2 h	-2.32	4.74	2.28	0.99591
±Δ	0.24	1.25	0.26	
■ 1 h	-0.97	2.66	0.90	0.93978
±Δ	0.19	1.53	0.17	

Figure 5. SRG depth dependence on film thickness at different exposures; Δd – SRG depth in micrometers; b – film thickness. Exponential fit is used.

Depth of recorded SRG depends on films thickness, too. SRG formation process versus film thickness is shown in Figure 5. Process is exponentially growing. To evaluate the effectiveness of each thickness, the ratio $\frac{\Delta d}{b}$ was calculated. Most efficient film thicknesses are from 3,5 μm to 5,5 μm, with value of ratio 0,21 (1 h exposure) – 4,05 (66 h exposure). For thicker films, as there is a lot of mass to transport, process is slower and they are efficient at long exposures. Deeper SRG can be recorded, but it takes more time. Example of recorded SRG and its profile are shown in Figure 6.

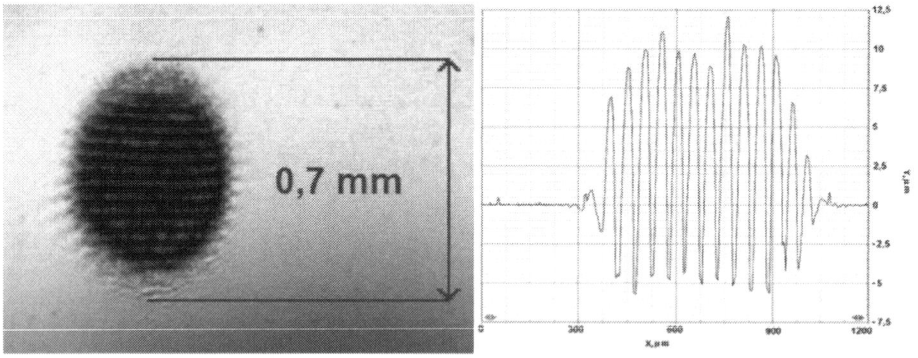

Figure 6. Picture and profile of 17 μm deep SRG recorded in 5, 5 μm thick film. Measured with profilemeter.

4. Conclusions

The new method is applicable for studying photo-induced mass transport in chalcogenide films. It is easier and cheaper way of recording SRG then holographic record. The quality of SRG is lower, but for research purposes this method is very good. For efficient recording S-P polarization is necessary. Recording beam intensity may vary but for thinner films ~1 μm it should be less then 3,77 W/cm^2 to prevent damaging of SRG. Optimal film thickness for SRG recording is ~ 4,5 μm.

Acknowlegements.
This work has been supported by ERAF project Nr. 2010/0204/2DP/2.1.1.2.0/10/APIAA/VIAA/010.

5. References

[1] Hisakuni H, Tanaka K, 1994, *Appl. Phys. Lett.* Vol. **65**, 23, 2925 – 2927 p.
[2] Kolobov A, Tanaka K, 2001, *in Handbook of Advanced Electronic and Photonic Materials and Devices,* Vol. **5**, H.S.Nalwa(Ed.),Academic Press, San Diego, p.47.
[3] Tanaka K, 1980 *J. Non-Cryst. Solids* **35&36**, 1073
[4] Roy A, Kolobov A V, Tanaka K 1998 *J.Appl.Phys.***83**, 4951
[5] Tanaka K, 2001, *in Handbook of Advanced Electronic and Photonic Materials and Devices,* Vol. **5**, H.S.Nalwa(Ed.),Academic Press, San Diego, p.120.
[6] Tanaka K, Terakado N, Saitoh A 2008 *J.of Optoelectron. Adv. Mater.* **10**(1), 124–130
[7] Hajito J, Janossy I 1983 *Phil. Mag. B* Vol. **47**, No.4, 347
[8] Gertners U, Teteris J, 2011, *J.of Optoelectron. Adv. Mater.* Vol. **13**, No. 11-12 , p.1462-1466
[9] Gertners U, Teteris J, 2012, *IOP Conf. Series: Mater. Science and Engineering*, **38**, 012026.

Luminescence of $LaBr_3$:Ce,Hf scintillation crystals under UV-VUV and X-ray excitation

V A Pustovarov[1], V Yu Ivanov[1], A N Razumov[1], D I Vyprintsev[2]
[1]Ural Federal University, Mira Street, 19, Yekaterinburg 620002, Russia,
[2]Stark Ltd Company, Marksa Street, 14, Obninsk 239035, Kaluga region, Russia

E-mail: pva@dpt.ustu.ru

Abstract. The present study was carried out by means of the low temperature time–resolved luminescence UV–VUV spectroscopy as well as the optical and thermally activation spectroscopy. The Ce^{3+} centers in regular lattice sites and located in the vicinity of the point defects of crystal structure were observed. Spectral and kinetic characteristics of the luminescence of these centers were defined. A photoluminescence of new point defects of the crystal structure are also manifested. We evaluated the energy of the interband transitions as $E_g \approx 6.2$ eV in $LaBr_3$. At the excitation energies E_{exc}>13 eV (higher than $2E_g$) the effect of the electronic excitations multiplication was detected.

1. Introduction

The $LaBr_3$:Ce scintillator manufactured by the company *Saint–Gobain*, in fact, is very attractive for both spectrometric applications and for medical imaging. Nowadays a mechanism of the scintillation pulse's formation in these crystals is well known due to the investigations of luminescence under different kinds of excitations as well as to research of the electron excitation energy transfer to Ce^{3+} impurity center [1–3]. Photoluminescence (PL) studies of d→f emission in Ce^{3+} ions and self–trapped excitons (STE) under excitation by photons in ultraviolet (UV) and vacuum ultraviolet (VUV) ranges in $LaBr_3$:Ce was reported in Ref. [1, 2] only. In these studies the values of the fundamental absorption edge and the interband transition energy E_g were determined approximately (E_g = 5.9 eV at T=10 K). In the later works the results presented in Ref. [1, 2] are quoted, however further similar investigations of $LaBr_3$ in VUV range were not carried out and this result seems to be ambiguous.

In our previous report [4] we investigated a nonlinearity of response light yield in $LaBr_3$:Ce in extremely ultraviolet (XUV) region (45–290 eV). Our first results of $LaBr_3$:Ce PL investigations under VUV excitation are presented in Ref. [5]. $LaBr_3$:Ce crystals are hygroscopic, therefore the method of growth of $LaBr_3$:Ce single crystals doped with hafnium was developed in Russia [6]. These $LaBr_3$:Ce,Hf crystals have low hygroscopicity and they are more suitable for practical applications. The main goal of present work is the studies of luminescence of the $LaBr_3$:Ce,Hf crystals produced in Russia. The present study was carried out by means of the time-resolved luminescence UV-VUV spectroscopy as well as the optical absorption spectroscopy.

2. Experiment

2.1. Methods

Time–resolved and time–integrated PL spectra (1.5–5.0 eV), time–resolved PL excitation spectra (3.7–32.0 eV) and PL decay kinetics were measured using SR at the SUPERLUMI station (beam line I, HASYLAB at DESY, Hamburg). The primary monochromator with Al or Pt gratings had a spectral resolution of 3.2 Å. The PL spectra were measured using a 0.3m monochromator model ARC Spectra Pro-308i and an R6358P photomultiplier tube (Hamamatsu) in two time windows Δt_1 = 11 ns (fast component) and Δt_2 = 30 ns (slow component), which were delayed relative to the start of the SR-excitation pulse by δt_1 = 3.0 ns and δt_2 = 36 ns respectively. The PL excitation spectra were normalized to the same number of SR–exciting photons using sodium salicylate. The time resolution of the whole detection system was 0.8 ns (FWHM). The temporal interval between SR pulses equaled 96 ns. The

400W deuterium discharge lamp and the prismatic DMR–4 monochromator were used to measure the PL excitation spectra in the low-energy range of 2.4–5.8 eV in Ural Federal University. The spectra of the X–rays induced luminescence (XRL) were studied by means of a BSV2:Cu X–ray tube (30 kV, 10 mA), a MDR–23 monochromator and a FEU–106 photomultiplier tube. The experimental details for XUV excitation are presented in Ref. [4].

2.2. Samples preparation

The examined single crystal with high optical quality (5 wt.% of cerium and 0.05–0.5 wt.% of hafnium in charge) have been grown using Bridgman process in the Russian company *Stark Ltd* [6]. The synthesized single crystals were certificated by a XRD and mass–spectrometry method with inductive–bound plasma, both the chemical and X-ray fluorescence analysis were also used. The introduction of hafnium into the melt reduces the hygroscopicity of grown crystals [9]. Because of the hygroscopic nature of $LaBr_3$, any contact with the atmospheric air leads to degradation of the sample surface that becomes cloudy due to the formation of crystalline hydrates. In all our experiments the samples (5x5x1 mm^3) were cleaved from crystal boule in an atmosphere of a dry hot air and immediately mounted in a cryostat with a fast vacuum pumping.

3. Experimental results

The luminescence spectra of $LaBr_3$:Ce,Hf crystals under X–ray and XUV excitations are very similar to those of $LaBr_3$:Ce crystals presented in Ref. [4]. The XRL spectra of $LaBr_3$:Ce,Hf crystals measured at different temperatures are shown in figure 1. These spectra contain emission bands at 356 and 382 nm. The time–resolved luminescence spectra are formed by the typical radiative d → f transitions in Ce^{3+} ions [2]. An additional broad band at 450 nm with microsecond PL decay kinetics appears at low temperatures. According to data [2], this band can be formed by self-trapped exciton (STE) emission. The PL spectra depend on the excitation energy (E_{exc}) significantly. The PL spectra of $LaBr_3$:Ce at T=8K under excitation by photons in the fundamental absorption region (E_{exc}=6.9 eV) corresponds well to the luminescence spectra under X–ray or XUV excitation. On the contrary, the PL spectra of $LaBr_3$:Ce,Hf shift to long wave region and contain narrow bands at 384 and 412 nm, as well as a broad band in the region of 500–600 nm with a microsecond PL decay kinetics (figure 2). The PL kinetics measured for the bands at 384 and 412 nm under excitation by photons with E_{exc}=3.7–5.0 eV have a single exponential decay with $\tau = 20 \pm 1$ ns. The value of the PL decay time is less than that in $LaBr_3$:Ce crystals (τ = 23–25 ns depending on E_{exc}, see Ref. [5]).

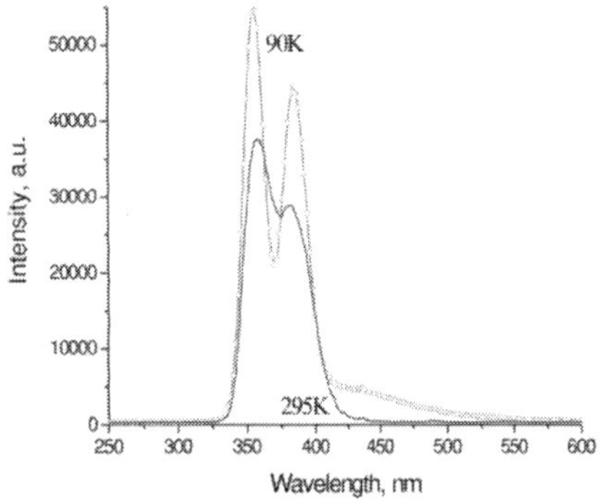

Figure 1. Luminescence spectra of $LaBr_3$:Ce,Hf crystals under X–ray excitation measured at 90 and 295 K.

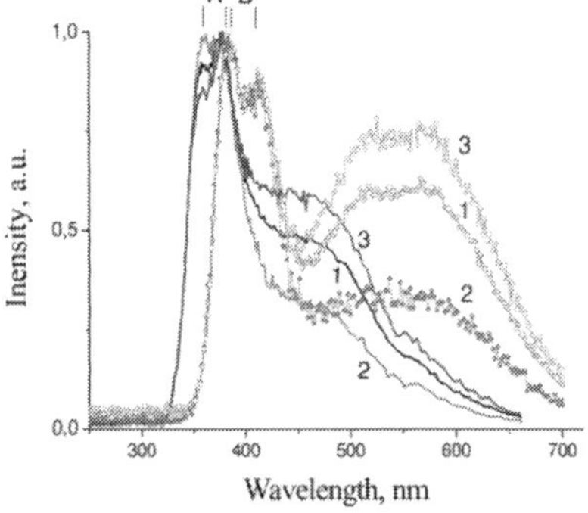

Figure 2. Time–resolved PL spectra of $LaBr_3$:Ce (solid lines) and $LaBr_3$:Ce,Hf (mugs–lines) under E_{exc}= 6.9 eV: time–integrated (1), fast (2) and slow (3) components, T=8 K.

Figure 3 shows the PL excitation spectra measured for 384 nm emission band and absorption spectrum of $LaBr_3$:Ce,Hf (5%). The absorption spectrum of $LaBr_3$:Ce (0.5%) calculated from transmission spectrum presented in Ref. [7] is also shown. The optical density in the region of 3.5–5.5 eV depends on the concentration of Ce^{3+} ions and its value increases dramatically at energies above 5.5 eV. The PL excitation spectra measured for 412 and 550 nm emission bands are represented in figure 4. Separate bands with maxima at 3.6 and 5.05 eV are observed.

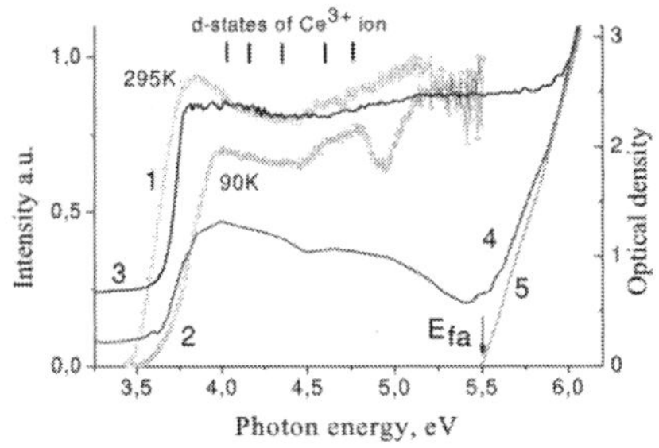

Figure 3. PL excitation spectra measured for 384 nm emission band (1, 2) in $LaBr_3$:Ce,Hf (5% of Ce), absorption spectra (T=295 K) of $LaBr_3$:Ce,Hf (5% of Ce) (3) and $LaBr_3$:Ce (0.5%) (4) from Ref. [7], (5) – linear fit. The energy positions of the d–states of Ce^{3+} ions in $LaBr_3$:Ce from Ref. [1] are shown.

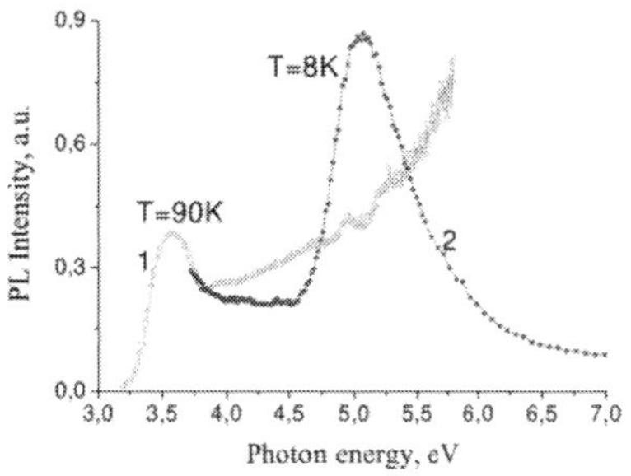

Figure 4. PL excitation spectra of $LaBr_3$:Ce,Hf measured for 412 (1), 540 or 600 nm (2) emissions.

Time-resolved PL excitation spectra of 384 nm Ce^{3+} emission band in the wide energy region presented on figure 5. The PL yield drops sharply at energies above 5.5 eV and increases at E_{exc}>13 eV. It should be noted that at E_{exc}>70 eV this spectrum has a sharp structure. We analyzed the PL excitation spectrum in XUV region in detail earlier in Ref. [4].

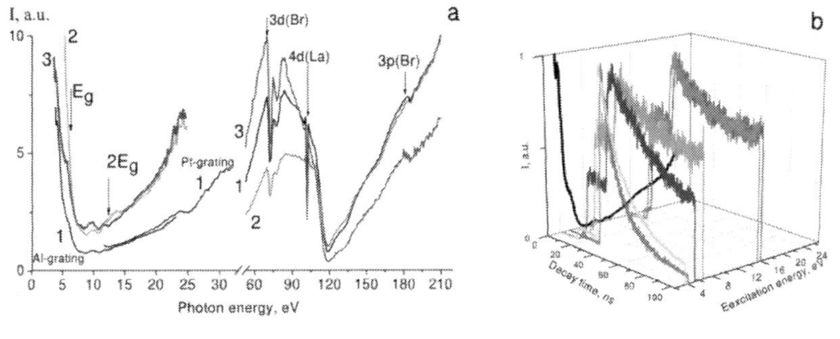

Figure 5. (a) – Time–resolved PL excitation spectra of 384 nm Ce^{3+} emission band in $LaBr_3$:Ce,Hf, T=295K: time–integrated (1), fast (2) and slow (3) components. Spectrum (1) was measured using the Al and Pt gratings. (b) – 3D image of PL excitation spectrum and PL decay kinetics.

4. Discussion

The analysis of presented spectra shows that Ce^{3+} ions in regular lattice sites (marked as the A–centers) and Ce^{3+} ions located in the vicinity of the defects of crystal structure (marked as the B–centers) are observed in $LaBr_3$:Ce,Hf. PL of the B–centers under intracenter excitation has a decay time of 20 ± 1 ns. PL spectrum is shifted to lower energies. Stokes shift does not change practically. PL was efficiently excited by 3.6 eV photons, however, f → d electronic transitions in the absorption spectrum do not appear in this spectral region. This fact indicates that the concentration of such B–centers is relatively low. Accordingly, in the XRL spectra only the A–centers appear.

New defects of the crystal structure were found in addition to the B centers in $LaBr_3$:Ce,Hf. The PL spectrum at T = 8 K has a broad band with maximum at 550 nm (2.25 eV, FWHM = 0.8 eV). The PL yield has its maximum at E_{exc} = 5.05 eV (Stokes shift is 2.8 eV) and it falls under excitation in the fundamental absorption edge and region of interband transitions. From this it follows that the intra center PL is observed, and that the efficiency of the energy transfer to these centers by excitons or band charge carriers is low. The structure of such defects is not clear.

The sharp increase of optical density at energies above 5.7 eV indicates the beginning of the fundamental absorption in these crystals. Linear fit shown in figure 3, allows us to estimate the energy of the edge of the interband transitions at E_{fa} = 5.5 eV. Slow microsecond components in the PL decay kinetics appear at PL excitation in this energy region (figure 5, curve 3). These facts indicate the participation of the band charge carriers in the formation of Ce^{3+} emission. Based on these data it is possible to estimate the energy of inter band transitions $E_g \approx 6.2$ eV in $LaBr_3$. It should be noted that our values E_{fa} and E_g differ from the data presented in Refs. [2, 5]. Apparently, the measurement of Urbach tail absorption in $LaBr_3$ crystals will determine a more precise value of E_g.

The efficiency of Ce^{3+} PL excitation drops in the fundamental absorption edge and region of interband transitions. The competition of capture of the band carriers on Ce^{3+} centers and on crystal structure defects, the formation of STE, and predominantly non–radiative annihilation of the band carriers on the crystal surface determines to this PL yield decrease. The PL yield stabilizes with a further increase of the kinetic energy of the band carriers. The PL yield increases monotonically at energies $E_{exc} > 13$ eV. Based on the studies of the luminescence excitation spectra in XUV region presented in figure 5–a (see in details Ref. [4]) we believe that at energies $E_{exc} > 13$ eV (higher than $2E_g$) the effect of electronic excitations multiplication is observed. This effect is an attribute of effective scintillation process. In crystals with a narrow valence band (E_v) of *p*-type the threshold energy for this effect should lie just in the region of (2–3)E_g [8]. We can reasonably assume that $E_v < E_g$ in $LaBr_3$ crystals, so the generation of secondary electron–hole pairs is due to the inelastic scattering of photoelectrons.

5. Conclusions

The present study was carried out by means of time–resolved luminescence UV–VUV spectroscopy as well as optical spectroscopy. Both Ce^{3+} centers in regular lattice sites and located in the vicinity of the point defects of crystal structure were observed. A PL of new point defects of the crystal structure was observed. We evaluated the energy of interband transitions as $E_g \approx 6.2$ eV in $LaBr_3$. At the energies $E_{exc} > 13$ eV (higher than $2E_g$) the effect of electronic excitations multiplication was detected.

The work was partially supported by HASYLAB (Projects I–20110050, II–20080119 EU) and Russian Ministry of Education and Science (grant No. 14.A18.21.0076).

References

[1] van Loef E V D, Dorenbos P, van Eijk C W E, Kramer K W and *Gudel H U* 2003 *Phys. Rev.* B **68** 045108
[2] Dorenbos P, van Loef E V D, Vink A P, van der Kolk E, van Eijk C W E, Kramer K W, Gudel H U, Higgins W M and Shah K S 2006 *J. of Lumin.* **117** 147
[3] Khodyuk I V and Dorenbos P 2010 *J. Phys.: Condens. Matter* **22** 485402
[4] Pustovarov V A, Ivanov V Yu, Vyprintsev D I and Shvalev N G 2012 *Techn. Phys. Lett.* **38** 784
[5] Pustovarov V A, Ivanov V Yu, Vyprintsev D I and Shvalev N G 2013 *Bulletin of the Russian Acad. of Science. Physics* **77** 217
[6] Vyprintsev D I *Patent of Russian Federation* 2426694, priority from 15.02.2010
[7] Chen H, Zhou C, Yang P and Wan J 2009 *J. Mater. Sci. Technol.* **25** 753
[8] Lushchik A, Feldbach E, Kink R, Lushchik Ch, Kirm M and Martinson I 1996 *Phys. Rev.* B **53** 5379

IOP Conf. Series: Materials Science and Engineering **49** (2013) 012048 doi:10.1088/1757-899X/49/1/012048

Mechanical and thermomechanical properties of radiation modified poly(ethylene-octene)/Ni-Zn ferrite nanocomposites

I Reinholds[1], V Kalkis[1], J Zicans[2], R Merijs Meri[2], I Bockovs[2], A Grigalovica[2] and G Muizzemnieks[1]

[1] Faculty of Chemistry of University of Latvia, 48 Valdemara Str., Riga, Latvia
[2] Institute of Polymer Materials, Riga Technical University, 14 Azenes Str. 14, Riga, Latvia

E-mail: ingars.reinholds@lu.lv

Abstract. Poly(ethylene-1-octene) copolymer (POE) composites filled with nickel-zinc ferrite nanoparticles have been modified by exposure to an electron beam at doses up to 500 kGy. The influence of radiation dose and ferrite content on mechanical properties has been investigated. Thermomechanical properties – thermorelaxation stresses formed in thermal heating and thermo residual stresses resulting in the process of full setting and cooling of materials have been investigated for radiation cross-linked oriented (extended up to 100%) composite samples. Increase of concentration of ferrite particles and increase of radiation dose affects a notable increase of elastic modulus and reduces the deformability in comparison to entire elastomer. Improvement of thermomechanical properties especially at low irradiation doses (100-150 kGy) have been detected for composites with increase of ferrite filler content up to 5 wt. %. It was found that gel content of POE increased up to 85% for pristine POE material with increase of irradiation dose up to 500 kGy due to the formation of cross-linked structure, increase of filler concentration up to 5 wt. % affect reduction in gel fraction due to uniform dispersion in amorphous (ethylene and substituted with hexyl branches) POE phases.

1. Introduction

Ethylene-octene copolymers (POE's) belong to a relatively new class of ethylene elastomers with high flexibility, rubber like mechanical properties and a processability of thermoplastics. Modification of cross-linkable polymer matrix like POE's by ionizing irradiation (γ-rays, accelerated electrons) can significantly improve durability of material [1]. Radiation modified cross-linkable thermoplastic polyolefins and their copolymers like metallocene-based POE's exhibit unique properties in comparison to unmodified materials: enhanced viscoelastic behavior above the melting point, increased thermal stability and improved mechanical, thermal and rheological properties as a reason of spatially formed cross-linked structure [1, 2].

There is importance from many industrial and engineering viewpoints due to the application of cross-linked POE in the manufacture of heat shrinkable materials (HSM), which are widely used in the cable and electric coatings, sealants and the other industrial elements as replacement of commonly known (HSM) based on low and high density polyethylene due to improved cross-linking efficiency of POE structures [2-4]. The previous studies have shown effect of small concentrations of liquid crystal polymers as reinforcement agents on mechanical and thermomechanical properties of HSM based on polyolefins [3]. It is predictable that inorganic filler particles are dispersing mainly in heterogeneous amorphous phase where the formation of cross-links occur that may affect increase of tensile and thermomechanical properties of polymer. The inorganic nanoparticles may act similarly by affecting reinforcement of tensile and thermomechanical properties of radiation modified POE composites. However, the incorporation inorganic nanofillers in polymer matrices possess improvement of mechanical properties such as hardness, elastic modulus and tensile strength at both low and high temperatures, as well as may affect change in dielectric, thermal and other properties [4]. Such materials filled with ferrite nanoparticles may have applications in engineering fields as microwave absorbers, special construction materials, soft magnets and shape memory materials [4-5].

In this research, POE composites with zinc-nickel ferrite nanofiller have been made. The composites have been irradiated with ionising radiation (accelerated electrons), and their tensile, structural and thermomechanical properties have been investigated. It should be noted that radiation modified POE composites with ferrite nanoparticles have not been previously described in literature.

2. Materials and investigation methods

2.1. Materials

Poly(ethylene-1-octene) copolymer with 1-octene content equal to 17 wt. % (*ENGAGE 8540*) supplied by the Dow Chemical Company (density 0.908 g/cm^3 and melt flow index 1.0 g/10 min) and nickel-zinc ferrite $Ni_{0.5}Zn_{0.5}Fe_2O_4$ nano particles with a diameter 10-30 nm, density - 4.8 g/cm^3 supplied by the Nanostructured & Amorphous Materials (USA) were used as the component materials. The composites of POE with concentrations of ferrite equal to 0, 2 and 5 wt. % were obtained by dual extrusion at 170-180°C and pressing at 190°C.

Irradiation of composite plates was performed by accelerated electrons (5MeV) in *LINAC LU4* (Latvia) in normal air at room temperature. The irradiation doses were 100, 150, 300 and 500 kGy.

2.2. Methods

The tensile tests of composite specimens were performed at 20±2°C on a Tinius Olsen H1KS UTM at a deformation speed of 1 mm/min for determination of elastic modulus E and at a deformation speed of 50 mm/min for stress-strain characteristics in accordance to ISO 527-2 standard. 7 parallel specimens have been tested.

The gel content of the irradiated composite samples was determined by evaluation of the content of insoluble fraction of cross-linked materials after solvent extraction in refluxing xylene for 48h. The extracted samples were then dried at 80°C for 24 h and weighted. Percent gel fraction was calculated as per the equation (1), where f_g is the percent gel fraction, m_i – the initial sample weight, m_g – the weight of sample after the extraction of soluble fraction. Results have been shown as the average of three parallel samples for each composition with average weight of sample 0.2 g.

$$f_g = m_i / m_g \cdot 100\% \qquad (1)$$

Thermomechanical behavior of the irradiated POE samples with 0 and 5 wt. % ferrite content, after their preliminary orientation by extension up to 100%, was investigated in the temperature range of 20 to 220°C. Specimens in the form of 5 x 5 x 2 mm strips were gradually heated and cooled under isometric conditions. The arising thermal relaxation σ_{TR} and residual thermal shrinkage σ_{RS} stresses were determined by a strain gage equipped with a 0.01 N sensitivity.

3. Results and discussion

3.1. Gel fraction

Figure 1 shows the effect of irradiation dose on gel content of pristine POE and POE composites with Ni-Zn ferrite content 2 and 5 wt. %. As a result of increase in spatial network formation (cross-linking extent), gel content increases gradually with increment of irradiation dose for pristine POE and reaches the average value of about 85% at the maximum irradiation dose (500 kGy). The values of gel content reduced for composites with increase of Ni-Zn ferrite content and with increase of irradiation dose. For example, gel fraction at 100 kGy decreased from 29 up to 24% with increase of ferrite content from 0 up to 5 wt. %, but at 300 kGy it decreased about 1.3 times – from 73 up to 56%.

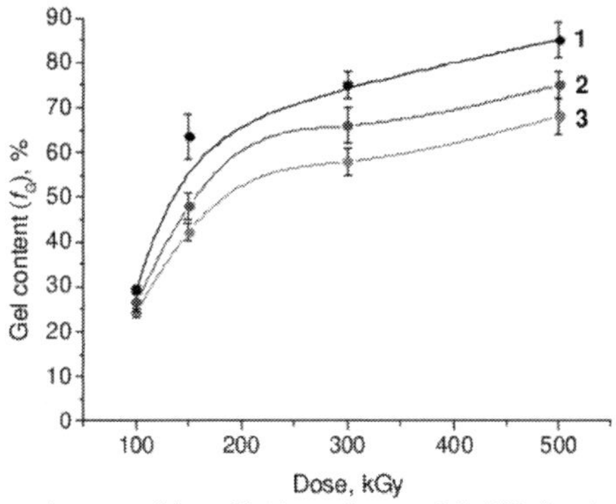

Figure 1. Gel fraction as a dependence of irradiation dose of POE-ferrite composites at ferrite concentrations 0 (1), 2 (2) and 5 (3) wt. %.

3.2. Tensile tests

The changes of tensile properties obtained from the stress-strain curves of unirradiated and irradiated up to 500 kGy POE composites are compared in Figure 2.

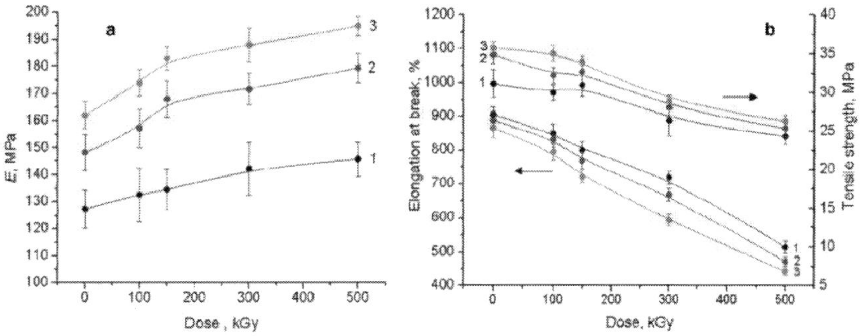

Figure 2. Elastic modulus (a), elongation at break and tensile strength (b) as a dependence of irradiation dose of POE-ferrite composites at ferrite concentrations 0 (1), 2 (2) and 5 (3) wt. %.

Incorporation of ferrite up to 5 wt. % in POE matrix affected increase of elastic modulus 1.27 times (elastic modulus increased from 127±6 MPa up to 162±5 MPa), increase of tensile strength 1.15 times and reduction of elongation at break 1.13 times.

Increment of elastic modulus with improvement of ferrite content up to 5 wt. % increased up to irradiation dose equal to 150 kGy, at which the modulus increased about 1.34 times. That is also seen in case of tensile strength and deformation at break that reach the optimal properties at 150 kGy, comparing composites with increment of ferrite content.

3.3. Thermomechanical properties

Thermomechanical curves of irradiated pristine POE and POE composites with ferrite content 2 and 5 wt. % are shown in Figure 3. The increase of thermorelaxation stress values coincides with increment of gel fraction due to increase of cross-linking. An improvement of thermorelaxation stresses σ_{TR} for irradiated up to 100 kGy composites with increase in ferrite content from 0 to 2 and 5 wt. %: the values of σ_{TR} increase from 0.33 to 0.4 MPa with increase of ferrite content up to 5 m.%.

The behavior of isometrical heating of thermomechanical curves at 100 kGy indicates an improvement of mechanical stresses with a maximum stress at orientation temperature (140°C) that increases with increment of ferrite content, a small increment of thermal shrinkage stresses also has been indicated. Increment of thermorelaxation stresses reduces with increase of irradiation dose comparing the pristine POE and it's composite with 5 wt. % that coincides with changes of mechanical properties.

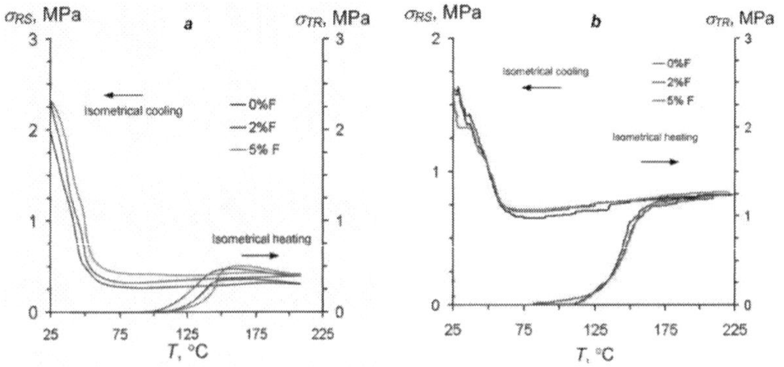

Figure 3. Thermomechanical curves of irradiated at 100 kGy and 300 kGy (b) oriented up to 100% POE-ferrite composites: dependence of thermorelaxation (σ_{TR}) and residual shrinkage (σ_{RS}) stresses at different ferrite content in composites

4. Conclusions

Structure properties of unirradiated and irradiated up to 500 kGy poly(ethylene-1-octene) composites filled with 0-5 wt.% Ni-Zn ferrite filler indicate a complex role of chain structure and crystallinity, irradiation dose and the content of ferrite particles on cross-linking efficiency of POE copolymers.

The mechanic study allow one to confirm first the reinforcement effect of Ni-Zn ferrite nanofiller for unirradiated POE elastomer composites and secondary an increase of cross-linking with increase of irradiation dose from 100 up to 500 kGy that affects increased stiffness, tensile strength and reduction of deformation at break, that even at dose as high as 500 kGy has relatively high values. Thermoshrinkable materials based on electron beam cross-linked POE composites with 0-5 wt. % Ni-Zn ferrite with improved mechanical and thermomechanical properties have been created after irradiation up to 100-150 kGy. A small effect of reinforcement of ferrite nanofiller on increase of thermorelaxation stresses has been reached for irradiated up to 100 kGy POE composites that reduces with increment of irradiation dose. That is explained with macromolecular cross-linking in POE's matrix interphase level at 300-500 kGy doses.

Acknowledgements: This work has been supported by the European Social Fund within the project "Support for Doctoral Studies at University of Latvia".

References

[1] Poongavalappil S, Svoboda P, Theravalappil R, Svobodova D, Danek M and Zatloukal M 2013 *Appl. Polym. Sci.* **128** 3026-3033
[2] Mishra J K, Chang Y W, Lee B C and Ryu S H 2008 *Rad. Phys. Chem.* **77** 675-679
[3] Kalkis V, Makimov R D and Zicans J 1999 *Polym. Eng. Sci.* **39** 1375-1382
[4] Biernholtz H, Kenig S, Dodiuk H 1992 *Polym. Adv. Tech.* **3** 125-131
[5] Buckley P R, McKinley G H, Wilson T S, Small W, Benett W J, Bearinger J P, McElfresh M W, Maitland D J 2008 *IEEE Trans. Biomed. Eng.* **53** 2075-2083

IOP Conf. Series: Materials Science and Engineering **49** (2013) 012049 doi:10.1088/1757-899X/49/1/012049

The influence of external weak magnetic field on anharmonic nanocontacts of Josephson type

D Sergeyev [1, 2] , K Shunkeyev[1], N Zhanturina[1, 3], S Shunkeyev[1]
[1]Aktobe State Pedagogical Institute, 030000, Aktobe, Kazakhstan
[2]Military Institute of Air Defense Forces, 030012, Aktobe, Kazakhstan
[3]Al-Farabi Kazakh National University, 050000, Almaty, Kazakhstan

E-mail: serdau@rambler.ru, shunkeev@rambler.ru

Abstract. The work considers the influence of weak external magnetic field on anharmonic Josephson's nanocontacts. The coordinate dependences of phase differences $\varphi(x)$, superconducting current $I_s(x)$ and magnetic field $H(x)$ anharmonic nanocontacts were found. It was defined the anharmonism parameter increase leads to phase difference $\varphi(x)$ decaying faster, and penetration of magnetic field decreases. It was shown the considered contacts react on external weak magnetic field in more passive way.

Introduction

In the recent years it became obvious that at some conditions the supercurrent dependence on phase differences in Josephson's nanocontacts contains the second harmonic, and a notable deviation from standard sinusoid dependence is observed $I_s(\varphi) = I_c \sin\varphi$ [1, 2]:

$$I_s(\varphi) = I_{c1}\sin\varphi \mp I_{c2}\sin 2\varphi, \qquad (1)$$

where I_s – supercurrent, I_{c1}, I_{c2} – the first and the second harmonics of critical current, φ – phase difference. Such objects are called anharmonic Josephson's nanocontacts. The dependence (1) is observed in high temperature superconductors due to d-wave symmetry [3] or due to bound Andreyev's states [4], in superconducting super lattices [5], in Josephson's structures on the base of exotic superconductors due to 4e current transport [6], and also in artificially created layered structures of Josephson's type [7]. The anharmonic supercurrent dependence on phase difference (1) is required for realization of φ-contact [8]. The interest to φ-contacts is explained with the wide use of such objects in modern micro- and nanoelectronics; that is why creation and research of anharmonic nanocontacts of Josephson' type is one of the urgent problems of the contemporary superconducting electronics. It, in turn, has important practical applications, for example, for the creation of the elements for a quantum computer [9].

---
[1] 34A Moldagulova avenue

IOP Conf. Series: Materials Science and Engineering **49** (2013) 012049 doi:10.1088/1757-899X/49/1/012049

The main equations

It is known the influence of magnetic field on Josephson's structures is estimated relative to the critical magnetic field [10]. If the effecting magnetic field is stronger than the critical value then Josephson junction is considered to be experienced strong magnetic field, and vice versa, less than the critical value – weak magnetic field. For the study the influence of weak external magnetic field on Josephson's nanocontacts we take Ferrel-Prange's equation (see for example [11]), describing the influence of weak field on Josephson's structures:

$$\frac{d^2\varphi}{dx^2}=\frac{1}{\lambda_J^2}\frac{I_s(\varphi)}{I_c}, \tag{2}$$

where $\lambda_J=\sqrt{\dfrac{c\Phi_0}{8\pi^2 j_c d}}$ – Josephson depth of magnetic field's penetration, Φ_0 – quantum of magnetic flux, $d=2\lambda+d'$ (λ – London's penetration depth, d' – depth of weak bound). We re-write equation (2) taking into account a model of current-phase dependence's anharmonism (1) introducing anharmonism parameter k $\left(I_s(\varphi)=I_c\sin\varphi-kI_c\sin 2\varphi\right)$ [12-14]:

$$\frac{d^2\varphi}{dx^2}=\frac{1}{\lambda_J^2}(\sin\varphi-k\sin 2\varphi). \tag{3}$$

The equation (3) can be presented as:

$$\frac{x\sqrt{2}}{\lambda_J}=\int_{\varphi_0}^{\varphi}\frac{d\varphi'}{\left(\frac{1}{2}k\cos 2\varphi-\cos(x)\right)^{1/2}}, \tag{4.1}$$

or in complex form:

$$\frac{x\sqrt{2}}{\lambda_J}=\int_{\varphi_0}^{\varphi}\frac{d\varphi'}{\left(\frac{1}{4}k\cos 2\varphi-\frac{1}{2}\cos(x)-\left(\frac{1}{4}k\sin 2\varphi\right)i+\left(\frac{1}{2}\cos(x)\right)i\right)^{1/2}}. \tag{4.2}$$

The solution of Ferrel-Prange's equation must characterize the distribution of phase difference along Josephson's structure. For the case of weak external magnetic field the following condition should be satisfied:

$$H_0\ll\frac{\Phi_0}{2\pi\lambda_J d}. \tag{5}$$

In this case weak currents will flow through Josephson's weak bound, and respectively, the phase difference $\varphi(x)$ will be little. As the qualitative aspect is important for us, we consider a modified Ferrel-Prange's equation (4.1), (4.2) for anharmonism of supercurrent, which may have approximate solution as:

$$\varphi(x)=\varphi(0)\exp\left(-\frac{x}{\lambda_J(1-2k)}\right), \tag{6}$$

or

$$\varphi(x) = 4\text{arctg}\left[\exp\left(-\frac{x}{\lambda_J(1-2k)}\right)\right]. \tag{7}$$

Further, inserting $\varphi(x)$ value in the following formulae we determine the distribution of supercurrent and magnetic field:

$$I_s(x) \sim \frac{d^2\varphi(x)}{dx^2}, \tag{8}$$

$$H(x) \sim \frac{d\varphi(x)}{dx} \text{ or } H(x) = \frac{\Phi_0}{2\pi d}\frac{d\varphi(x)}{dx}. \tag{9}$$

The figures 1, 2, 3 show the distribution of phase differences, supercurrent and magnetic field at different values of k anharmonism parameter.

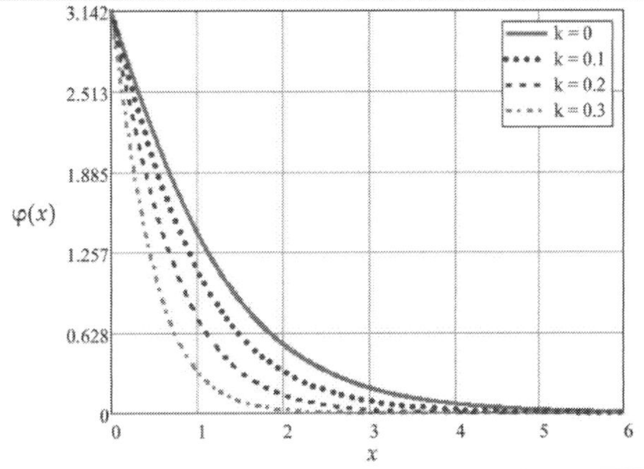

Figure 1. The distribution of phase difference at different values of anharmonism.

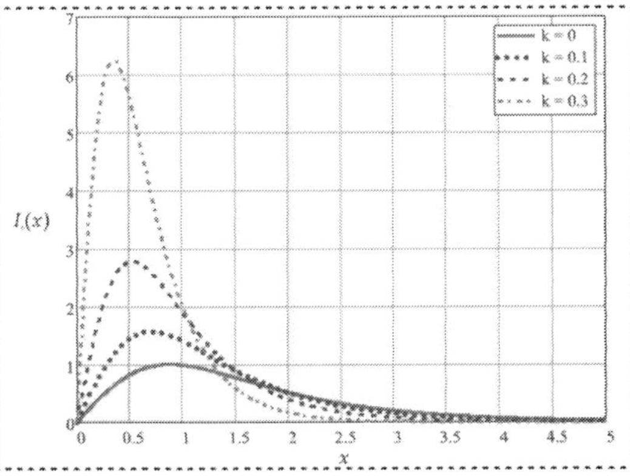

Figure 2. The distribution of supercurrent at different values of anharmonism.

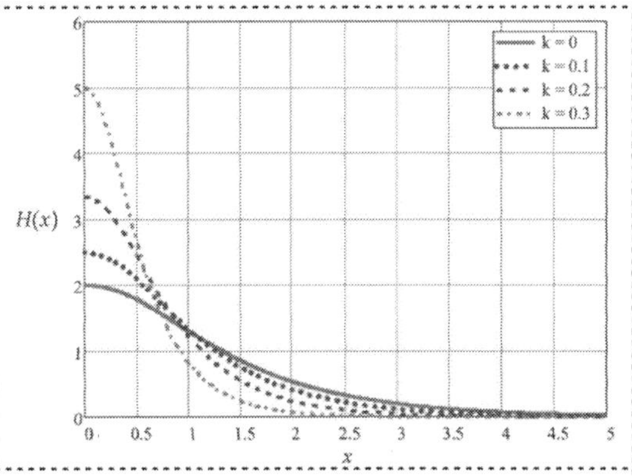

Figure 3. The distribution of magnetic field at different values of anharmonism.

Discussion

The figure 1 shows the anharmonism parameter sufficiently influences on the distribution of phase difference depending on coordinates $\varphi(x)$. At $k=0$ the change of $\varphi(x)$ is shown for harmonic case. With the increase of anharmonism parameter in Josephson's nanosystem the phase difference decay takes place faster. This happens due to the weakness of the currents flowing through Josephson's transition in anharmonic case.

The next figures 2 and 3 show the weak external magnetic field impact on anharmonic Josephson contacts more passive than in the same structures with harmonic ratio supercurrent-phase (at $k = 0$). The weakening of magnetic field's penetration into Josephson nanocontacts with the increase of anharmonism parameter is explained with the prevalence of Cooper's pairs at $T \ll T_c$. This means the value of screening supercurrent in the system increases with the anharmonism parameter, and with the increase of supercurrent (fig. 2), displaced magnetic field enlarges, respectively.

Conclusion

Thus, the present work determined the distributions of $\varphi(x)$, $I_s(x)$ and $H(x)$ on coordinates. It was discovered that weak external magnetic field impact on anharmonic Josephson's nanocontacts in more passive way compared to harmonic ones. Such Josephson's nanocontacts can be used for the creation of low cubits that interact with external magnetic fields more passive.

Acknowledgements

This work was supported by the grant funding of Ministry of Education and Science of the Republic of Kazakhstan No 1955/GF3, 1332/GF-OT.

References

[1] Klenov N V, Pugach N G, Sharafiev A V, Bakurskiy S V and Kornev V K 2010 *Phys. Solid State* **52** 2246
[2] Kornev V K, Karminskaya T Y, Kislinskii Y V, Komissinski P V, Constantinian K Y and Ovsyannikov G A 2006 *J. of Phys.: Conf. Series* **43** 1105
[3] Löfwander T, Shumeiko V S and Wendin G 2001 *Supercond. Sci. Technol.* **14** R53
[4] Komissinski P V, Il'ichev E, Ovsyannikov G A, Kovtonyuk S A, Grajcar M, Hlubina R, Ivanov Z, Tanaka Y, Yoshida N and Kashiwaya S 2002 *Europhys. Lett.* **57** 585
[5] Il'ichev E, Zakosarenko V, IJsselsteijn R P J, Schultze V, Meyer H-G and Hoenig H E 1998 *Phys. Rev. Lett.* **81** 894
[6] Gogadze G A and Omelyanchouk A N 1996 *Fiz. Nizk. Temperatur* **22** 648
[7] Sergeyev D M 2012 *Russian Phys. J.* **55(1)** 84
[8] Bakurskiy S V, Klenov N V, Karminskaya T Yu, Kupriyanov M Yu and Golubov A A 2013 *Supercond. Sci. Technol.* **26** 015005
[9] Ioffe L B, Geshkenbein V B, Feigel'man M V, Fauchere A L and Blatter G 1999 *Nature* **398** 679.
[10] Likharev K K 1986 *Dynamics of Josephson Junctions and Circuits* (New York: Gordon and Breach Science Publishing)
[11] Ferrell R and Prange 1963 *Phys. Rev. Lett.* **10** 479
[12] Goldobin E, Koelle D, Kleiner R and Buzdin A 2007 *Phys. Rev.* B **76** 224523
[13] Sergeyev D M 2013 *Solid State Phenomena* **200** 272
[14] Sergeyev D M and Shunkeyev K Sh 2009 *Russian Phys. J.* **8/2** 556

Dielectric properties of $BaTiO_3$ based materials with addition of transition metal ions with variable valence

D Sitko[1], W Bąk[1], B Garbarz-Glos[1], A Budziak[2], C Kajtoch[1] and A Kalvane[3]

[1]Institute of Physics, Pedagogical University of Cracow, Poland
[2]The H. Niewodniczanski Institute of Nuclear Physics PAN, Cracow, Poland,
[3]Institute of Solid State Physics, University of Latvia

E-mail: sitko.dorota@gmail.com

Abstract. Dielectric properties of $BaTiO_3$(BT), $BaTiO_3$+0.1wt.%MnO_2(BTMn) and $BaTiO_3$+0.1 wt.%Fe_2O_3 (BTFe) solid solution was investigated. The effects of substitution on the microstructure and on the electric properties of $BaTiO_3$ samples were studied by performing X-ray diffraction and electric measurements. X-ray diffraction analysis of the samples shows the formation of single-phase compound with a tetragonal structure at room temperature. SEM microphotographs exhibit the uniform distribution of grains. The electric properties (a real part of electrical permittivity(□ε') and an imaginary part of electric modulus (M")) were investigated in the temperature range from 150K to 600K and in the frequency range from 0.1 Hz to 10 MHz. It was found that a Fe and Mn ions -doping influence the electric properties of the investigated polycrystalline materials.

1. Introduction

Doped ceramics allows the control of electric, electromechanical and electro-optic properties of ferroelectric functional materials such as $BaTiO_3$. By a proper substitution attractive materials for capacitors and transducers with a wide range of applications from power engineering to medical-device technology [1, 2]. In many recent papers [3, 4] the influence of substitutions on electric properties and the phase transition temperature of BT ceramics have been extensively investigated.

A large number of different dopants can be accommodated in the $BaTiO_3$ lattice. The ionic radius of dopants is the parameter which mainly determines the substitution site. According to Tsur for ionic radii, r ≤0.87Å, the dopants occupy the B-site. For ions of the transition metals such as a Mn^{3+} (0.78Å) and Fe^{3+} (0.64 Å) ions it is well established that they preferentially were substituted on the Ti^{3+}(0.74Å) site [5-7]. Additionally, the dopants, event at small concentration, can cause lattice deformations [8-11] as well as changes in the charge balance of the material. The oxygen vacancies are responsible for a charge compensation , involves the creation of a stoichiometric amount of oxygen vacancies, so they behave as acceptors of dopants [12]. The oxygen vacancies could be easily created by a loss of oxygen from a crystal structure during sintering at high temperature. The aim of this work is clarify the influence of Mn and Fe ions on the dielectric properties of $BaTiO_3$.

2. Experimental procedure

The $BaTiO_3$+ 0.1 wt.% x (x=MnO_2, Fe_2O_3) polycrystalline samples were prepared by a solid state synthesis [13] in the Institute of Solid State Physics University of Latvia.

The X–ray studies were performed with the use of an X'Perrt PRO (PANalytical) diffractometer using the $Cu_{K\alpha}$ radiation ($\lambda = 0.154178$ nm) and a graphite monochromator. The patterns were obtained at the rising temperature. To analyse and fit the spectra, the program FULLPROF based on Rietveld method was used. The investigations of the microstructure of specimens were performed using the Hitachi S4700 SEM (Scanning Electron Microscope) with field emission and the Noran Vantage EDS (Energy-Dispersive X-ray Spectroscopy) system. The X-ray microanalysis EDS was applied to investigate the homogeneity of the composition and the EPMA (Electron Probe Microbeam Analysis) was applied to analyse the distribution of elements on the sample surface. The investigations of chemical composition were performed using the Noran - Vantage microanalyses which is a part of the Hitachi scanning electron microscope. All the electric measurements of the samples were performed with the use of an Alfa – AN modular measurement system with a temperature control system Quatro Krio 4.0 range from 0.1Hz to 10 MHz.

3. Results and discussion

Surface SEM investigations were performed on doped (BTMn, BTFe) and undoped (BT) samples. The parameters of the elementary cell of all samples were calculated. From the obtained results was stated the elementary cell parameters are: a = b = 3.9943Å, c= 4.0354Å, a = b = 3.9996Å, c = 4.0339Å and a = b = 3.9995Å, c = 4.0338Å for BT, BTFe and BTMn respectively. The space group of BT, BTFe and BTMn samples at room temperature is P4mm. The SEM images of $BaTiO_3$ show that the microstructure consists of intergranular pores and grains of various size. Moreover, one can observe that substitutions have an inhibiting effect on the grain growth process and, consequently, a relative homogeneous microstructure, which reveals a higher amount of an intergranular porosity and the lower size of grains than those ones in a non-modified sample. This may be due to the nucleating behavior of Mn, Fe ions in the $BaTiO_3$ ceramics. Figures 1 indicates that the addition of transition metal ions decreases the size of grains. The EDS investigations showed that BTMn and BTFe ceramics contained Ba, Ti, O, Mn and Fe elements near their surfaces and no other impurity elements were detected in the spectrum. The homogeneity of element distribution in the samples was investigated by an EPMA method with use of an X-ray microprobe. The obtained "mappings" confirmed the qualitative composition of the examined samples.

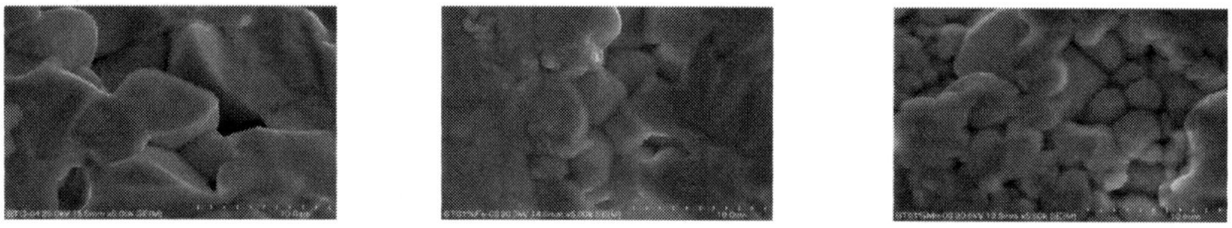

Figure 1. The SEM images of BT, BTFe and BTMn ceramics (magnification 5 000x).

A dielectric spectroscopy is a convenient and widely used method for the electrical properties characterization of ceramic ferroelectrics. In particular, a.c. measurements in a wide frequency range (10^{-1}Hz ÷ 10^{7}Hz) give reliable characteristics of both: a bulk (grain) and a grain boundary dielectric polarization dynamic. The dependence of real part of electric permittivity (ε') on temperature at the constant value of the frequency (10 kHz) electric field of applied for the investigated samples was studied. The maxima of electric permittivity ($\varepsilon'(T)$) seen in the figure 2 are observed at the temperatures corresponding to the temperatures of phase transitions. In the case of BT – ceramics, these temperatures are well correlated with the temperature obtained by calorimetric measurements (not shown here). For BTMn material, the value of ε' is damped down in the whole temperature range but the temperatures of phase transitions are the same as for those BT. For the BTFe material the temperature of the paraelectric – ferroelectric transition is almost the same as for BT. The low temperature transitions posses a diffusion character with the maximum value of ε' occurs at a temperature of 253K. The frequency dependencies of real part of electric permittivity ($\varepsilon'(\nu)$) at 600K for the investigated samples are presented in figure 3. The measured frequency dependence of ε' consists of two characteristic regions: one of frequency above and the other below $\sim 10^4$ Hz. For the low frequency region, the ε' values of all investigated samples are high values and they are frequency dependent.

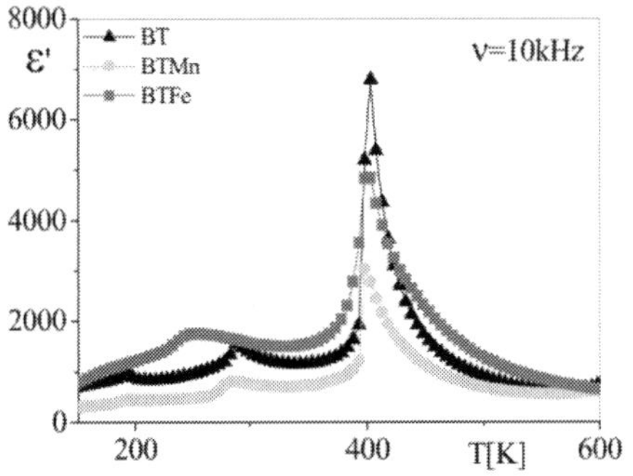

Figure 2. The temperature dependence of ε' for BT, BTFe and BTMn.

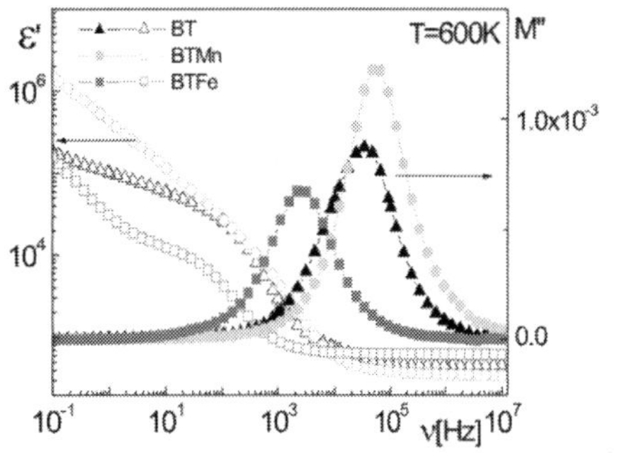

Figure 3. The frequency dependence of ε' and M" for BT, BTFe and BTMn.

For frequencies higher than $\sim 10^4$ Hz, the ε' values observed in the same scale show the lack of frequency dependence. The same two frequency regions can be defined also for an electric modulus (imaginary part of M"). The values of M" indicate the weaker polarization processes which were identified as connected with a bulk polarization. The electric modulus M* and complex permittivity ε* are related by the following formula: $M^*=1/\varepsilon^*=M'+iM''$. The dependence of imaginary part of the electric modulus (M") on frequency at selected temperature (423K, 473K, 523K and 573K) is presented in figure 4(a). The increase in temperature cause two effects: the first one is the increase of the M" peak values up to about 5 times, and the second one the shift of the curves in the figures towards higher frequencies. The characteristics of M"(ν,T) at temperature: 273K, 323K, 373K and 393K, for the same sample are presented in figure 4(b). At the lower temperatures the M" behaviour is not a regular function of temperature and frequency and the M" values are of one order of magnitude smaller than those in figure 4(a). On the basis of the data presented in figure 4 the relaxations times were calculated using the $\omega\tau = 2\pi\nu_m\tau = 1$ equation, where ν_m – value of frequency for the M" peaks at fixed temperature. The collected results are presented as Arrhenius plots.

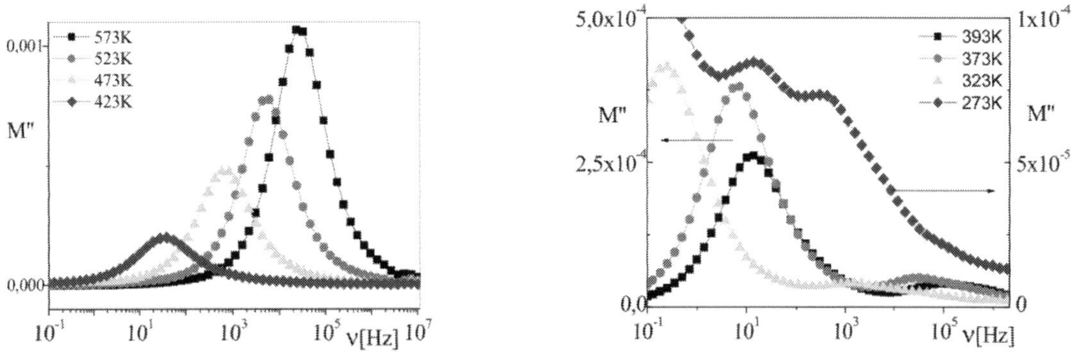

(a) (b)

Figure 4. The frequency dependence of MO for BTMn at the temperature: (a) 423K, 473K, 523K and 573K (b) 273K, 323K, 373K and 393K.

IOP Conf. Series: Materials Science and Engineering **49** (2013) 012050 doi:10.1088/1757-899X/49/1/012050

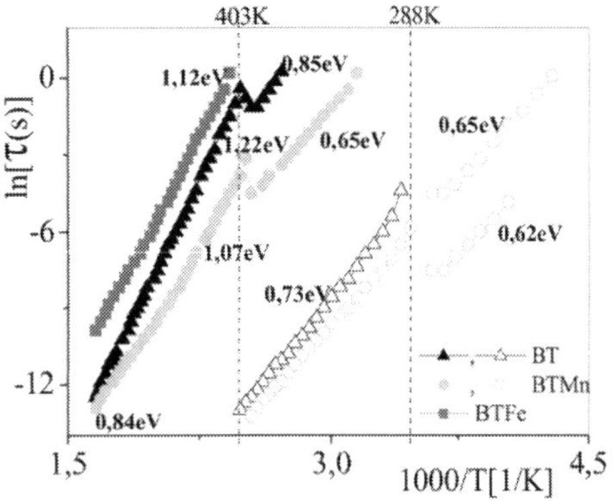

Figure 5. Arhenius plot of relaxation times for BT, BTFe and BTMn samples (open symbols –grain, solid symbols – grain boundary).

The thermal activation energies for all investigated materials are shown in figure 5. The visible breaks seen in figure 5 occur at the temperature of phase transitions for all samples. The Fe dopant shifts the curves towards higher temperatures compared with the BT sample. The Mn dopant causes opposite shift and appearance of next values of thermal activation energy. These values of the activation energies may be related to the electron excited process from the oxygen vacancy.

4. Conclusions

The structural and electric properties of titanate ceramics prepared by a solid state synthesis were investigated. The space group (P4mm) and the lattice parameters were obtained. EDS and EPMA measurements were used for a microstructure and chemical analysis. SEM investigation reveals a fine grain and dense microstructure. No significant impurities were detected in an EDS spectrum and the samples are in good stoichiometric ratio. The obtained dielectric data suggest that, the substitution of transitions metal ions has a significant influence on the value of real part of electric permittivity (ε') and character of the electric modulus M*. The thermal activation energies for investigated samples which characterized the electrical properties of grains and grain boundaries were calculated.

5. References

[1] Setter N and Waser R 2000 *Acta Mater.* **48** 151.

[2] Heartling G H 1999 *J. Am. Ceram. Soc.* **82** 797.

[3] Garbarz-Glos B, Bak W, Molak A, and Kalvane A 2012 *Phase Transition* **1-9** iFirst.

[4] Garbarz-Glos B, Bormanis K and Sitko D 2011 *Ferroelectrics* **424**, 36

[5] Tsur Y, Dunbar T D and Randall C A 2001 *J. of Electroceramics* **7**, 25

[6] Hagemann H J and Hennings D 1981 *J. Am. Ceram. Soc.* **64** (10) 590

[7] Pontin R G, Slade E F and Ingram D J E J 1969 *Phys.* **C2** 1146.

[8] Kirianov A, Ozaki N, Ohsato H, Kohzul N, and Kishi H 2001 *Jpn. J. Appl. Phys.* Part 1 **40**, 5619.

[9] Keith G M, Rampling M J, Sarma K, Alford N Mc, and Sinclair D C 2004 *J. Eur. Ceram. Soc.* **24** 1721

[10] Hennings D F K 2001 *J. Eur. Ceram. Soc.* **21** 1637.

[11] Buscaglia M T, Buscaglia V, Viviani M, Nanni P and Hanuskova M 2000 *J. Euro. Ceram. Soc.* **20**, 1997

[12] Prades M, Maso N, Beltran H, Cordoncillo E, West A R 2008 *Am. Ceram. Soc.* **91**, 2364

[13] Sitko D, Bąk W, Garbarz-Glos B, Antonova M and Jankowska-Sumara I 2012 *Ukr. J. Phys. Opt.* **V13** 34

Optical properties of natural topaz

. Skvortsova, N. ironova- lmane, L. rinkler, G. C ikvaidze

Institute of Solid State Physics, University of Latvia

e-mail: vera@cfi.lu.lv

Abstract. The results of investigation of infrared, Raman and UV-Visible absorption spectra of natural topaz crystals from Ukraine before and after fast neutron irradiation are presented. We assume that the ~ 620 nm band in topaz crystals is associated with the presence of Cr^{3+}, Fe^{2+} and Mn^{2+} impurities. The broad band with maxima at 650 cm^{-1} observed in Raman spectra for topaz irradiated by fast neutrons may be connected with lattice disorder. Exchange interaction between radiation defect and impurity ions during neutron irradiation leads to appearance of additional absorption band in UV-VIS spectra and bands broadening in infrared and Raman spectra of investigated crystals.

1. Introduction

The study the optical properties of topaz can be used to determine the quality of jewelry topaz, for the successful use of these crystals in thermoluminescence dosimetry [1] as well as for evaluation of physical and chemical conditions of formation of mineral [2]. Topaz is an aluminium fluorosilicate with a fairly constant chemical composition $Al_2SiO_4(OH,F)_2$. The only major variation found in different samples is related to the [OH]/[F] concentration ratio. The structure of topaz consists of SiO_4 groups linking octahedral chains of $Al[O_4(F,OH)_2]$ in a zigzag fashion parallel to the crystalline c-axis. Four of six anions surrounding Al^{3+} ion belong to SiO_4 tetrahedron and the remaining two anions are F^- or OH^- group. Topaz crystallizes in the orthorhombic system, space group Pbnm (D^{16}_{2h}) [3, 4] with four molecules per unit cell, but the OH/F substitution turns its symmetry into triclinic (P1) [5]. The lattice parameters are a=4.6499 A°, b=8.7968 A° and c=8.3909 A° and is normally found as well-developed prismatic crystals with pyramidal terminations.

Usually in nature topaz occurs in the form of colorless crystals. Coloration of topaz crystals is possible through the incorporation of transition metal impurities or by irradiation. Blue topaz now is one of the gem industry's most commercialized gemstones. The origin of the blue color in topaz is not well understood.

The paper presents study of infrared, Raman and UV-Visible absorption spectra of natural topaz crystals from Ukraine before and after fast neutron irradiation.

2. aterials and et ods

Natural colorless and blue topaz from Ukraine was investigated. The sample thickness was in the range of 1.8 –3.4 mm.

The neutron irradiation was performed at the Latvian 5 MW water-water research reactor. The fluence of fast neutrons with energy > 0.1 MeV was in the range $10^{14} – 5{\cdot}10^{17}$ cm^{-2}. A cadmium filter

IOP Conf. Series: Materials Science and Engineering **49** (2013) 012051 doi:10.1088/1757-899X/49/1/012051

was used for absorption of thermal neutrons. The sample temperature did not exceed 40°C during the neutron irradiation.

Infrared spectra were measured at room temperature using VERTEX 80v vacuum FT-IR spectrometer with the actively aligned UltraScan™ interferometer, which provides peak spectral resolution. Raman spectra were measured by Raman Spectrometer Advantage 785 (USA) with laser 785 nm.

The standard technique to measure absorption spectra has been based on a "Specord 210" (Analytikjena) double-beam spectrophotometer operating in the spectral region of 190-1100 nm. Optical measurements before and after irradiation were carried out at the temperature 300 K.

3. Results and Discussion

The absorption spectra of topaz single crystals before irradiation (curve 1), after fast neutron irradiation (curve 2) and natural blue topaz from Volodarsk Volynskii, Ukraine (curve 3) are given in figure 1. Before fast neutron irradiation in absorption spectra of colorless topaz there was observed only one band at 230 nm. After irradiation the absorption band 230 nm intensity increases and there appear bands with maxima at 305, 430 and 620 nm and crystals obtain blue color. In natural blue topaz there was identified an intensive band at 230 nm and less intensive bands at 430 and 620 nm. Two lines around 230 and 305 nm belong to a single defect such as silanone (=Si=O) as it was suggested in [6] basing on similar lines observed in spodumene. Blue color is connected with a broad absorption band in the red part of the spectrum generated by the so-colled R-centers (two F-vacancies with two trapped electrons) [7]. It was previously concluded that the optical absorption band centered at 620 nm is closely correlated with an O^- center interacting with two Al ions of the topaz structure. This O^- center is produced by the irradiation in the hydroxyl sites which substitutes for fluorine in the topaz structure [8, 9]. As suggested Priest V. at al [10] the blue defects in neutron damaged topaz is the doubly occupied dangling silicon bond. The calculations performed in [11] using $AlCrFH_7O_{11}Si$ model molecule with unrestricted CIS method to simulate UV and visible spectra shows that the theoretical spectrum of the model is in a considerably good agreement with experimental one. Excitation spectra of Cr^{3+} investigated by Tarashchan A N at al. [12] in topaz crystal show the same lines. We assume that the band ~ 620 nm includes not only the above mentioned bands, but also the bands associated with the presence of impurities of Cr^{3+}, Fe^{2+} and Mn^{2+} ions. Band 430 nm also belongs to impurity Cr^{3+}ion.

The IR spectra of the topaz sample before and after fast neutron irradiation are shown in figure 2.

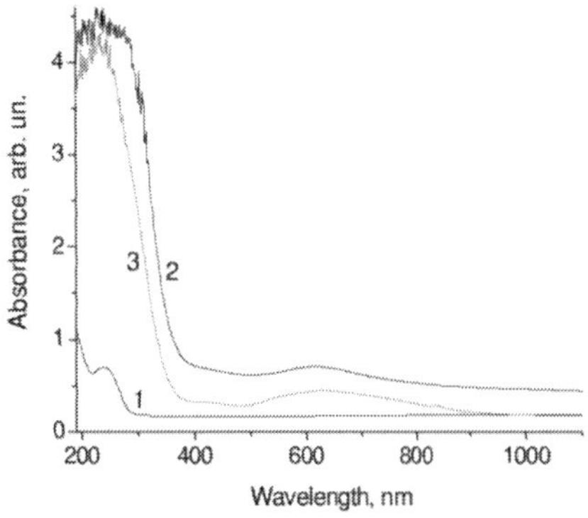

igure 1. Topaz absorption spectra before (1) and after neutron irradiation (2), (3) natural blue topaz from Volodarsk Volynskii, Ukraine.

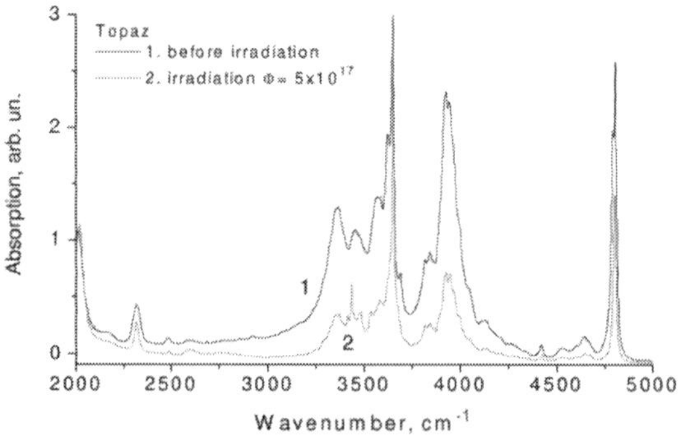

igure 2. Infrared spectra of topaz crystals:

1. before fast neutron irradiation, 2. after irradiation $\Phi = 5 \cdot 10^{17}$ cm^{-2}.

The hydroxyl ions are normally configured into the mineral assembly as crystallization water or as hydroxyl ion replacing other ions of the structure. Depending on its origin topaz always contains hydroxyl groups which substitute for fluorine atoms. The ratio OH/(OH+F) varies from almost zero up to thirty percent. When the OH content becomes sufficient, the structure is distorted to triclinic or monoclinic symmetry. OH- stretching mode is observed on wave number around 3650cm^{-1}, with a strong asymmetry around 3622 cm^{-1}. The observed OH stretching vibrations are in good agreement with the position of approximately 3650 cm^{-1} for band observed in [13, 14] and calculated by Churakov [15]. The band at 3650 cm^{-1} was decomposed in two components at 3639 and 3650 cm^{-1}. These two peaks were assigned to two vibrations (stretching modes) of the OH molecules substituting for F [16]. Additionally, a broader and weaker band was observed around 3450 cm^{-1} assigned earlier due to anomalous OH group [17]. Before irradiation of colourless topaz in infrared spectra there were observed two bands around 4809 and 4787 cm^{-1}, which are the combination of main (OH) stretching and Al-OH bending modes (1163 cm^{-1}) as suggested in [18]. Fast neutron irradiation reduces this band intensity and band became symmetrical with maximum 4802 cm^{-1}. Bands around 3950 cm^{-1} were observed. After fast neutron irradiation bands intensities decrease, and only the 3650 cm^{-1} band intensity increases. Neutron irradiation produces the electron and hole defects in topaz crystals. They probably interact with impurities and hydroxyl ions and reduce the local symmetry

Raman spectra of natural blue topaz, colourless topaz and irradiated by fast neutron at fluence $\Phi = 10^{17}cm^{-2}$ and $5x10^{17}$ cm^{-2} are shown in figure 3. Spectra of natural blue and colourless topaz analysed in this work are comparable to those found in the literature. The bands observed around 276-293 cm^{-1}, 561 cm^{-1}, 858cm^{-1}, 929cm^{-1}, 1006cm^{-1} are assigned to the various Si-O vibration modes in SiO_4 group based on the work of Griffith [19], Beny and Piriou [16] and Souza [20], the bands between 235 and 380 cm^{-1} are considered to be caused by the symmetric Si-O ring deformation, and the bands between 300 and 400 cm^{-1} are satellite bands [13]. The series of sharp bands in the 330 - 500 cm^{-1} range are assigned with stretching and bending modes of [AlO_6] octahedral coupled with the bending modes of [SiO_4] tetrahedra. The band observed at 325cm^{-1} is assigned to Al-F stretching modes [13].

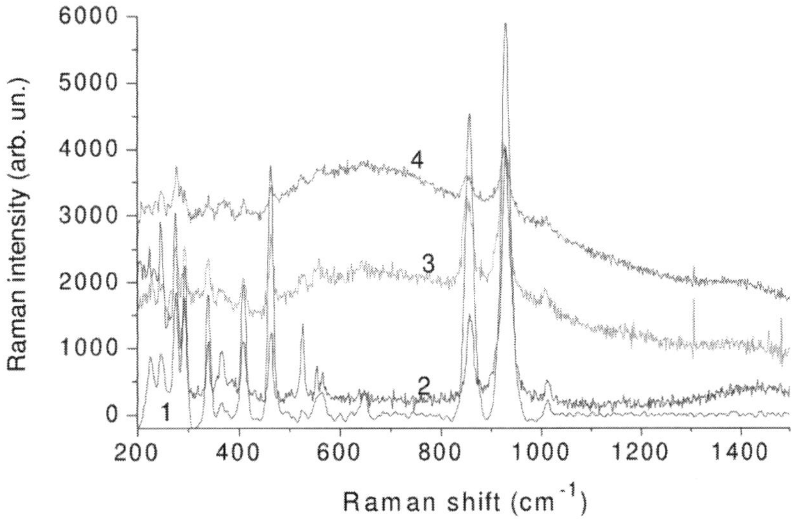

igure 3. Raman spectra of topaz crystals: 1. natural blue, 2. before and 3. after fast neutron irradiation $\Phi = 10^{17}$ cm^{-2}, 4. $\Phi = 5\cdot10^{17}$ cm^{-2}.

The Raman spectrum of colourless topaz (figure 3) before irradiation is the same as that of the natural blue topaz. Intensities of some bands for these two crystals slightly differ. After fast neutron irradiation the broad band with maxima at 650 $cm^{-1}$are observed. The narrow lines intensities decrease (figure 3, curve 3 and 4) with fast neutron fluence increasing. The narrow lines intensities decrease and broad band appearance may be connected with topaz lattice disorder.

4. Conclusion

The infrared, Raman and UV-Visible absorption spectra of natural topaz crystals from Ukraine before and after exposure to a fluence of fast neutrons up to $5\cdot10^{17}cm^{-2}$ (E>0.1 MeV) are investigated. We assume that the band ~ 620 nm in topaz crystals is associated with the presence of impurities of Cr^{3+}, Fe^{2+} and Mn^{2+} ions. The broad band with maxima at 650 cm^{-1} observed in Raman spectra for irradiated by fast neutron topaz may be connected with lattice disorder. Exchange interaction between radiation defect and impurity ions during neutron irradiation leads to appearance of additional absorption band in UV-VIS spectra and bands broadening in infrared and Raman spectra of investigated crystals.

5. Ackno ledgments

This work was supported by Latvian Science Council Grant No 402/2012 and ERDF project Nr.2DP/2.1.1.2.0/10/APIA/VIAA/010.

References

[1] Magalhaes C M S, Macedo Z S, Valerio M E G, Hernandes A C and Souza D N 2004 *Nuclear Instruments and Methods in Physics esearch* **218** 277
[2] Ivanova O A and Korovkin M V 2001 *I vesti a Toms ogo olitehniches ogo universiteta* **3 4** 85 (in Russian)
[3] Ribbe P H and Gibbs G V, 1971 *Am. Mineral.* **5** 24
[4] Northrup P A, Leinenweber R and Parise J B 1994 *Am. Mineral.* **9** 401
[5] Pinheiro M V B, Fantini C, Krambrock K, Persiano A I C, Dantas M S S and Pimenta M A 2002 *Phys. e .* ***65***, 104301
[6] Bonventi W Jr, Isotani S and Albuquerque A R P L 2012 *Advances in Condensed Matter Physics* **Article ID 8 38 4**. doi:10.1155/2012
[7] Platonov A N, Taran M N and Balyatskii V S 1984 *Nature of color gems* (Moscow: Nedra) (in Russian).
[8] Da Silva D N, Guedes K J, Pinheiro M V B, Spaeth J-M and Krambrock K 2005 *Phys. Chem. Minerals* **32** 436
[9] Leal A S, Krambrock K, Ribeiro L G M, Menezes M A B C, Vermaercke P and Sneyers L 2007 *Nuclear Instruments and Methods in Physics esearch A* **58** 423
[10] Priest V, Cowan D L, Reichel D G and Ross F K 1990 *. A lied Physics* **8** 3030
[11] Goto H, Niwa A, Greenhidge D C, Kato N, Ida T, Mizuno M, Endo K and Tada T 2005 *. of Surface Analysis* **12** 249
[12] Tarashchan A N, Taran M N, Rager H and Iwanuch W 2006 *Phys. Chem.Minerals* **32** 679
[13] Kloprogge J T and Frost R L 2000 *S ectrochim. Acta A* **5** 501
[14] Wunder B, Rubie D C, Ross C R, Medenbach O, Seifert F and Schreyer W 1993 *Amer. Mineral.* **8** 285
[15] Churakov S V and Wunder B 2004 *Phys. Chem. Minerals* **34** 131
[16] Beny J M and Piriou B 1987 *Phys. Chem. Minerals* **15** 148
[17] Aines R D and Rossman G R 1986 *Amer. Mineral.* **1** 1186
[18] Prasad P S R and Gowd T N 2003 *eological Society of India* **1** 202
[19] Griffith W P 1969 *Chem. Soc. A* 1372
[20] Souza D N, Lima J F, Valerio M E G, Fantini C, Pimenta M A, Moreira R L and Caldas L V E 2002 *Nuclear Instruments and Methods in Physics esearch* **191** 230

IOP Conf. Series: Materials Science and Engineering **49** (2013) 012052 doi:10.1088/1757-899X/49/1/012052

Dielectric properties of lithium niobate single crystals doped with gadolinium

M N Palatnikov[1], N V Sidorov[1], V A Sandler[2], K Bormanis[3], I Smeltere[3]
[1]Institute of Chemistry, Kola Science Centre RAS, Apatity, Murmansk Region, Russia
[2]Ivanovo State University, Ivanovo, Russia
[3]Institute of Solid State Physics, University of Latvia, Riga, Latvia

E-mail: ilze.smeltere@cfi.lu.lv

Abstract. In this paper dielectric properties of $LiNbO_3$:Gd single crystals at temperatures ~ 290–490 K at narrow frequency range (0.5–10^6 Hz) are presented. The influence of growth conditions was also investigated. The anomalies on dielectric plots depend on the development of the micro- and nano-domain structure.

1. Introduction

Lithium niobate $LiNbO_3$ (LN) crystals have been widely used for tailoring optical and piezoelectric devices. They are used in many comercial applications however the intrinsic defects asociated with the structure and growth conditions influence the properties of this material. A number of anomalies of conductivity, optical, dielectric and pyroelectric properties of lithium niobate crystals within a temperature range of 300—400 K has been observed by different authors [1-8]. A super-structural sub-lattice of clustered defects formed at the cation sub-lattice of rare earth elements REE-modified LN crystals is observed to have steps of the size of a few lattice translation periods [9]. There are clusters formed close to the defects (Nb_{Li}) having a pattern with periodic nano-size fractal structures of steps between 1 – 2 nm [10].

In the present work the influence of gadolinium Gd on the observed dielectric anomalies in $LiNbO_3$ single crystal was investigated.

2. Results and discussion

LN single crystals were grown under stationary and non-stationary conditions. The dielectric dispersion of $LiNbO_3$:Gd single crystals was investigated. Measurements were performed at temperatures ~ 290–490 K within narrow frequency range (0.5–10^6 Hz). Short periodical temperature oscillations at crystal – fusion phase boundary within the crystal growth promotes the forming of contrary polarized domains [11-17]. Domain boundaries are characterized by a complex structure. The examination of lithium niobate crystals containing gadolinium ($LiNbO_3$:Gd) has revealed formation of regular micron-scale domain structures (RDS) characteristic for this kind of crystals (Fig. 1).

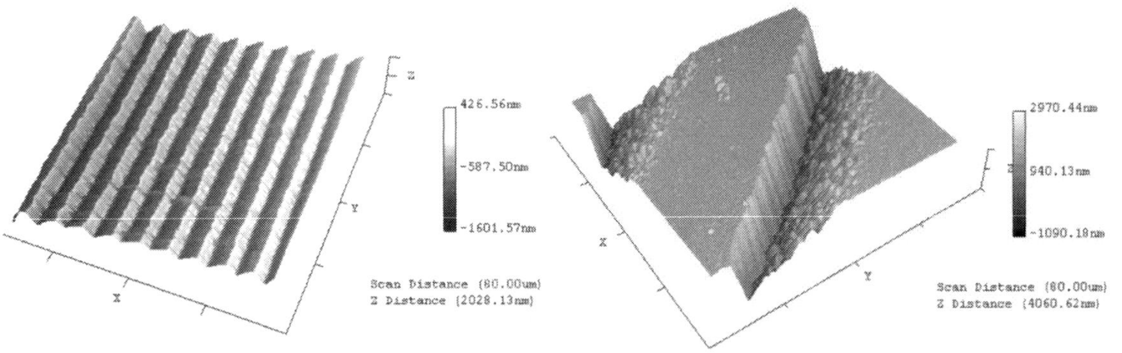

Fig. 1. Regular domain structure in non-stationary grown $LiNbO_3$:Gd [Gd=0.44 wt%] single crystal. Image is taken with Atomic force microscope Nano R2[TM]

There is an anomaly on the dielectric permittivity plot $\varepsilon'_{33}(T)$ observed at around 330-380 K. Dielectric permittivity decreases with increasing frequency and disappears at $f \geq 10$ kHz. Dielectric dispersion at low frequencies follows Debye equation. Fig. 2 shows the Cole-Cole diagram for $LiNbO_3$:Gd single crystal.

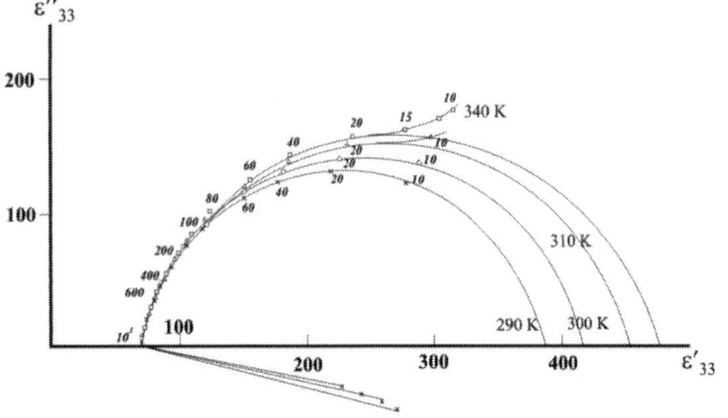

Fig.2. Cole-Cole diagram for $LiNbO_3$:Gd [Gd=0.44 wt%, z-orientation] single crystal in the temperature range from 290 to 400 K. Frequencies are shown in Hz at curves

Monodispersion is located at temperatures 290-340 K in the frequency range 10 Hz to 10 kHz and can be described with an equation (1):

$$\varepsilon^*(\omega) = \varepsilon' - i\varepsilon'' = \varepsilon_\infty + \frac{\varepsilon_0 - \varepsilon_\infty}{1 + (i\omega\tau)^{1-\alpha}} \qquad (1)$$

where ε^* is the complex dielectric permittivity, ε' and ε'' are he real and imaginary parts of the dielectric permittivity, ω – the angular frequency, τ – relaxation time.

The dielectric dispersion nature is changed with increasing temperature. These changes appear as linear areas on the Cole-Cole diagram at low frequencies. These linear areas become more pronounced and extended with increasing temperature. Relaxation becomes longer in time. Dielectric dispersion following the equation (1) is suppressed above 410 K (Fig. 3). It means that there are two dispersion processes present at 290~410 K: Debye processes (I) and the linear area at lower frequencies (II). When temperature exceeds 490 K only the process (II) appears on Cole-Cole diagram.

The depth of dispersion does not change essentially while increasing temperature from 290 K to 340 K. Further increase in temperature above 340 K (even 4 K) drastically decreases the depth of dispersion.

$LiNbO_3$:Gd crystal is associated with one Debye relaxation process in temperature range from 290 K to 340 K with a relaxation time $\tau \sim 2.5 \cdot 10^{-2}$c at room temperature.

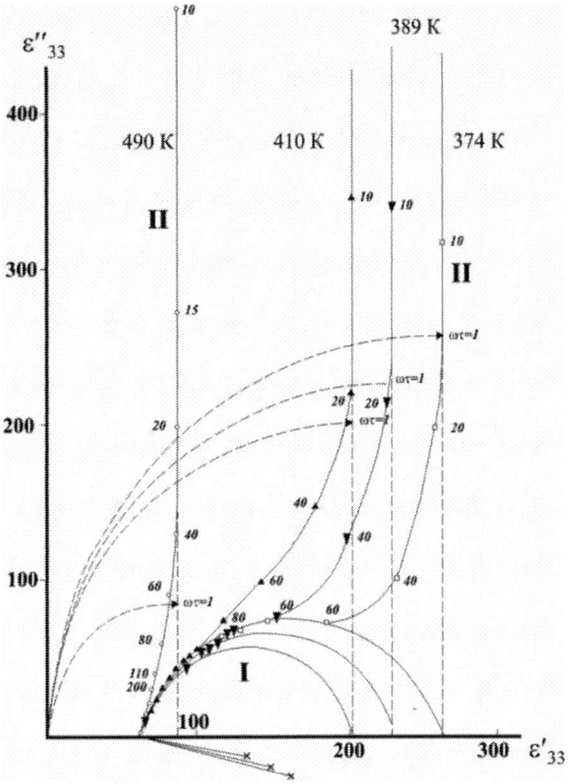

Fig.3. Cole-Cole diagram for $LiNbO_3$:Gd [Gd=0.44 wt%, z-orientation] single crystal in the temperature range from 374 to 490 K. Frequencies are shown in Hz at curves

The Debye process (I) disappears after holding the sample at 340 – 350 K for 4 hours. The investigation of dielectric dispersion under different voltage (0 – 10 kV/cm) in bias field shows decrease of dispersion depth. At the same time there is no evidence for the Debye nature change.
The observations mentioned above allow us to assume that the annomalies are caused by the interaction of spontaneous polarization relaxation and defect relaxation (associated with impurity of Gd^{3+}) with periodic domain or nanostructure boundaries. Constant electric field leads to domain growth and consolidation of domains causing the decrease of dispersion depth.
Stationary grown $LiNbO_3$:Gd (0.52 wt%) without developed micro- and nano- domain structure was also examined. Figure 4a shows two relaxation processes at T=290 K in frequency range of 1 Hz to 1 kHz. The first is Debye relaxation, the second emerges as a straight line associated with the static conductivity of the sample. The Debye process disappears instantly with a slight rise in temperature (Fig. 4b). Numerically dispersion depth is much higher for samples grow non-stationary. Smaller dispersion depth is the evidence of the unipolarity of a crystal.

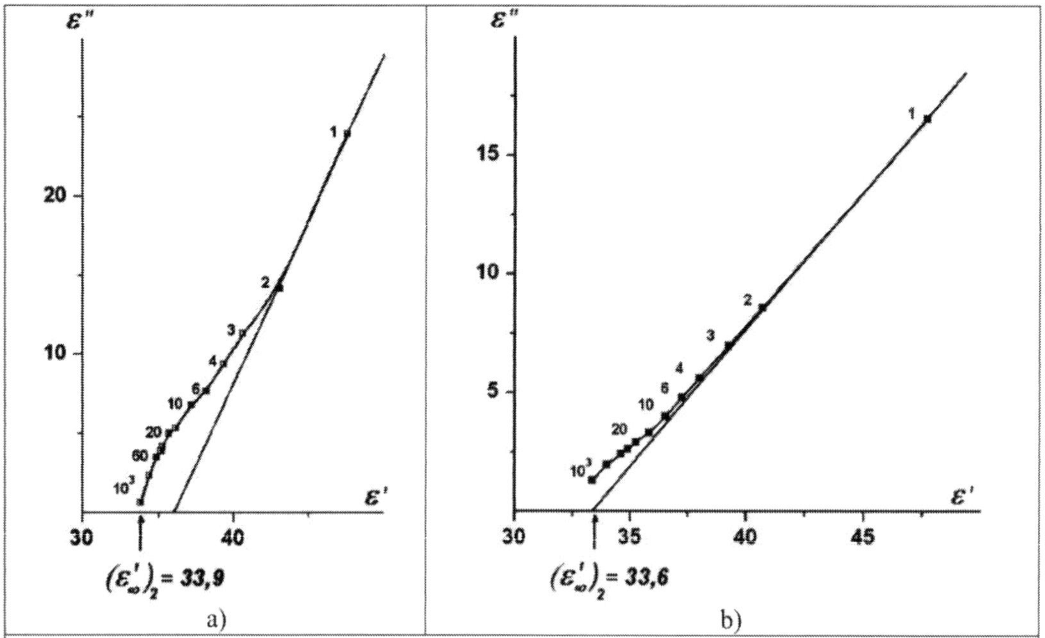

Fig.4. Cole-Cole diagram for stationary grown $LiNbO_3$:Gd (Gd=0.52 wt%): a) measured at 305 K; b) 314 K

3. Conclusions

Dielectric properties of stationary and non-stationary grown $LiNbO_3$:Gd single crystals at temperature ~ 290–490 K in narrow frequency range (0.5–10^6 Hz) were investigated. Interaction of spontaneous polarization relaxation and defect relaxation with periodic domain or nanostructure boundaries causes the low-frequency dielectric dispersion. The dispersion depth is defined by evolution of micro and nano domain structure.

4. References

[1] Kuzminov Y.S. 1987 *Moscow Pub*: *Science* 264
[2] Kamencev V P, Nekrasov A V, Pedko B 1983 *Physics Series USSR Press* **47** 791
[3] Rotberg M B, Novik V K, Gavrilova N D 1969 *Crystallography* **14** 938
[4] Pedko B, Lebedev E V, Kislova I L, Volk T L 1998 *SPP* **40** 337
[5] Blistanov A A, Makarevskaya E V, Geraskin V V 1978 *SPP* **20** 2575
[6] Bagdasarov H S, Bodanov M J 1987 *SPP* **29** 2380
[7] Belabaev K G, Gabrielan V T, Sarkisov B H 1973 *Crystallography* **18** 198
[8] Voskresensky V M, Starodub O R, Sidorov N V, Palatnikov M N, 2011 *Crystallography* **56** 26
[9] Pedko B, Rudyak V M, Shabalin A L 1990 *Physics Series USSR Press* **54** 1154
[10] Palatnikov M N, Sherbina O B, Sidorov N V, Bormanis K 2010 *Crystallography* **55** 859
[11] Palatnikov M N, Sandler V A, Sidorov N V, Guryanov A V, Kalinnikov V T 2000 *SPP* **42** 1456
[12] Palatnikov M., Shcherbina O, Biryukova I, Sidorov N 2008 *Ferroelectrics* **374** 41
[13] Palatnikov M., Shcherbina O, Kazakov A A 2008 *Inorganic materials* **44** 360
[14] Naumova I I 1994 *Crystallography* **39** 1119
[15] Naumova I I, Glinko O A 1996 *Crystallography* **41** 749
[16] Naumova I I, Evlanova N F, Gilko O A 1997 *J.Cryst.Growth.* **181** 160
[17] Ming N, Hong J, Feng D *J.* 1982 *Mater. Sci.* **17** 1663

IOP Conf. Series: Materials Science and Engineering **49** (2013) 012053 doi:10.1088/1757-899X/49/1/012053

Chemical and physical modification of hemp fibres by steam explosion technology

Anna Sutka [1,2], Silvija Kukle[1], Janis Gravitis[2], Agris Berzins[3]
[1]Riga Technical University, Institute of Textile Materials Technologies and Design, Azenes St. 14, LV – 1048, Riga, Latvia
[2] LSIWC Laboratories of Biomass Eco-Efficient Conversation, Dzerbenes St. 27, Riga LV-1006, Latvia
[3] Department of Chemistry, University of Latvia, Kr.Valdemara St. 48, Riga, LV-1013, Latvia

E-mail: anna.sutka@rtu.lv

Abstract. In current research attempt has been made to analyse hemp fibres treated with steam explosion (SE) technology. Disintegration of hemp fibres separated from non-retted, dew-retted and dried stems of hemp ('Purini')[1] by alkali treatment and steam explosion (SE) were investigated. An average intensive SE in combination with the hydro-thermal and alkali after-treatment allows decreasing the diameter of hemp fibres and reduce the concentration of non-celluloses components, among them hemicelluloses, lignin, pectin, waxes and water [1;2].

1. Introduction

Hemp bast fiber is Latvian renewable resource, which is rich in cellulose. High mechanical strength of crystalline cellulose pays the attention of scholars, how to extract from it the cellulose micro- and nanofibres. Due to low density, biodegrability, interesting thermal, mechanical, acoustic and aseptic properties of hemp fibres, low cost and eco-friendly raw material, and this natural fibrous material is used as a (partial) replacement of synthetic fibres, such as glass, carbon or metallic fibres [3,4]. Therefore is carried out intensive studies of pulp processing in the form of micro- and nanofibres. Acquisition of micro– and nanofibres from cellulose bast fibers poses a problem because cellulose is insoluble/soluble difficult in practice-use solvents. Besides cellulose macromolecules are integrated into microfibrils in the bast fibers, which in merged in to macrofibrils, fibers, fiber bundles, technical fibers and each of the entities in addition to crystalline cellulose contains amorphous areas/cover ligno-cellulosic, lignin, pectin, and/or in wax form that need to remove before further processing using alkali treatment and steam explosion technique .

2. Materials and Methods

Dew-retted hemp fibres of local variety 'Purini' grown on the experimental fields of the Latgalian Agriculture Research Center LLZC (Latvia, district Vilani) and NaOH (Commercial grade) are used in this research. Hemp fibres were prepared by cutting into uniform size of approximately 2 mm length. This size of fiber allows steam explosion process taking place in the chemical and physical processes to penetrate deeper into the fibres in the inner layers.

Table 1. Sample variants under investigation

Sample		Pretreatment	SE, bar	Water extraction	0,4% NaOH extraction
A	Non-retted	-	-	-	-
B	Dew-retted	-	-	-	-
C	Dew-retted	-	32	+	+
D	Dew-retted	4 wt. %NaOH	-	+	+
E	Dew-retted	4 wt. %NaOH	16	+	+

During the SE process, fibres are chemically modified and mechanically defibrillated. The biomass is subjected to the treatment with high pressure saturated steam and rapid decompression resulting in substantial breakdown of the lignocellulosic structure, hydrolysis of hemicelluloses, depolymerisation of lignin components and defibrillation [5] .The process of which includes fast impregnation of the plant material with saturated steam in a closed reactor. The treatment proceeds at moderate temperature and pressure for a desired period of time (from some seconds to some minutes). The steam explosion process gives clean fibers with rough surfaces and increased crystallinity index.
SEM micrographs of fibres surface were taken using a scanning electron microscope VEGA Tescan 5136M (Czech Republic – UK). Prior to SEM evaluation, the samples were coated with gold by means of a plasma sputtering apparatus.
X-ray XRD patterns were determined on a Bruker D8 Discover diffractometer using copper radiation (CuKα) at a wavelength of 1.54180 Å. The tube voltage and amperage were set to 40 kV and 40 mA. The divergence slit was set at 0.6 mm, and the anti-scattering slit was set at 8.0 mm. The diffraction pattern was recorded using a scanspeed 0.5s/0.02° from 5–45° in 2θ scale and LynxEye position sensitive detector.
The crystallinity index (CI), which is a measurement of the amount of crystalline cellulose with respect to the global amount of amorphous materials, was evaluated using Segal empirical method. The CI has frequently been determined by means of the empirical Segal equation [6]:

$$C,\% = \frac{I\text{crystalline}}{I\text{crystalline} - I\text{amorphous}} \times 100$$

3. Results and discussion

As seen from micrograph B (Fig. 1), untreated elementary fibers of dew-retted hemp are still fastened in bundles with amorphous matrix; its surface is uneven, with protuberances of different size. After 4 wt. % NaOH pre-treatment, SE, water extraction and extraction by 0.4 wt. % NaOH (Fig. 1, E) the fibrous hemp parts are separated into elementary fibers the surfaces being clean, insoluble substances are not seen the substances other than cellulose being evaporated (13,8% loss of weight) during the SE treatment (Table 1). Disintegration and/or weakening of the binding of fibers in bundles is decisive for further separation of the fibers in subsequent operations.

Figure 1. SEM micrographs of hemp fibres without pre-treatment, dew-retted (F) and after various treatments (Table 2)

Evaporation intensities corresponding to pressure 16 bar (Table1) increase with the pressure up to 32 bars. After SE treatment at pressure of 32 bars and following water treatment all the water soluble components are removed leaving 75, 3 % residue after drying at room temperature to constant mass. The ultimately recovered total mass of extracted components decreases with increasing the pressure indicating that part of the components soluble in NaOH have been converted into evaporable and water-soluble substances during the SE treatment and removed by water treatment (Table 3).

Table 2. Modes of dew-retted hemp fibre samples STEX treatment and after-treatment

Sample	Residue, %	Water solub., %	Resid.after wat., %	Alk.solub.,%	Resid.after Alk.extr., %
E	93.3	5	88.4	2.8	85.5
C	86.2	10.9	75.3	7.1	68.3

The crystallinity of non-retted (A) hemp fibres was calculated to be 70,9% from the X-ray diffraction patterns and increase with the retting (71,8%; B), alkaline treatment (75,9%; D); SE (79,7 %; C) and SE combination with alkaline treatment (79,9%; E) (Table 3, Figure 2). Figure 2 presents the crystallinity obtained by Segal calculation method. These values show that treatments with SE and NaOH increase the crystallinity index. Alkali treatment hydrolyses the amorphous parts of cellulose present in fibres and after treatment the material contains more crystalline cellulose [7]. The Segal method is used for empirical measurements to allow rapid comparison of cellulose samples. It is useful for comparing the relative differences between samples.

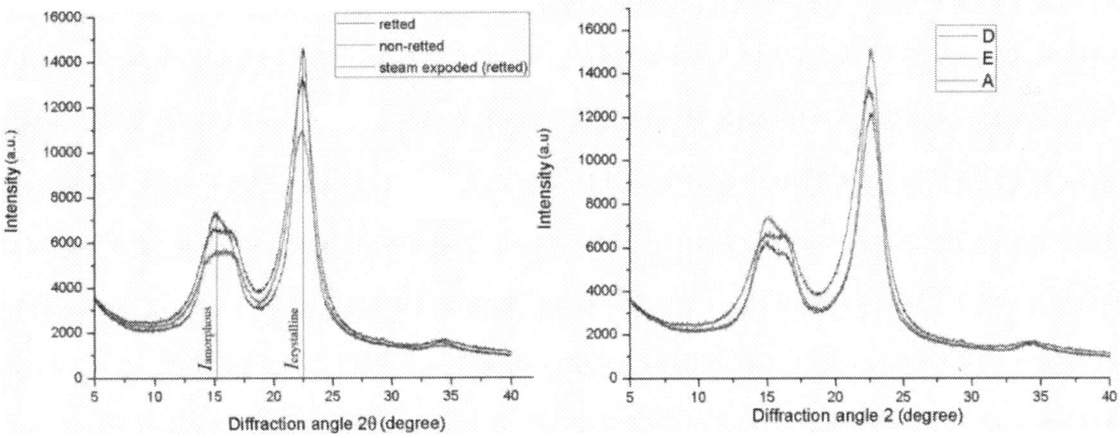

Figure 2. Determination of the crystalline cellulose content in the hemp fibres

Table 3. Crystallinity of untreated and treated hemp fibres

Sample	Crystallinity, %
A	70,9
B	71,8
C	79,7
D	75,9
E	79,9

4. Conclusions

SEM and X-ray methods were used for identification of changes in the chemical and physico-chemical properties of treated hemp fibres in comparison to untreated fibres. Results of chemical modifications of hemp fibres confirmed the changes of fibres associated with removing of impurities and pectin, hemicelluloses, lignin and waxes. After alkaline treatment, the fiber diameter decreases but the surface appearance improves. Removing of these substances depends on treatment. Scaning electron microscopy confirmed the changes in surface morphology of treated hemp fibres. The Segal method is used for empirical measurements to allow rapid comparison of cellulose samples. It is useful for comparing the relative differences between samples.

Acknowledgments
This work has been supported by the European Social Fund within the project «Support for the implementation of doctoral studies at Riga Technical University». The authors are grateful to the staff of Laboratory of Biomass Eco-Efficient Conversion, Latvian State Institute of Wood Chemistry.

References

[1] Putnina A., Kukle S., Gravitis J. Effect of Steam Explosion Treatment on Hemp Fibres Microstructure. 12th World Textile Conference AUTEX 2012 "Innovative Textile for High Future Demands": Book of Proceedings, Croatia, Zadara, 13.-15. June, 2012, 871 -874
[2] Kukle S., Gravitis J., Putnina A. Processing Parameters Influence on Disintegration Intensity of Technical Hemp Fibres. Journal of Biobased Materials and Bioenergy. Vol.6, No.4, 2012, 440-448
[3] Siqun Wang, Qingzheng Cheng, T. G. Rials and Seung-Hwan Lee, Cellulose Microfibril/Nanofibril and Its Nanocompsites, Proceedings of the 8th Pacific Rim Bio-Based Composites Symposium, 301-308.
[4] M. Fan Elementary hemp fibres and strength. BioResources 5 (4), 2010., 2307-2322
[5] M. Heitz, E. Capek-Monard, P. G. Koeberle, J. Gagno, E. Chornet, R. P. Overend, J. D. Taylor, E. Yu. Fractionation of Populus tremuloides at the Pilot Plant Scale: Optimization of Steam Pretreatment Conditions using the STAKE II Technology. Bioresource Technology 35 (1991) 23-32
[6] Segal L, Creely JJ, Martin Jr, Conrad CM. An empirical method for estimating the degree of crystallinity of native cellulose using the Xray diffractometer. Text Res J 1959;29:786–94.
[7] M. L. Troedec, et al., "Influence of Various Chemical Treatments on the Composition and Structure of Hemp Fibres," Composites Part A: Applied Science and Manufacturing, Vol. 39, No. 3, 2008, pp. 514-522. doi:10.1016/j.compositesa.2007.12.001

IOP Conf. Series: Materials Science and Engineering **49** (2013) 012054 doi:10.1088/1757-899X/49/1/012054

Atomic and electronic structure of hydrogen on ZnO $(1\bar{1}00)$ surface: *ab initio* hybrid calculations

A.B. Usseinov[1*], E.A. Kotomin[2], Yu.F. Zhukovskii[2], J. Purans[2], A.V. Sorokin[2], A.T. Akilbekov[1]

[1]L.N. Gumilyov Eurasian National University, Astana, Kazakhstan
[2]Institute of Solid State Physics, University of Latvia, Riga, Latvia
*E-mail: useinov_85@mail.ru

Abstract. Hydrogen atoms unavoidably incorporated into ZnO during growth of bulk samples and thin films considerably affect their electrical conductivity. The results of first principles hybrid LCAO calculations are discussed for hydrogen atoms in the bulk and on the non-polar ZnO$(1\bar{1}00)$ surface. The incorporation energy, the atomic relaxation, the electronic density redistribution and the electronic structure modifications are compared for the surface adsorption and bulk interstitial H positions. It is shown that hydrogen has a strong binding with the surface O ions (2.7 eV) whereas its incorporation into bulk is energetically unfavorable. Surface hydrogen atoms are very shallow donors, thus, contributing to the electronic conductivity.

1. Introduction

Understanding of the atomic and electronic structure of defective/doped ZnO is of great importance for improving performance of electrodes in optoelectronic devices based on transparent conducting oxides, *e.g.,* LED displays, *etc.* A study of a role of hydrogen impurities penetrating into ZnO thin films from plasma is of particular interest. Hydrogen atoms could occupy interstitial positions (H_i) or substitute O ions ([1–4] and references therein).

Previous theoretical studies mainly focused on hydrogen in ZnO bulk, which is found to act as a shallow donor [3,4]. However, the effect of hydrogen on the structural and electronic properties of different ZnO *surfaces* remains unclear: only a few relevant studies have been published so far [5,6], although this is important for improving technologies of ZnO thin film production.

In the current study, we have performed *ab initio* calculations of the atomic hydrogen adsorption on the non-polar ZnO$(1\bar{1}00)$ surface characterized by quite low surface energy (as described below) and studied earlier experimentally by means of ellipsometry and electronic energy loss spectroscopy [7]. To draw conclusions on the hydrogen effect on the electronic conductivity, the total and projected densities of one-electron states (DOS, PDOS) have been calculated and accompanied by analysis of the electronic charge redistribution, lattice relaxation, hydrogen adsorption energy at surface and its incorporation into the bulk. The paper is organized as follows: details of calculations are discussed in Section 2, main results for H incorporated in ZnO bulk and upon ZnO$(1\bar{1}00)$ surface are analyzed in Section 3, and main conclusions are summarized in Section 4.

2. Computational details

Large-scale *ab initio* calculations have been performed using the method joining LCAO (linear combination of atomic orbitals) basis set and hybrid exchange-correlation Perdew-Burke-Ernzerhof (PBE0) functional [8] as implemented in *CRYSTAL2009* computer code [9]. Such hybrid functionals allow us to perform very accurate calculations of the band gap, unlike standard LDA or GGA-type functionals. The atomic basis sets for Zn and O atoms have been taken from Ref. [10], and that for the hydrogen atom from Ref. [11]. Within the SCF procedure, the accuracies (tolerances) 10^{-7}, 10^{-7}, 10^{-7}, 10^{-7}, 10^{-14} have been chosen for calculations of Coulomb and exchange integrals. Effective charges on atoms have been estimated using Mulliken population analysis. The integration over Brillouin zone in the reciprocal space has been performed with a 6×6×4 Pack-Monkhorst grid. When modeling hydrogen absorbed in the bulk, the (2×2×2) supercell (periodic) model with 32 atoms has been chosen.

For $(1\bar{1}00)$ surface modeling, we have considered the slab model unit cell of finite thickness repeated along the chosen z axis. When simulating adsorption of isolated hydrogen atom on the $(1\bar{1}00)$ surface, the aforementioned slab extended in *CRYSTAL* calculations as 2×2 in the x and y directions correspond to the chosen periodic distribution of surface hydrogen atoms (Fig. 1). We have considered several hydrogen adsorption positions: above the surface atoms Zn (1), O (2) and in the "*hollow*" position (3) as shown in Figure 1 (b). In order to save computation time, we have performed partial geometry optimization, in which the relaxation has been provided for three upper layers.

The energy of hydrogen adsorption has been determined using the following expression: $E_{ads} = E_{tot}^{\mathrm{ZnO}} + E_{tot}^{\mathrm{H}} - E_{tot}^{\mathrm{H/ZnO}}$, where $E_{tot}^{\mathrm{H/ZnO}}$, E_{tot}^{ZnO} and E_{tot}^{H} are the total energies of slab with the adsorbed hydrogen atom, pure slab, and isolated hydrogen atom in the ground state, respectively. If the adsorption energy is positive, the process is energetically favorable.

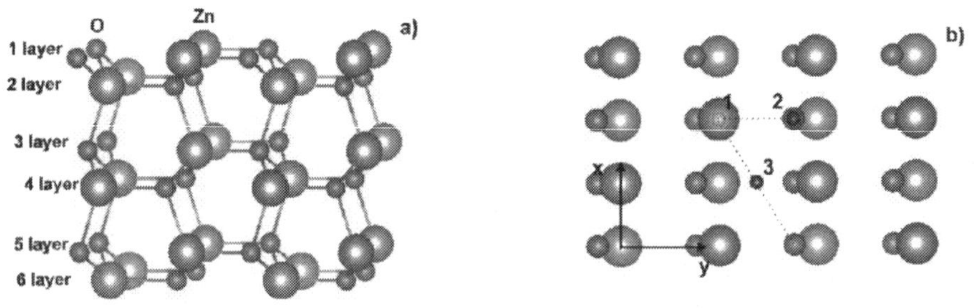

Figure 1. Unrelaxed 6-layer slab model of the ZnO $(1\bar{1}00)$ surface (a) and top views of three different positions of the hydrogen atom upon the surface (b).

3. Results and analysis

3.1 Hydrogen in the ZnO bulk

The analysis of the calculated density of states (DOS) in Fig. 2 shows the two effects: (*i*) narrowing of the band gap (E_g = 2.72 eV) as well as (*ii*) H contribution in a wide energy spectrum below and within the conduction band. Appearance of such a broad H band can be explained by a relatively high defect concentration used in our calculations (1/16 ≈ 6%) and a proximity of H atom energy level to the bottom of the conduction band. This is in qualitative agreement with previous conclusion about H_i as a shallow donor [1]. (The hybrid B3LYP calculations for 400-atom supercell have predicted H energy position only 0.1 eV below the bottom of conduction band [4] whereas O–H bond length is 0.98 Å, very close to our result – 0.982 Å). The strong interaction between H and O is also observed in the differential electron density map (Fig. 3a), accompanied by a polarization of surrounding atoms.

The calculated energy of H incorporation is found to be 1.8 eV, in qualitative agreement with the value of 1.0 eV obtained in the GGA calculations when using the energy of H atom in a free H_2 molecule [2] as the reference state. For a comparison, our energy has to be reduced by half the binding energy in a molecule (2.26 eV) which gives 0.67 eV. Similar calculations for the larger (3×3×2) supercell have gave a slightly larger value of 1.88 eV, which means that the interaction of periodically distributed H atoms is small, while in the 2×2×2 supercell it is large enough.

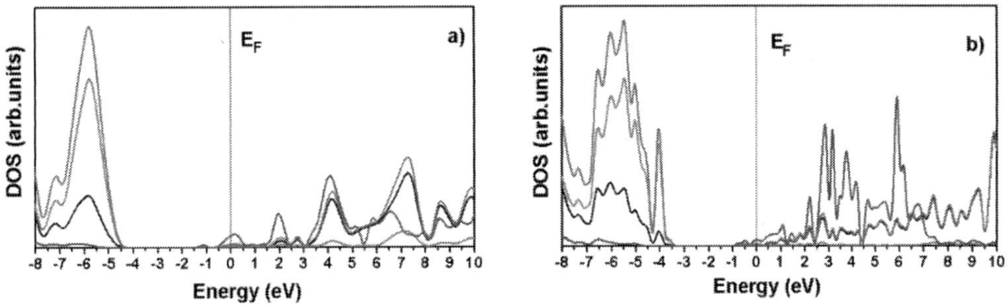

Figure 2. DOS projected onto all orbitals on Zn (black), O (red), H (blue), and the total density of states (cyan): a) hydrogen in (2×2×2) supercell ZnO; c) hydrogen adsorbed on the $(1\bar{1}00)$ surface of ZnO. For the hydrogen atom, PDOS is increased by a factor of 50. E_F denotes the Fermi level.

3.2. Hydrogen upon the surface

Similar calculations have been performed for hydrogen adsorbed on the nonpolar $(1\bar{1}00)$ surface (Fig.1). The $(1\bar{1}00)$ surface energy varies only slightly with increase of the number of layers. Therefore, 6-layer slab model has been selected for calculations of the relaxation of surface geometry

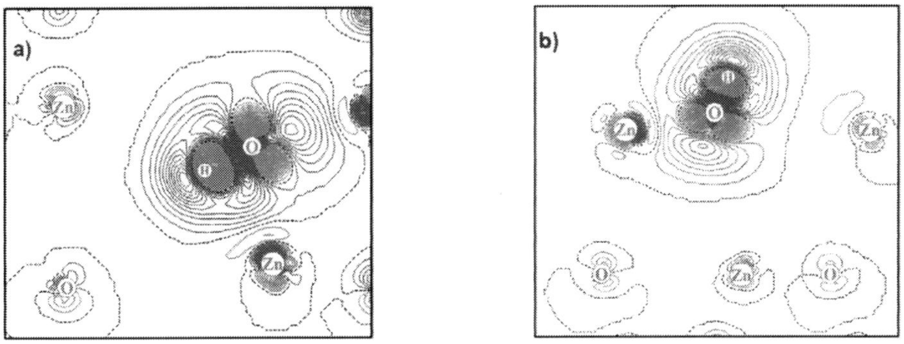

Figure 3. The differential electron density plots are drawn for the hydrogen atom positioned a) in (2×2×2) ZnO bulk supercell; b) upon the ZnO$(1\bar{1}00)$ surface. The solid (red) and dotted (blue) isolines correspond to positive (excess) and negative (deficiency) electron density, respectively. Increment for isolines is $0.001e$ within the range from $-0.1e$ to $0.1e$.

and the electronic properties of a pure surface. The analysis of calculations on the absolute and relative displacements of surface ions shows that after slab relaxation the surface Zn ions are displaced inwards, to the slab center, whereas O ions are shifted outwards, which is typical for oxide surfaces. This leads to a relatively small *surface rumpling*.

The adsorption energies of H atom have been obtained for three possible surface sites (Fig. 1 and Table 1): energetically the most favorable is the one atop the surface oxygen ions (E_{ads} = 2.71 eV). (It is also energetically favorable for adsorption of H atoms in the *hollow* positions.) As mentioned above, if hydrogen atoms before adsorption are produced by dissociation of H_2 molecules, the adsorption energy with such the reference state will by smaller by half of its binding energy. This means that only adsorption atop surface O atoms would be energetically favorable, with the energy of 1.58 eV, whereas H adsorption in other configurations mentioned in Table 1 turn out to be unfavorable.

The calculated DOS (Fig. 2b) shows that hydrogen atoms adsorbed on ZnO surface act as very shallow donors, while H contribution in the energy range close to the bottom of conduction band is much higher by amplitude than in the bulk. The electronic density redistribution also confirms

IOP Conf. Series: Materials Science and Engineering **49** (2013) 012054 doi:10.1088/1757-899X/49/1/012054

formation of a strong O–H bonding (Fig. 3b). Thus, adsorbed H atoms are expected to contribute strongly to the electronic conductivity.

Table 1. Adsorption energy E_{ads} (eV) of the hydrogen atom in different positions (Fig. 1b), the effective charge of surface ions q_{eff} (e), their vertical displacements Δz (Å) from defect-free lattice sites, as well as the distance d (Å) between the adsorbed H atom and a surface ion.

Site	E_{ads}	q_{eff}	Δz	$d_{Zn(O)\text{-}H}$
O	2.71	–0.79	0.071	0.96
Zn	–4.42	0.85	–0.31	1.61
Hollow	0.41	–0.18[a]	1.074[b]	1.91[c]

[a] Effective charge of H atom (e);
[b] Hydrogen distance from the surface (Å);
[c] Distance between the hydrogen atom and surface Zn ion.

4. Conclusions

As a result of first principles hybrid LCAO calculations, we have obtained the atomic and electronic properties of ZnO:H in the bulk and on the non-polar $(1\bar{1}00)$ surface. The interstitial anti-bonding hydrogens act as shallow donors, and this is the more so on the $(1\bar{1}00)$ surface. Incorporation energy of hydrogen atoms from plasma into interstitial positions is quite large (1.8–1.9 eV), whereas their adsorption atop surface O ions is energetically strongly favorable (2.7 eV). Thus, one could expect a high concentration of plasma-induced hydrogen on ZnO surfaces with a considerable contribution to the electronic conductivity.

Acknowledgements

This research was partly supported by the Latvian ERAF Project No. 2010/0272/2DP /2.1.1.1.0/10/ APIA/ VIAA/088 and Jülich Supercomputing Center (Project HSS13).

References

[1] McCluskey M D and Jokela S J 2009 *J. Appl. Phys.* 106 071101
[2] Van de Walle C G 2000 *Phys. Rev. Lett.* 85 1012
[3] Oba F, Togo A, Tanaka I, Paier J, and Kresse G 2008 *Phys. Rev.* B 77 245202
[4] Gallino F, Pacchioni G, and Valentin C 2010 *J. Chem. Phys.* 133 144512
[5] Ozawa K and Mase K 2010 *Phys. Rev.* B 81 205322
[6] Valtiner M, Todorova M, and Neugebauer J 2010 *Phys. Rev.* B 82 165418
[7] Matz R and Lüth H 1979 *Appl. Phys.* 18 123
Dorn R, Lüth H, and Büchel M 1977 *Phys. Rev.* B 16 4675
[8] Perdew J P, Ernzerhof M, and Burke K 1996 *J. Chem. Phys.* 105 9982
[9] Dovesi R, Saunders V R, Roetti C, *et al* 2009 *CRYSTAL09 User's Manual* University of Torino, Italy. URL: http://www.crystal.unito.it
[10] Jaffe J E and Hess A C 1993 *Phys. Rev.* B 48 7903
[11] Gatti C, Saunders V R, and Roetti C 1994 *J. Chem. Phys.* 101 10686

IOP Conf. Series: Materials Science and Engineering **49** (2013) 012055 doi:10.1088/1757-899X/49/1/012055

Photovoltaic properties of glass forming pyranyliden derivatives in thin films

R.Grzibovskis, A.Vembris, J.Latvels

Institute of Solid State Physics, University of Latvia, Riga, Latvia

E-mail: aivars.vembris@cfi.lu.lv

Abstract. Organic solar cells based on solution processable materials could provide new low cost production method. Most popular materials are polymer **P3HT** and fullerene derivative **PCBM**. There is active study to find effective low molecular weight materials. Instead of planar structure, most of the research is made on bulk heterojunction samples as they are providing better efficiency. In our work we have studied three newly synthesized materials consisting 2-tert-butyl-6-methyl-4H-pyran-4-ylidene fragment. Each compound has different electron acceptor group. From photoconductivity experiments its threshold values were obtained. For photovoltaic experiments bulk heterojunction samples containing investigated compounds mixed with **PCBM** were made. Best results with Incident Photon to Charge Carrier Efficiency (IPCE) larger than 0.4% at wavelength between 390 and 570 nm were obtained for system containing compound with malononitrile as electron acceptor group. Still, efficiency could be greatly improved as IPCE spectral dependence is not repeating absorption spectrum which show that generated charge carriers extraction from the sample was limited.

1. Introduction

Extensive researches of organic solar cells have been carried out in various scientific groups. First studies involved relatively expensive and low-efficient methods as evaporation in vacuum [1, 2]. In the last years new concept has been employed – solution processed solar cells [3, 4]. It provides possibility to use low-cost solvent-based printing or spin-coating methods. Successful approach is so called bulk heterojunction solar cell where two materials are mixed together. One of the most popular objects for studies is polymer-fullerene bulk-heterojunction solar cell where poly(3-hexylthiophene) (**P3HT**) is mixed with [6,6]-phenyl C61-butyric acid methylester (**PCBM**). This structure typically has efficiency of 3-4% [5]. Investigation of new organic compounds for organic photovoltaic still continues to increase their efficiency and spectral range.

In this paper photoelectrical and photovoltaic properties of three original organic compounds are shown. Ability to form glass-like structure in thin films is main advantage of these compounds. All three compounds have the same electron donor group but different electron acceptor group.

2. Materials

Chemical structure of the investigated compounds is shown in figure 1. All compounds consist of 2-tert-butyl-6-methyl-4H-pyran-4-ylidene fragment as a backbone of the molecule. Each of them contains various electron acceptor groups. Indandione group is in **ZWK-1TB** molecule, barbituric acid – **JWK-1TB** and malononitrile – **DWK-1TB**. Synthesis of the investigated compounds is described elsewhere [6]. Bulky trityloxyethyl group in the investigated compounds supports formation of amorphous state in thin films from organic solvents.

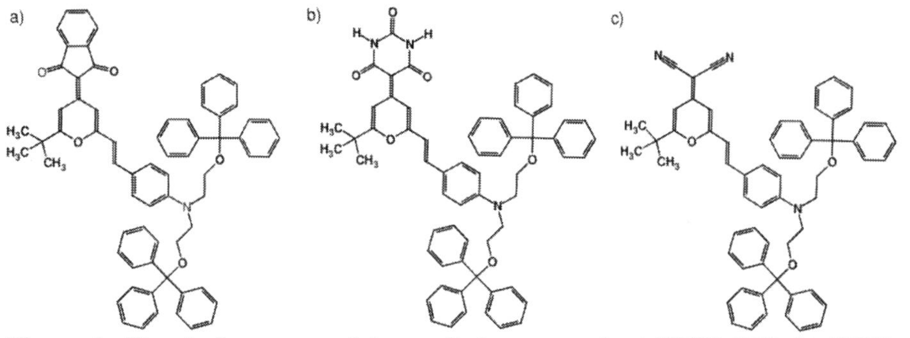

Figure 1. Chemical structure of the studied compounds a) **ZWK-1TB**, b) **JWK-1TB**, c) **DWK-1TB**.

3. Experimental

3.1. Sample preparation

For photoconductivity measurements sandwich type samples with structure ITO/organic layer/Al were made. Organic compounds were dissolved in chloroform with 40 mg/ml concentration. Active layer with the thickness of 500-600 nm was spin-coated on ITO-coated glass at 400 rpm speed for 40 seconds. Then sample was heated at 120°C for 15 minutes. Semitransparent aluminum electrodes with the thickness of 20 nm were deposited by thermal evaporation in vacuum (EDWARD AUTO 306) at pressure of $6*10^{-6}$ mBar.

Samples for photovoltaic experiments were made in sequent steps. PEDOT:PSS layer was spin-coated on ITO-coated glass at 2500 rpm for 80 seconds and dried for 30 minutes at 160°C. Then blended solution was made dissolving our compounds and PCBM in mass ratio 1:1 in chloroform. The solution was spin-coated on PEDOT:PSS layer at 400 rpm for 40 seconds and dried at 120°C for 15 minutes. Finally BaF and aluminum was deposited on top of the active layer by thermal evaporation in vacuum (EDWARD AUTO 306) at pressure of $6*10^{-6}$ mBar. Obtained structure of the samples was ITO/PEDOT:PSS(50nm)/organic compound:PCBM 1:1 (500nm)/BaF(1nm)/Al(70nm).

3.2. Photoelectrical and photovoltaic measurements

Xenon lamp was used as light source in the photoconductivity and photovoltaic measurements. Interference filters in spectral range from 400 nm to 700 nm with step 10 nm and half-width 10 nm were placed in front of the lamp. Average illumination intensity was around 1 mW/cm^2. The samples were illuminated from ITO and Al electrodes while 1 V positive or negative voltage was applied in photoconductivity experiment. In the figure 2 and 3 it is shown as ITO+, ITO-, Al+ and Al-. Current-voltage characteristics were measured while ITO side was illuminated in photovoltaic experiment. Measurements of the photoconductivity and photovoltaic spectral dependence were performed in vacuum ($1\cdot10^{-5}$mbar).

3.3. Determination of photoconductivity threshold energy

Charge carriers are strongly localized in organic semiconductor which means that their surrounding becomes polarized. In this way electronic polaron is created [7]. Difference between molecular polaron levels M_p^- and M_p^+ is called adiabatic energy gap. These polaron levels M_p^+ and M_p^- are actual charge conduction levels in organic material. Adiabatic energy gap value can be estimated from intrinsic photoconductivity quantum efficiency as threshold energy E_{th} [8]. Spectral dependence of photoconductivity quantum efficiency could be used to determinate E_{th} [9].

4. Results an discussion

Photoconductivity quantum efficiency spectrum for **JWK-1TB** sample is shown in figure 2. Similar spectra were observed for all the compounds. Minimum of the quantum efficiency at the maximum absorption position indicates charge generation mainly in the volume of the sample [10]. Rising the

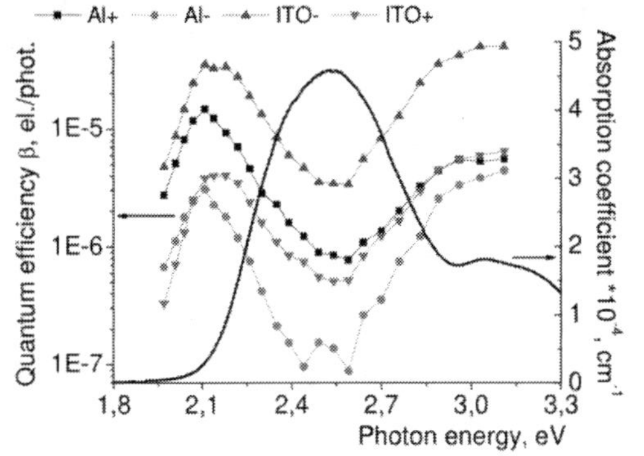

Figure 2. Photoconductivity quantum efficiency and absorption spectra of **JWK-1TB** sample. Abbreviations explained in section 3.2.

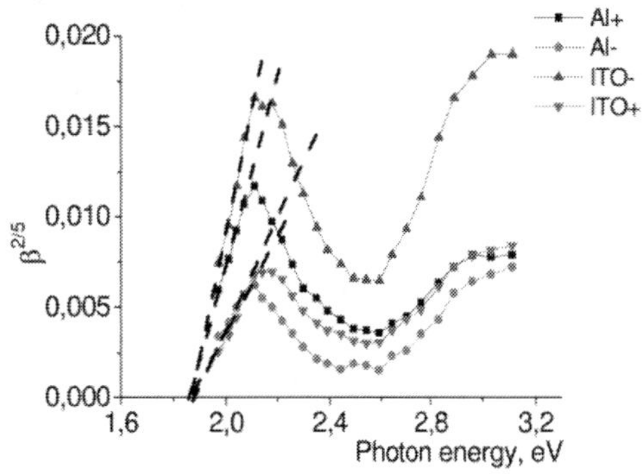

Figure 3. Determination of the threshold energy of compound **JWK-1TB**. Abbreviations explained in section 3.2.

photoconductivity quantum efficiency values to the power of 2/5 (see figure 3) threshold energy value was obtained. These values are 1.85±0.03 eV and 1.95±0.03 eV for compounds **JWK-1TB** and **DWK-1TB**, respectively. It shows that these compounds cover almost all visible light range.

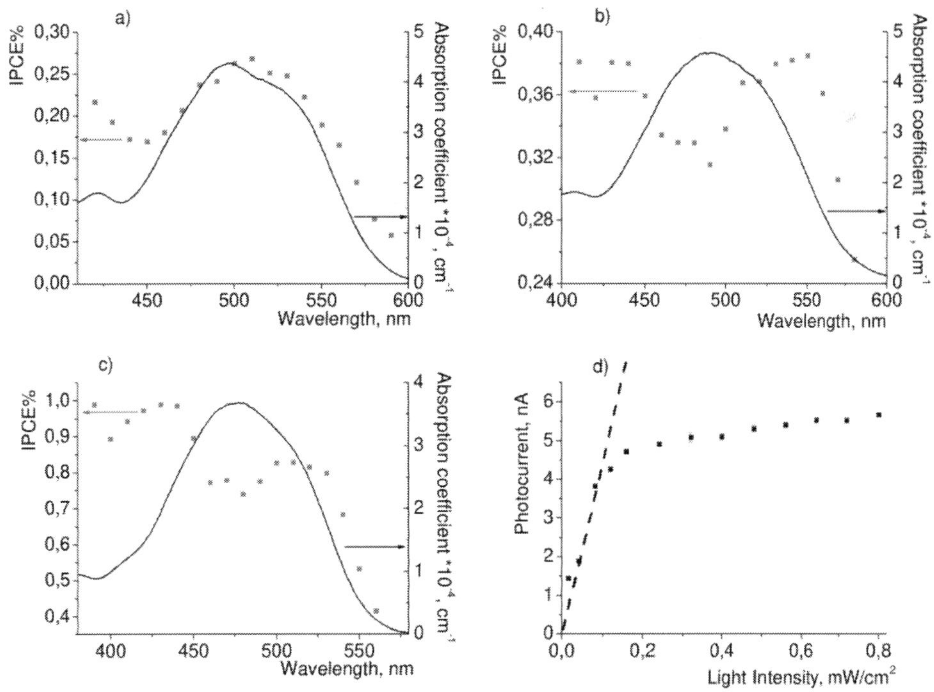

Figure 4. Absorption spectra of the pure film (line) and IPCE spectra of the blend with **PCBM** (squares) of a) **ZWK-1TB**, b) **JWK-1TB** and c) **DWK-1TB**. d) photocurrent of **JWK-1TB:PCBM** under 480nm wavelength illumination at various light intensities.

Determination of the threshold energy for the sample with **ZWK-1TB** was impossible due to very low photocurrent.

Incident Photon to Charge Carrier Efficiency (IPCE) was calculated form obtained current-voltage characteristics at each wavelength. Obtained IPCE spectral dependences of the organic compound-fullerene samples are shown in figure 4 a,b,c. Correlation between the absorption and IPCE spectra was observed for **ZWK-1TB:PCBM** which indicates good exciton separation at the interface and charge carrier extraction from the sample. In the case of **JWK-1TB** and **DWK-1TB** mixed with **PCBM** anti-correlation was observed. ICPE has minimum where absorption has maximum. To explain this phenomena photocurrent dependence on incident light intensity at 480 nm (it corresponds to maximum absorption) was measured for **JWK-1TB:PCBM** (see figure 4 d). Linear dependence was acquired only at low light intensity. Photocurrent rapidly saturates at higher intensity which means that charge carrier extraction is limited. The IPCE efficiency could be greatly increased by overcoming this issue. One of the possible solutions could be thinner active layer.

Maximum short circuit current, open-circuit voltage, IPCE and fill factor is presented in table 1. Almost all parameters are higher for **DWK-1TB:PCBM** system in comparison with other systems. Only fill factor is lower than for **ZWK-1TB:PCBM** but it could be increased by optimizing the sample.

Table 1. Photoconductivity threshold energy E_{th}, short circuit current I_{sc}, open-circuit voltage U_{oc}, Incident Photon to Charge Carrier Efficiency IPCE and fill factor FF of the prepared systems. Number in brackets shows incident light wavelength at which this value was obtained.

Compound	E_{th}, eV	I_{sc}, $\mu A/cm^2$	U_{oc}, V	IPCE max, %	FF max
ZWK-1TB	---	0.76 (490nm)	0.17 (490nm)	0.3 (510nm)	0.32 (530nm)
JWK-1TB	1.85	1.10 (470nm)	0.40 (500nm)	0.4 (550nm)	0.21 (560nm)
DWK-1TB	1.95	2.50 (490nm)	0.60 (480nm)	1.0 (390nm)	0.29 (540nm)

5. Conclusions

The threshold energy of photocurrent indicates that glass forming original organic molecules with 2-*tert*-butyl-6-methyl-4*H*-pyran-4-ylidene fragment as a backbone can be used to convert visible light to electricity. Photovoltaic effect was observed in all investigated organic compound:PCBM systems. The best results of short circuit current, open-circuit voltage and IPCE were obtained for the compound with malononitrile as electron acceptor group. These results could be improved because generated charge carrier extraction is limited. It could be overcome by preparing thinner active layer.

Acknowledgments
This work has been supported by the ERAF Project No. 2010/0252/2DP/2.1.1.1.0/10/APIA/VIAA/009 and the European Social Fund within the project «Support for Doctoral Studies at University of Latvia»

References
[1] Peumans P, Bulović V and Forrest S R 2000 *Appl. Phys. Lett.* **76** 2650
[2] Rostalski J and Meissner D 2000 *Sol. Energy Mater. Sol. Cells* **61** 87
[3] Shaheen S E, Radspinner R, Peyghambarian N and Jabbour G E 2001 *Appl. Phys. Lett.* **79** 2996
[4] Dyakonov V 2002 *Physica E* **14** 53
[5] Deibel C and Dyakonov V 2010 *Rep. Prog. Phys.* **73** 096401
[6] Zarins E, Siltane K, Misina E, Kokars V, Lazdovica K, Vembris A, Kampars V, Muzikante I and Rutkis M 2012 *Proceedings of SPIE* **8435**, 84351Q
[7] Silinsh E A, Bouvet M and Simon J 1995 *Mol.Mat.* **5** 1
[8] Silinsh E A and Capek V 1994 *Organic Molecular Crystals. Interaction, Localization and Transport Phenomena.* (New York: AIP Press)
[9] Grzibovskis R, Latvels J and Muzikante I 2011 *IOP Conf. Series: Materials Science and Engineering* **23** 012021
[10] Indrikova M, Latvels J, Muzikante I and Turovska B 2011 *Materials Science* **17** 125

IOP Conf. Series: Materials Science and Engineering **49** (2013) 012056 doi:10.1088/1757-899X/49/1/012056

TSL and fractional glow study of Ge-doped α-quartz.

A. Zolatarjovs, A.N. Trukhin, K. Smits, D. Millers
Institute of Solid State Physics, University of Latvia

E-mail: alexeyzolotarjov@gmail.com

Abstract. Crystalline α-quartz doped with 0.1wt% and 0.9wt% germanium was studied using TSL and FGT equipment. Sample was chosen because previously it is known that Ge in quartz is efficient trap for electrons, therefore it could be used for detection of hypothetic self-trapped hole in α-quartz. However previous investigations of ODMR and TSL shows that in α-quartz the hole is still mobile and trapping occurs only on defect states. The activation energies for both TSL peaks are found by fractional glow and Hoogenstraaten method. The TSL distribution changes depending on Ge concentration and also on irradiation type. The TSL peaks below 70K in quartz doped with Ge could belong to hole trapped on Ge.

1. Introduction

The Ge center in α-quartz is described by several researchers [1 – 8] and the thermostimulated luminescence (TSL) was also observed [7, 8]. However the formation of the excited states responsible for TSL is not clear. On the one hand is known that in Ge doped α-quartz efficient electron trapping takes place at low temperatures [4, 5]. On the other hand, there must be traps for holes also since TSL was observed. However the evidence is that the hole in α-quartz remains mobile even at low temperatures and could be trapped only on defects [6, 7]. Since only one kind of dopant namely Ge is incorporated, it is not known if it is responsible for trapping electrons or holes. Additional information for understanding these processes is needed. Therefore we modified our equipment so that two different thermostimulated methods – classical TSL and fractioned glow technique (FGT) – could be used and, additionally, the luminescence spectra of glow peaks could be monitored. The testing of the modified equipment showed that the TSL peaks of Ge-doped α-quartz can be well reproduced and are very close to those described in [7] Additionally, the thermal activation energies were extracted from FGT results. These experimental data were used for clearing up some details of thermostimulated recombination in Ge doped α-quartz.

2. Samples and methods

2.1. Samples

The investigated samples were 0.1 wt.% and 0.9 wt.% Ge-doped synthetic crystalline α-quartz, grown by hydrothermal method. To achieve good thermal contact, samples were polished and pressed between two copper plates with holes in the centre for irradiation and luminescence measurement. The temperature was measured by a transducer diode fastened to these copper plates.

2.2. Equipment

The TSL measurement equipment consists of sample irradiation source, temperature control and luminescence measurement units. For the sample irradiation at low temperature two different sources were used: the X-ray tube with W target operating at 40 kV, 10 mA and 7.9 eV F_2 laser (PSX-100-2) with pulse energy 0.5 mJ and pulse duration 5 ns. The luminescence spectra were recorded with spectrometer (Andor Shamrock B303-I) coupled with a CCD camera (Andor DU- 401A-BV) at exit port. The 150 lines per inch grating covered 295-865 nm spectral range in one measurement. This allowed measuring both TSL intensity and spectral distribution at each temperature point. For temperature control, the closed-cycle helium cryostat (Sumitomo HC-4) with temperature controlling unit (Lakeshore 331) was used. The temperature control range was 8-300K. All controlling devices, as well as data acquiring devices were connected to a PC and controlled using various TSL/FGT programs (ISSP UL).

2.3. TSL and FGT methods

For TSL measurements, the classic TSL method was applied: irradiation at low temperature followed by heating with a constant heating rate and monitoring of luminescence in dependence of temperature. For determination of activation energies 2 methods were used – initial rise method [9] and Hoogenstraaten method [10]. For the first method, only a fraction of a total glow is required, which leads to a non-linear temperature change, while the second method is based on the peak shifts resulting from different heating rates in classic TSL.

3. Results and discussion

First of all, the classic TSL was measured and the result obtained was compared with that found in [7]. Two TSL glow-curve peaks were observed at 51K and 75K (β=0.1 K/s, X-ray), which coincide well with the previous observations [7]. In addition, we measured TSL glow-curve when the sample was excited with UV laser. Results are shown in Fig.1. Excitation with UV laser resulted in a peak shift to higher temperatures, and the shift of a first peak is greater, than that of the second, so this process can not be related to a heating rate difference (remained constant) or temperature inaccuracy of any kind. Also, when exciting with UV laser with longer wavelength (266 nm), the second peak is lower in comparison to the first, compared to the case of the 157 nm excitation. Under X-ray irradiation at 8 K the luminescence intensity increases and reaches saturation after 20 min. To verify the impact of irradiation dose, an experiment with two different irradiation doses were made (5 min and 60 min). The irradiation time increased the TSL intensity; however the irradiation time/dose neither had an impact on peak positions nor on the peak ratio Fig.2. (inserted graph).

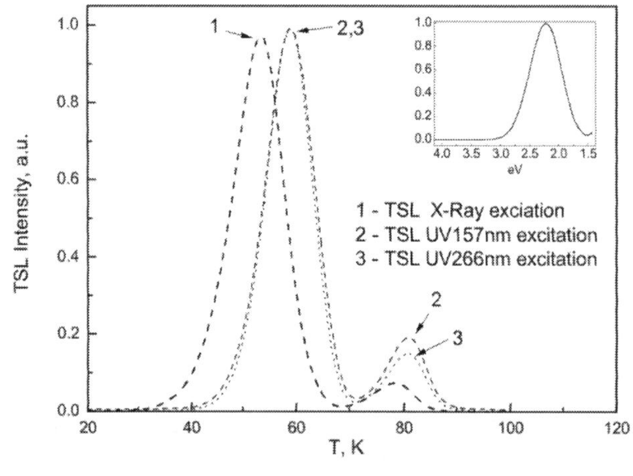

Figure 1. TSL glow- curve for X-Ray and UV excitation (heating rate β=0,1 K/s) Insert shows luminescence spectral distribution.

The luminescence spectra for all excitations are the same (Fig1. insert). Moreover, the photoluminescence and cathodoluminescence spectra measured at 8 K are the similar too, indicating that the recombination center in all cases is the same, however, the traps differ. Previously the TSL peaks were associated with Ge luminescence center which corresponds to self-trapped exciton (STE) on Ge and is found in Ge-doped samples. We verified that the luminescence spectral properties are the same in two samples with different Ge concentrations, however TSL spectra slightly differs, especially the intensity and position of second TSL peak.

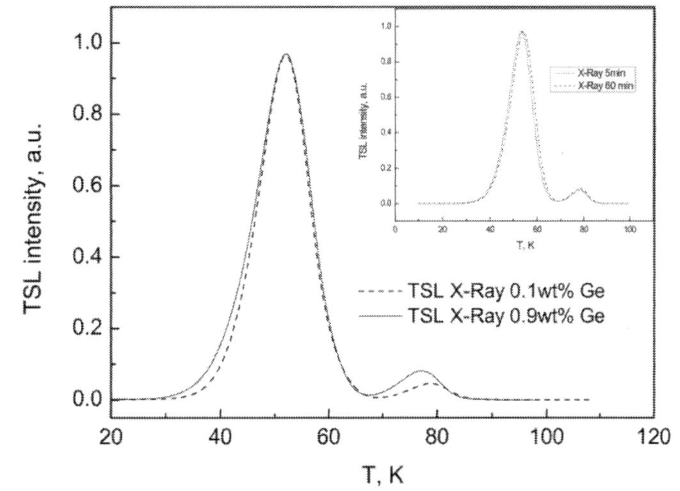

Figure 2. TSL glow- curve for SiO_2 crystal with 0.1wt Ge and 0.9wt% Ge. The insert shows the TSL for sample with 0.9wt% Ge irradiated for 5min and 60min.

As expected, the luminescence intensity increased with increase of the Ge doping concentration. The first TSL peak was similar for both samples; however, the second was slightly different (Fig.2). The ratio between the intensities of the first and second TSL peaks (after X-ray irradiation) for sample with 0.1wt% Ge was 25, but for sample with 0.9 wt% Ge the ratio was 11. With the decrease of Ge concentration the second peak decreases. The TSL peaks obtained under UV irradiation shows opposite dependence. The second TSL peak is even more intense and the respective intensity ratios were found 5 for 0.9 wt% Ge and 1.8 for 0.1wt% Ge samples. Not only the peak intensity ratio changes, also the TSL peak positions of UV irradiated samples slightly changed. This indicates that the mechanics behind this is more complex than previously assumed and not only STE but also hole traps are involved.

The FGT method was used to determine the activation energies for both peaks. The activation energy 0.065 eV (0.07eV under UV irradiation) was found for TSL peak at 50 K and this energy is close for both kinds of excitation. The activation energy for 75 K peak was 0.13 eV for X-ray excited sample and slightly lower for UV excited sample (Fig.3). Additionally TSL spectra with six different heating speeds were recorded and the Hoogenstraaten method was used to calculate the activation energies for both peaks. The calculated activation energy values 0.057eV and 0.123eV are slightly lower to these found by FGT method (Fig.3). This could indicated that the peak are not completely first order, but Hoogenstraaten method is applicable for first order TSL peaks. The TSL measurements for UV irradiated samples shows peak shift to larger energies, however the FGT energies under UV irradiation for second peak shows even lower calculated energies.

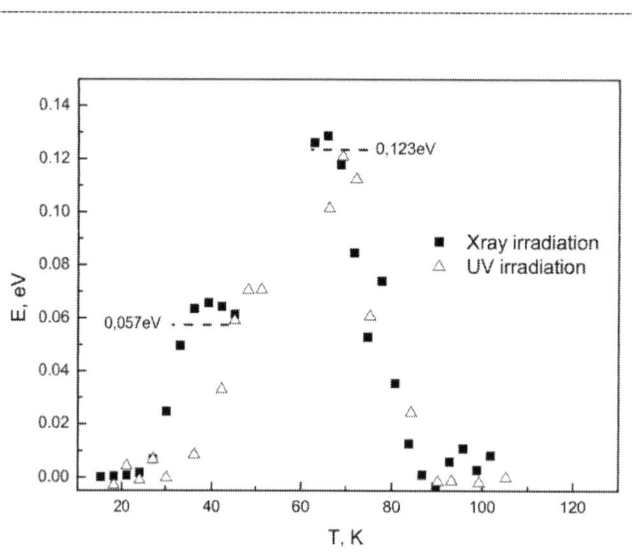

Figure 3. FGT Energy graph as a function of temperature. Dash lines represent data calculated from Hoogenstraaten method.

4. Conclusions

Two TSL peaks were detected in the range 8 – 100 K in crystalline SiO_2 doped with germanium having intensity roughly proportional to the Ge concentration. It is found that the peak positions are changed for different kind of excitation. For the case of X-ray excitation the peaks are situated at 50K and 75 K. For excitation with 157 nm excimer laser they are situated at 60 K and 80 K. The difference is explained as trap structure modification. Activation energies were determined for both excitation types. The activation energies found by FGT and Hoogenstraaten method were close; the small differences could be explained that Hoogenstraaten method is applicable for first order TSL peaks.

5. Acknowledgments

The work was supported by LZP project 2013.10-5/014 and
ERAF Nr.2010/0253/2DP/2.1.1.1.2.0/10/APIA/VIAA/079

6. References

[1] Trukhin A N 1992 *J.Non-Cryst. Solids* **149** 32-45
[2] Trukhin A and Poumellec B 2004 *Sol. St. Commun.* **129** 285-289
[3] Trukhin A N Boukenter A Ouerdane Y Girard S 2011 *J.Non-Cryst. Solids* **357** 3288-3291
[4] Halperin A and Ralph I E 1964 *J. Chem. Phys.* **39,1** 63-73
[5] Weil I A 1975 *Radiation Effects* **26,1** 261-265
[6] Hayes W and Jenkin J T 1985 *J. Phys. C: Solid State Physics* **18** L849-L853
[7] Trukhin A N 1986 *Sov. Solid State Physics* **28** 1460 – 1464
[8] Godmanis I Truhins A 1981 *LPSR ZA Vēstis Fiz.Teh.zin.* **6** 26-29
[9] Chen R and McKeever S W S 1997 *Theory of Thermoluminescence and Related Phenomena* 110-112
[10] Hoogenstraaten W 1958 *Philips Res. Rept.* **13** 515

IOP Conf. Series: Materials Science and Engineering **49** (2013) 012057 doi:10.1088/1757-899X/49/1/012057

Structural, electrical and optical characteristics of Al-doped zinc oxide thin films deposited by reactive magnetron sputtering

M. Zubkins, R. Kalendarev, K. Vilnis, A. Azens, J. Purans

Institute of Solid State Physics, University of Latvia, Riga, Latvia
E-mail: zubkins@cfi.lu.lv

Abstract. ZnO:Al (AZO) thin films on glass were deposited by DC reactive magnetron sputtering at approximately 300°C substrate temperature. Structural, electrical and optical properties were investigated as a function of oxygen flow. XRD data shows that AZO thin films are polycrystalline with pronounced c-axis orientation and the grain size increasing with the oxygen flow. The lowest achieved resistivity within the deposited set of samples was $7.6 \cdot 10^{-4}$ Ωcm. The transmittance of AZO films was above 80 % at 550 nm with the optical band gap between 3.4 and 3.8 eV.

1. Introduction

ZnO is a direct wide band gap (3.4 eV) semiconductor with large exciton binding energy (60 meV) and wurzite-type hexagonal structure [1-3]. As-grown ZnO films are n-type conductors due to sub-stoichiometric composition and omnipresent hydrogen impurity which acts like a shallow donor [3]. Al-Doped ZnO films are a lower cost alternative to expensive In_2O_3:Sn (ITO), because of virtually unlimited availability of source material. ZnO thin films are characterized by high chemical and thermal stability, and non-toxicity [4]. The most often used ZnO based composition is Al doped ZnO (AZO). However, industrialy perspective technologies of DC reactive magnetron sputtered Al-doped ZnO films have not yet reached the performance of ITO in terms of the combination of electrical conductivity and optical transmittance, as well as stability of the process [4]. Therefore, the deposition techniques must still be developed as well [5].

Physical vapour deposition techniques such as sputtering, pulsed laser deposition, evaporation etc. are commonly used for ZnO thin films deposition.

This paper reports on characterisation of structural, electrical and optical properties of AZO thin films deposited by reactive magnetron sputtering upon varied amount of oxygen in the sputtering atmosphere.

2. Experimental details

AZO thin films were deposited on glass substrates by reactive magnetron sputtering from a metallic (Zn=98%, Al=2%) target in an Ar+O_2 atmosphere. The substrates were washed in acetone before deposition. The chamber was pumped to base pressure below $1 \cdot 10^{-5}$ Torr by a turbo-molecular pump. During the deposition the substrate was kept at about 300°C temperature, the sputtering was conducted at 20 mTorr working pressure and constant 100 W sputtering power. The target to substrate separation was 5 cm. Argon gas flow was kept constant at 50 sccm, oxygen gas flow was varied to cover the range from partly to fully oxidized target. The flows of both gases were controlled by MKS mass flow controller. Deposition time was 20 minutes. The process was controlled by Plasma emission monitoring (PEM), carried out by a high resolution spectrometer (Ocean Optics "HR4000") connected to the chamber view port through a fibre optic cable. A set of samples was deposited in the oxygen flow range between 2 and 4.5 sccm. The setpoint was approached from the metallic (low-oxygen) side.

Films thickness was determined by profilometer (CART "Veeco Dektak 150"). Thicknesses for different films were in the range from 400 to 500 nm, with the corresponding deposition rate between 20 and 25 nm/min. Resistivity, electron concentration and Hall mobility were measured by Hall effect using van der Pauw configuration (HMS5000). Structural analysis was carried out by X-ray diffraction, XRD (X'Pert Pro MPD). Films transmittance in the range from 200 to 1100 nm was determined by a double-beam spectrophotometer (Analytik Jena AG "Specord 210"). All measurements were performed at room temperature.

3. Results and discussion

3.1 AZO structural analysis

XRD data shows typical textured polycrystalline hexagonal ZnO structure (Fig. 1). The (002) peak has the highest intensity indicating that AZO films are strongly oriented with the c-axis perpendicular to the substrate surface. The diffraction angles are slightly shifted compared to the standard (002) angle of 34.34° [6] due to non-stoichiometry and aluminium doping. For the films deposited at lower oxygen content, low intensity (101) peaks are present as well. No phases of Zn, Al, Al_2O_3 or $ZnAlO_4$ were detected. At oxygen flows below 2 sccm, the films were X-ray amorphous or nanocrystalline and therefore not presented in the paper. Based on the FWHM values, the crystal size (Fig. 2) increases from 20 to 55 nm with the increased oxygen flow.

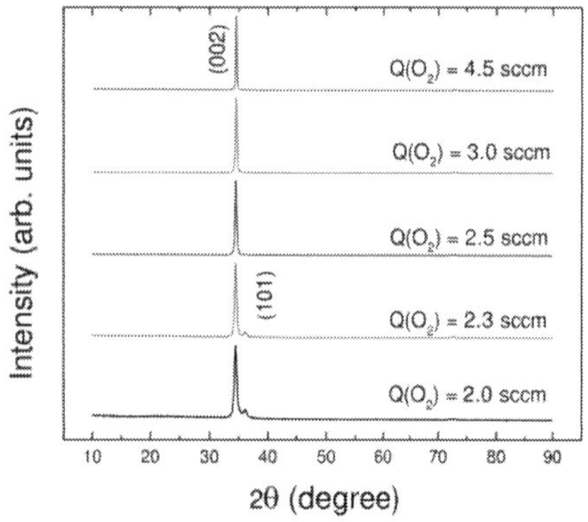

Figure 1. XRD spectra for the films deposited with different oxygen content in the sputtering atmosphere.

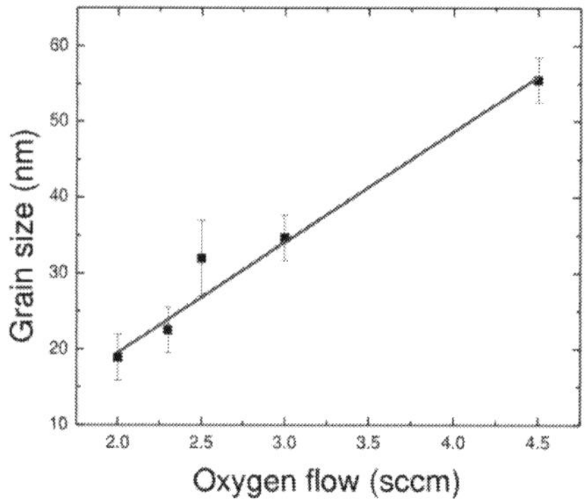

Figure 2. Grain size dependence on the oxygen content in the sputtering atmosphere.

3.2. Electrical properties

Hall effect of AZO films was investigated as a function of oxygen flow in the sputtering atmosphere. The films exhibit n-type conductivity. Within the studied range, the lowest resistivity of $7.6 \cdot 10^{-4}$ Ωcm was obtained for the film deposited at 2.3 sccm oxygen flow (Fig.3a). For this film, the carrier concentration and Hall mobility are $4.6 \cdot 10^{20}$ $1/cm^3$ and 18.8 $cm^2/(Vs)$, respectively.

Material resistivity is determined by the charge carrier concentration and mobility. For the studied set of films, the carrier concentration was found to decrease with the oxygen flow (Fig.3b). A similar trend has been observed for films of pure ZnO [7], Indium – Zinc oxide [8], and ITO [2], suggesting that the number of carriers is related to the deviation from stoichiometry. Interstitial oxygen may be present in ZnO [9], which would act as an acceptor in the films deposited at high oxygen content.

Based on the increased grain size, a higher Hall mobility could be expected for the films deposited at higher oxygen content. However, a mobility decrease with the oxygen flow is seen in Figure 3c, indicating that the grain boundary scattering is not the only mechanism affecting the mobility. It may not be ruled out that interstitial oxygen may contribute to charge carrier scattering as well. Oxygen acting as a scattering centre has been suggested for the mobility decrease observed for films deposited at different pressure [10]. Further research is needed to confirm the reason for the mobility decrease.

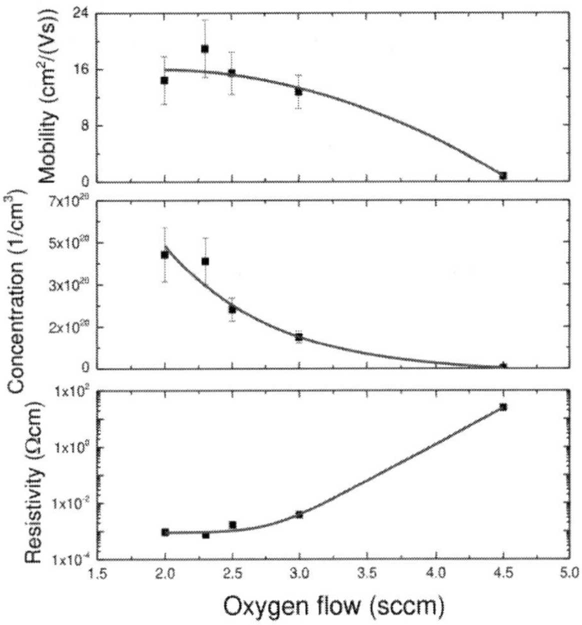

Figure 3. Hall effect as a function of oxygen flow: (a) resistivity, (b) carrier concentration, (c) Hall mobility.

3.3. Optical properties

For the studied set of films, the optical transmittance is approximately 85% at 550 nm. Transmittance curves shown in Fig. 4 are normalized with respect to the transmittance of bare glass substrate. Films deposited at lower oxygen content are less transparent in the NIR region, which can be caused by higher charge carrier concentration.

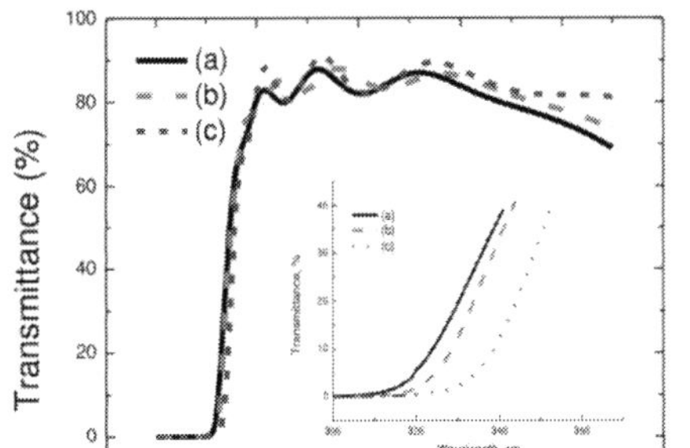

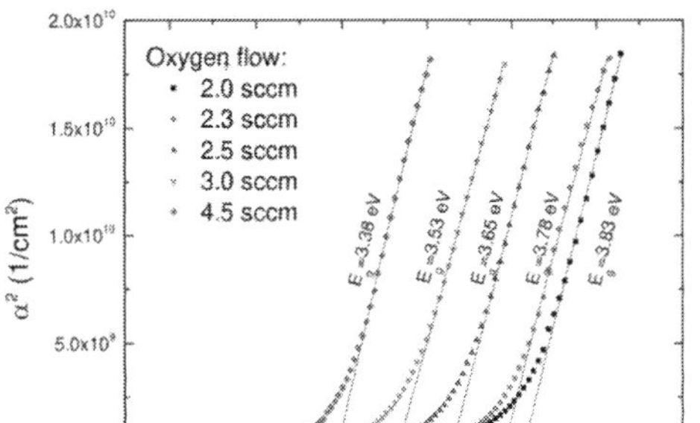

Figure 4. Transmittance of AZO films deposited at different oxygen flows: (a) 2.0 sccm, (b) 2.3 sccm, (c) 2.5 sccm.

Figure 5. Square of the AZO films absorption coefficient α as function of photon energy. Linear extrapolation of α^2 shows that the optical gap depends on the oxygen flow.

The optical absorption coefficient was measured using transmittance data (*T*) and films thickness (*d*): $\alpha = \frac{1}{d}\ \ln(\frac{1}{T})$. Fig. 5 shows α^2 as a function of photon energy. The optical band gap can be obtained by extrapolating the linear part of the curve towards the *x*-axis. The optical band gap of the studied films was found to change from 3.4 to 3.8 eV depending on the oxygen flow.

Degenerate semiconductor absorption edge shift in the shorter wavelength direction with increasing electron concentration is known as Burstein-Moss effect. Since zinc oxide is heavily doped with aluminium, optical band gap increases as the lower conduction band states become occupied and unavailable for excitation.

4. Conclusions

AZO thin films were deposited by DC reactive magnetron sputtering technique with varied oxygen/argon content in the sputtering atmosphere. The lowest achieved resistivity was $7.6 \cdot 10^{-4}$ Ωcm at approximately 85% optical transmittance and E_g=3.78 eV. Oxygen content in the sputtering atmosphere was found to significantly affect the films structure and thus the electrical and optical properties.

Acknowledgements

The work has been performed within ERDF project 2010/0272/2DP/2.1.1.1.0/10/APIA/VIAA/088. The authors would like to thank Dr. A. Plaude for XRD measurements and Dr. L. Trinklere for transmittance measurements.

References

[1] Ozgur U 2005 *J. Appl. Phys.* **98** 041301
[2] Minami T 1999 *J. Vac. Sci. Technol.* A **17 (4)** 1765-1772
[3] Bjorheim T S 2012 *The Journal of Physical Chemistry C.* **116** 23764-23772
[4] Fortunato E 2007 *MRS bulletin* **32** 242-247
[5] Castaneda L 2011 *Materials and Applications* **2** 1233-1242
[6] Park H 2011 *Thin Solid Films* **519** 6910-6915
[7] Subramanyam T K 1999 *Phys. Stat. Sol* **173** 425
[8] Jung Y S 2003 *Thin Solid Films* **445** 63–71
[9] Knutsen K E 2012 *Phys Rev B.* **86** 121203(R)
[10] Charpentier C 2011 *EPJ Photovoltaics* **2** 25002

IOP Conf. Series: Materials Science and Engineering **49** (2013) 012058 doi:10.1088/1757-899X/49/1/012058

Mechanical properties and XRD of Nafion modified by 2-hydroxyethylammonium ionic liquids

V Garaev[1], S Pavlovica[1], I Reinholds[1] and G Vaivars[1,2,*]

[1] Department of Chemistry, University of Latvia, Kr. Valdemara 48, Riga, LV-1013, Latvia
[2] Institute of Solid State Physics, University of Latvia, Kengaraga 8, Riga, LV-1063, Latvia

E-mail: [*] Guntars.Vaivars@cfi.lu.lv

Abstract. In this work, the Nafion 112 membrane impregnated with 2-hydroxyethylammonium carboxylate ionic liquids have been investigated. The used ionic liquids were 2-hydroxyethylammonium formate [HEA]F, acetate [HEA]A and lactate [HEA]L. Prepared composite membranes Nafion/ionic liquid are characterized by mechanical testing, such as tensile test and creep test. It is found that ionic liquids decrease elastic modulus and creep compliance, but do not have significant effect on the tensile strength. Also, composite membranes were studied by wide angle X-ray diffraction. All ionic liquids shift the peak maximum to the lower angle. In this work, only biodegradable ionic liquids were used for composite preparation.

1. Introduction

Fuel cells are one of the promising alternative sources of clean energy. Proton exchange membrane fuel cell (PEMFC) is currently selected from many different types of fuel cells for energy back-up systems. A key component of this type of fuel cell is proton exchange membrane (PEM). Nafion membrane is the commonly used as PEM. High mechanical strength, excellent thermal and chemical stabilities are Nafion membrane advantages over others PEM. However, Nafion membrane has high cost and its proton conductivity is decreasing at high temperatures. Therefore, Nafion membrane needs to be modified. Nowadays, there are many ways how membrane can be modified, e.g. incorporating titanium oxide [1], zirconium phosphate [2], etc. Last trend is to use ionic liquids (ILs) due to their high conductivity, good thermal stability and wide electrochemical window [3,4]. There are two main methods how to modify membrane with an IL: solution casting method [5] and IL-swollen method [6]. The IL-swollen method was used in this work.

In this study, 2-hydroxyethylammonium carboxylates were used for the modification of Nafion membrane. These ILs are non-toxic and highly biodegradable [7,8]. The impact on conductivity and thermal stability with the same ILs has been studied in our previous research [9]. However, for practical applications the mechanical properties of composite membrane are very important. 2-hydroxyethylammonium lactate already was used for the modification of other type of PEM [10].

IOP Conf. Series: Materials Science and Engineering **49** (2013) 012058 doi:10.1088/1757-899X/49/1/012058

2. Experimental

2.1. Materials

Nafion 112 (equivalent weight 1100 g/mol, thickness 50.8 μm) was purchased from Ion Power. The ILs 2-hydroxyethylammonium formate [HEA]F, acetate [HEA]A, lactate [HEA]L (for chemical structures see figure 1) were synthesized in reaction of 2-hydroxyethylamine with formic, acetic or lactic acids by the method described elsewhere [7,8]. 2-Hydroxyethylamine and formic acid (90%) (*Aldrich*) were distilled before use. Others carboxylic acids were analytical grade commercial products (*Aldrich*). The ionic liquids purity is ≥98% for lactates and ≥99% for formates and acetates (by HPLC/MS).

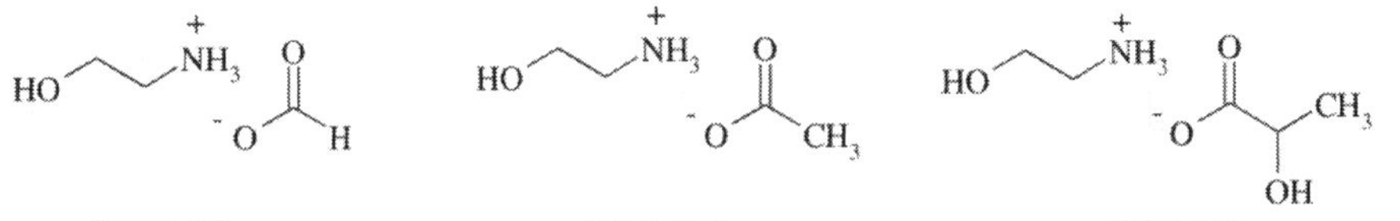

Figure 1. Chemical structures of used 2-hydroxyethylammonium carboxylate ionic liquids

2.2. Impregnation with ionic liquids

Nafion membrane samples were pretreated by sequential immersion in boiling 0.5 mol/L H_2SO_4 solution and then in boiling deionized water. After that membrane was blotted with a tissue to remove any residual liquid and thereafter dried at 80 °C in oven for 1 h. Impregnation with ionic liquids was performed by immersing the membrane samples into the corresponding IL at 50 °C temperature for 90 min. After impregnation the membranes surface was blotted with a tissue. Membranes were afterwards dried at 80 °C in oven until constant weight.

2.3. Characterizations

Stress-strain curves of the membrane samples were measured by a Tinius Olsen H1K-S UTM material testing machine. The membranes were cut into "dogbone" shaped specimens with the gauge length and width of 22 mm x 5 mm, respectively. Measurements were carried out at room temperature in air atmosphere with a constant deformation speed of 50 mm/min.

Creep test was performed at homemade apparatus at room temperature in air atmosphere. The membranes also were cut into "dogbone" shaped specimens with the same dimensions as in the previous experiment. Membranes specimens were clumped in the apparatus and a constant weight was hung from the bottom of the "dogbone" for 30 min. Elongation was recorded by a MarCator 1086 digital indicator.

Wide angle X-ray diffraction (XRD) patterns were recorded on Bruker D8 ADVANCE diffractometer. Diffraction spectra were measured using a Cu K_{α} source ($\lambda K_{\alpha 1}$ = 1.54180 Å) operated at 40 kV and 40 mA. Scans were made from 5° to 60° 2θ. The scattering angle was increased by 0.02° 2θ steps, with a time per step equal to 0.1 s.

3. Results and Discussion

3.1. Mechanical properties

The tensile stress at break of dry Nafion 112 membrane and its composites with ILs are shown on figure 2. It was found that the tensile strength is ~15 MPa for dry Nafion 112 membrane. Impregnated Nafion 112 membranes have the similar values as dry membrane. IL with acetate anion slightly increases tensile strength, but with lactate decreases it. The elastic moduli of dry and impregnated Nafion 112 membranes are shown on figure 3. The dry membrane has the highest elastic modulus (~250 MPa). The incorporation of ILs into the membrane decreases the elastic modulus value. The biggest effect on the elastic modulus has [HEA]L. Composite with this IL has 2.5 times lesser elastic modulus (~84 MPa) than a pure membrane. Elastic modulus of composite with [HEA]F also is several times smaller than a pure membrane. At the same time [HEA]A does not significant decreases elastic modulus of the Nafion 112 membrane.

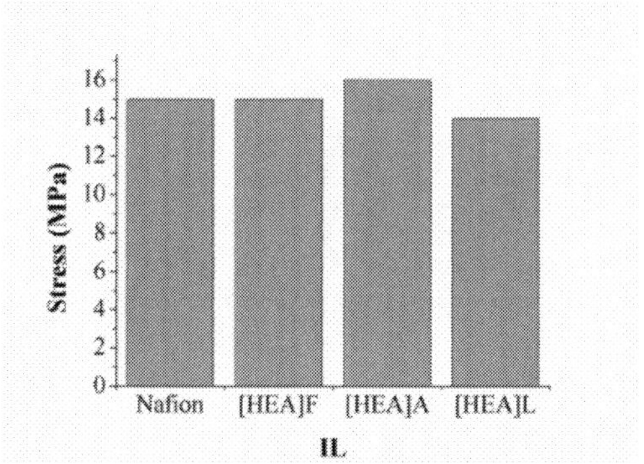

Figure 2. Tensile stress at break of Nafion 112 and impregnated Nafion membranes.

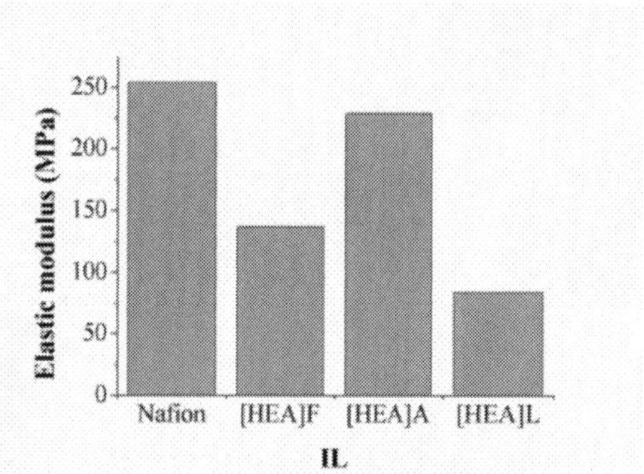

Figure 3. Elastic modulus of Nafion 112 and impregnated Nafion membranes.

Figure 4 shows the short-time creep test. Since the membrane has the permanent pressure on its surface in PEMFC, it is important to know how the constant stress will affect on the membrane in time. Nafion 112/[HEA]L composite has the biggest creep compliance as compared with the others composites. The creep compliance of Nafion 112/[HEA]F is lower than composite with [HEA]L. The dry membrane and impregnated with [HEA]A have the similar creep compliances. The creep test results show the good matching with elastic modulus values from figure 3.

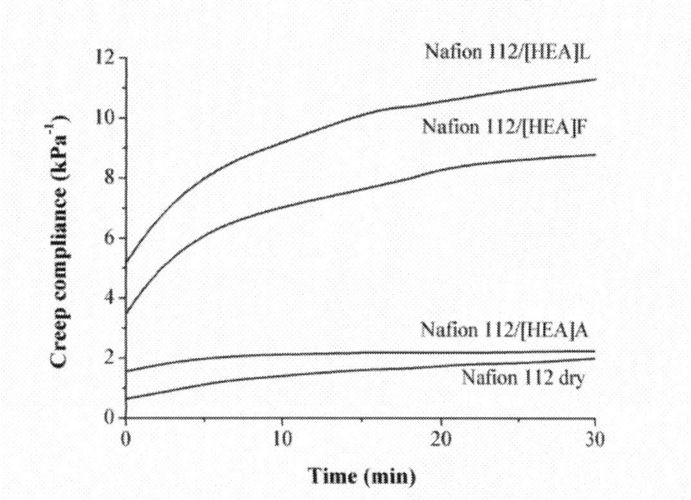

Figure 4. The short-time creep test data of dry Nafion 112 and impregnated Nafion membranes.

3.2. XRD measurements

The results of XRD measurements of dry Nafion 112 and its composite with ILs are shown on figure 5. The first maximum is attributed to the distance between two PTFE segments which form clusters in the membrane [2]. The dry Nafion 112 membrane has the peak with the maximum at ~18.1°. Composite membranes have the peak maximum at the similar 2θ, but these maximums are slightly shifted to the lower 2θ. This means that the distance increases between two PTFE segments.

Composites do not have the same shift value. The biggest shift have Nafion 112/[HEA]A and Nafion 112/[HEA]L – the peak maximum at ~17.8°, but Nafion 112/[HEA]F at ~17.9°.

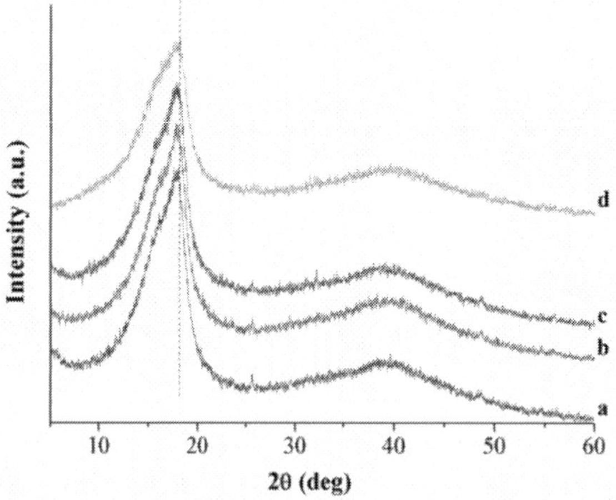

Figure 5. The XRD data of Nafion/[HEA]F (a), Nafion/[HEA]A (b), Nafion/[HEA]L (c) and dry Nafion (d).

4. Conclusions

Composite membranes were prepared from Nafion 112 and three different biodegradable 2-hydroxyethylammonium ionic liquids: [HEA]F, [HEA]A and [HEA]L. After impregnation into membrane the used ionic liquids do not decrease the mechanical strength, but at the same time ionic liquids act as plasticizers and decrease the elastic modulus as well as increase the creep compliance. Also, all ionic liquids increase the distance between two PTFE segments according to XRD spectra. The Nafion/[HEA]A has potential to be used in PEMFC according to its good mechanical properties. The PEMFC working temperature is about 80 °C, so the mechanical properties at higher temperatures need more investigation, because mechanical properties depend on glass transition temperature.

Acknowledgments

Valery Garaev is grateful to the European Social Fund (ESF) for the financial support of this research (Nr. 2009/0162/1DP/1.1.2.1.1/09/IPIA/VIAA/004).

References

[1] Satterfield M B, Majsztrik P W, Ota H, Benziger J B and Bocarsly A B 2006 *J. Polym. Sci., Part B: Polym. Phys.* **44** 2327
[2] Chabé J, Bardet M and Gébel G 2012 *Solid State Ionics* **229** 20
[3] Martínez de Yuso M V, Neves L A, Coelhoso I M, Crespo J G, Benavente J and Rodríguez-Castellón E 2012 *Fuel Cells* **12** 606
[4] Yang J, Che Q, Zhou L, He R and Savinell R F 2011 *Electrochim. Acta* **56** 5940
[5] Cho E, Park J-S, Sekhon S S, Park G-G, Yang T-H, Lee W-Y, Kim C-S and Park S-B 2009 *J. Electrochem. Soc.* **156** B197
[6] Schmidt C, Glück T and Schmidt-Naake G 2008 *Chem. Eng. Technol.* **31** 13
[7] Zicmanis A, Pavlovica S, Gzibovska E, Mekss P and Klavins M 2010 *Latvian J. Chem.* **3** 269
[8] Pavlovica S, Zicmanis A, Gzibovska E, Klavins M and Mekss P 2011 *Green Sus. Chem.* **1** 103
[9] Garaev V, Kleperis J, Pavlovica S and Vaivars G 2012 *IOP Conf. Ser.: Mater. Sci. Eng.* **38** 012064
[10] Oliveira L, José N M, Boaventura J, Iglesias M and Mattedi S B 2011 *AIP Conf. Proc.* **1400** 149

IOP Conf. Series: Materials Science and Engineering **49** (2013) 012059 doi:10.1088/1757-899X/49/1/012059

Luminescent materials based on nanoparticles of silicon dioxide for solar cells

[1]T N Nurakhmetov, [1]K A Kuterbekov, , [2]T A Schmedake, [1]A Zhanbotin, [1]A Kainarbay, [1]J M Salihodja, [1]A M Zhunusbekov, [1]S Pazylbek, [1]K Bekmyrza, [1]D Daurenbekov, [1]A Baitemirova

[1] Physical Technical Department, L.N. Gumilyov Eurasian National University, 2 Mirzoyana Str., Astana, Republic of Kazakhstan, 010008

[2] Chemistry Department, University of North Carolina at Charlotte, 9201 University City Blvd, Charlotte, NC 28223, USA

kainarbay_azh@enu.kz

Abstract. The report considers luminescent mesoporous silica dioxide particles of activated RE ions and lead sulfide nanocrystals of different sizes. For the purpose of assessing the possibility of converting by lead sulfide nanoclusters of solar radiation into the red-orange light, the following methods will be drawn to the attention: mesoporous silica particles sol-gel synthesis, additional light scattering method as an additional approach for the determination of particle size. The element composition of a particle size of 300 ± 40 nm was identified using the scanning electron microscope with the system of energy disperse analysis. Particle images were obtained using the field scanning electron microscope. Zeta Potential Analyzer was used to measure the surface charge of the particles. Particles in solution were excited by near ultraviolet and visible light. The intensity of intrinsic emission was registered in the range of 400-550 nm. The diagnosis made visible the electron induced energy transmission from the matrix to the Eu^{3+} and Tb^{3+}. Emission of Eu^{3+} and Tb^{3+} has been registered at 300K. It is assumed that the identical matrixes could work as converters of solar emission spectrum, to which some solar cells are sensitive.

1. Introduction

Luminescent silica (LS) sol-gel materials grade promising environmentally friendly class of phosphors resulting from the carboxylic acid catalyzed condensation of alkoxysilane or aminopropyltriethoxysilane precursors [1]. It is also possible to use luminescent silica as solar emission converters [2, 3]. Conversion of solar emission into electrical energy is performed by semiconductor photovoltaic converter (PVC). Efficiency of conversion is about 25% in the laboratory and 10-15% in the production conditions, which is much inferior to conventional energy sources. Solar silicon cells are sensitive only to a certain spectral range of the continuous spectrum of solar radiation. It amounts 20-30% of the entire spectral range of the solar spectrum. The increase of efficiency based on silicon is carried out follows: luminescent nanocrystalline films of various sizes are placed on the surface of silicon. Luminescence of different energy that occurs during the solar radiation can be collected from each film layer. This increases the overall efficiency of PVC with a set of nanocluster films. Luminescent materials that convert ultraviolet and visible solar radiation into the red light are placed on the silicon surface in the form of a film. This way the efficiency of PVC based on silicon can be increased.

To produce luminescent silica we used the same synthesis technique as described in previous letter [4]. The paper provides additional information about physical properties of luminescent silica material.

2. Experimental conditions

All photoluminescence measurements of luminescent silica spheres were performed in 3 g of a 30/70 ethanol/toluene mixed solvent. The solvent ratio was chosen to minimize scattering effects with the spheres by refractive index matching. We used LMCS spheres after 6 hours of 200°C calcination. This calcination terms leads to most significant photoluminescence of the LMCS particles relatively to another calcination conditions [4]. Particles concentration was 0.3 mg/ml. 1/10 (lanthanide/LMCS) mass of Eu or Tb chloride was doped into silica spheres suspended in ethanol/toluene solution using ultrasonic bath.

Photoluminescence spectroscopy data were performed with the CM2203 spectrofluorometer. Monochromator slit widths for both the emission and excitation were set to 2 nm band-pass.

For SEM images, LMCS solution was deposited on a doped silicon wafer and the solvent was evaporated. SEM imaging performance assisted with Raith 150 SEM/Leo Gemini Column and JOEL-6460LU with the EDAX system. 10 kV voltage acceleration was applied. Brookhaven Corp. light scattering technique was used as an additional method for the determination of particle size. Brookhaven Corp. Zeta Potential Analyzer was used to measure the surface charge of the particles.

3. Results and discussion.

Mesoporous colloidal silica spheres possess the desirable properties of low polydispersity at the sub-micron level and very high surface area resulting from the mesoscale porosity. The LMCS spheres used in this manuscript had an average diameter of 377 nm and respective standard deviation of 3% (Figure 1).

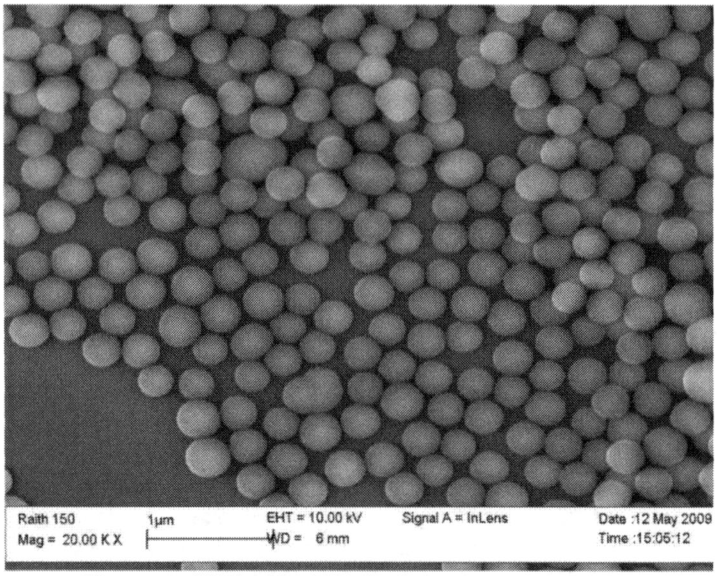

Figure 1. SEM of LMCS (TAP6) after calcination.

Energy dispersive analysis results of LMCS particles are in good agreement with the theoretical expectation. The ratio of the intensities of the peaks for silicon (Si) and oxygen (O) is 1:2, as expected for a SiO_2 compound. The index shows peaks of carbon and nitrogen impurities. These atomic weights of LMCS based on k(α)-lines intensities are given in Table 1.

Table 1. The atomic content of the elements in LMCS

	atomic %			
Sample#	C k(α)	N k(α)	O k(α)	Si k(α)
S1*	16.56	3.67	51.09	28.69
S2	29.19	6.00	41.57	23.24
S3	27.10	4.92	40.05	27.86
S4	20.55	4.61	48.77	26.08

*S1 - TEOS, S2 -TAP noncalcined, S3 -TAP6, S4 -TAP48 [4].

Dynamic light scattering method for TEOS particles [4] gives their sizes with very wide range. Figure 2 shows standard distribution for sizes of given particles. However, SEM analysis demonstrates monodisperse-like distribution. A discrepancy in the data should be attributed to the partial aggregation of the particles in ethanol/toluene solution.

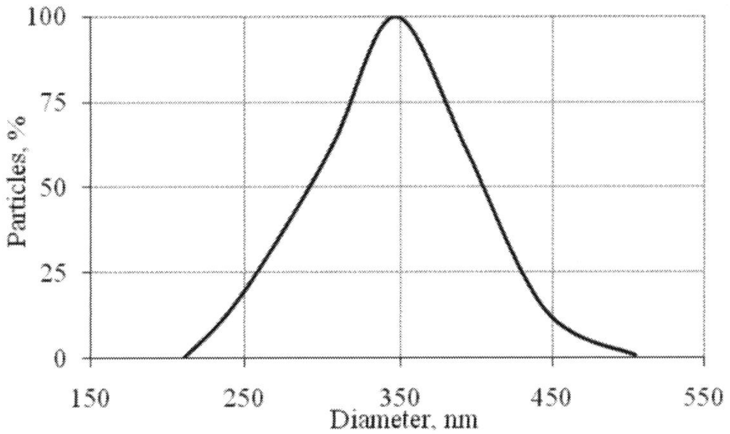

Figure 2. The statistical distribution of TEOS particles sizes.

Potentiometric measurements indicate the presence of a positive charge on the particle surface. For noncalcined particles, the charge value is 43.5 ± 0.0216 mV, for 6 hours at 200°C calcined particles, the value is 48.52 ± 0.0167 mV. The positive charge of the noncalcined particles can be attributed to the presence of NH^{3+} groups on the particle surface. After doping lanthanides chlorides to the calcined particles, the surface charge increases up to 56.45 ± 0.0272 mV.

Photoluminescence spectroscopy data have been collected for different excitation wavelengths. The excitation was fixed at 314, 320, 360, and 378 nm according to luminescence excitation spectra of calcined LMCS (Figure 3).

Figure 3 shows most significant excitation wavelength for calcined particles at 378 nm. As expected, LMCS particles give highest luminescence when exited at 378 nm (Figure 4). Comparing the emission spectra obtained under excitation at 320 nm [4] and at 378 nm, the maximum intensity of the last one is ~7 times higher.

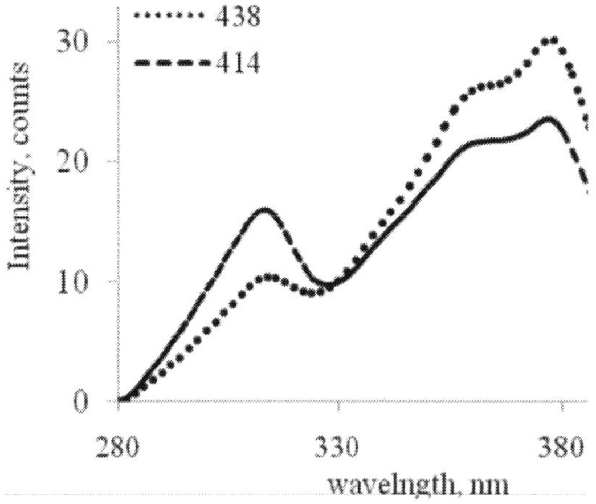

Figure 3. Excitation spectra of calcined LMSC measured for the 414 and 438 nm emission wavelengths.

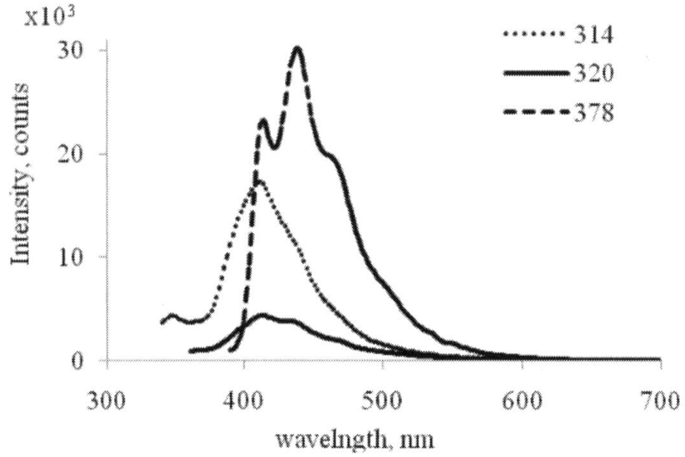

Figure 4. Luminescence spectra of calcined LMCS excited at 314, 320, 378 nm.

It is important to note that the high emission intensity of silica dioxide system provides matrix-to-lanthanide energy transfer. Thus, next step of the current research is to provide photoluminescence measurements of LMCS particles doped with lanthanides and excited at optimum excitation energy for matrix.

This research was sponsored by Ministry of Education and Science of the Republic of Kazakhstan, Award #511, from 05.04.2012.

References

[1] Green WH, Le KP, Grey J, Au TT, Sailor MJ. Science 1997;276(5320):1826-8.
[2] K Bothe, K Ramspeck, D Hinken, C Schinke, J Schmidt, S Herlufsen, R Brendel, J Bauer, J-M Wagner, N Zakharov, and O Breitenstein. 2009 J. Appl. Phys. 106, 104510
[3] L Desmet, J M. Ras, D K G de Boer, and M G Debije. 2012 Optics Letters, Vol. 37, Issue 15, pp. 3087-3089
[4] A Zhanbotin, R B Soriano, A K Ikonen, T N Nurakhmetov, T A Schmedake. 2010 Materials Letters 65 10–12

IOP Conf. Series: Materials Science and Engineering **49** (2013) 012060 doi:10.1088/1757-899X/49/1/012060

Preparation and characterization of nanostructured Fe-TiO_2 thin films produced by electrophoretic deposition

Ineta Liepina[a], Gunars Bajars[a], Andrejs Lusis[a], Gundars Mezinskis[b], Martins Vanags[a]

[a]Institute of Solid State Physics, University of Latvia, Latvia
[b]Institute of Silicate Materials, Riga Technical University

liepina.ineta@gmail.com

Abstract. Fe-TiO_2 thin films have been produced by a sol-electrophoretic deposition method on metallic Ti foil substrates. X-ray diffractograms and Raman spectra confirms the anatase structure of deposited and annealed Fe-TiO_2 films. Obtained Fe-TiO_2 thin films demonstrate photoactivity under visible light radiation due to the doping with Fe^{3+} ions. Open circuit potential results shows that increase of the film thickness and surface area improves the photoactivity of Fe-TiO_2.

1. Introduction

In recent years nanostructured titanium dioxide (TiO_2) thin films have been focused great interest due to its unique optical and electrical properties, as well as low cost, durability and chemical inertness. TiO_2 thin films have a wide range of practical applications, such as photocatalysts, solar cells, optical filters, antireflective coatings, planar wave-guides and gas sensors. The wide band gap (E_g = 3.21 eV for anatase and E_g = 3.00 eV for rutile [1]) limits application of TiO_2 only in UV irradiation therefore modifications are necessary in order to gain photoactivity in visible light region of the spectrum (~ 40 % of solar radiation). Generally accepted way of extending the spectral response of TiO_2 to the visible light is doping with transition metal ions like Mn(II), Cu(II), Cr(II), Fe(III), Cr (III), V(V), Mo(V) [2, 3, 4].

In this case Fe^{3+} (0.64 Å) is a favorable choice since its radius is close to that of Ti^{4+} (0.68 Å). Furthermore iron ions act like hole-electron pair traps and reduce their recombination rate because the energy level of Fe^{2+}/Fe^{3+} is close to that of Ti^{3+}/Ti^{4+} [5]. The content of Fe^{3+} ions in TiO_2 for optimal photoactivity has been a subject of discussion for some time, however some studies find that 0,1 mol% increases photocatalytic properties most significantly [6, 7].

Photoactivity of the doped TiO_2 substantially depends on the dopant ion and its concentration, as well as preparation method and treatment. Electrophoretic method has been widely applied to obtain TiO_2 thin films [8]. Advantages of electrophoresis over other coating methods are the homogenity of thin films on large substrates, as well as better control of particle size and shape.

The purpose of current work was to electrophoreticaly obtain Fe-TiO_2 thin films that have enhanced photoactivity in visible light. Characterization of these films include determination of TiO_2 phase and particle size (X-Ray diffraction, Raman spectroscopy), as well as surface morphology and microstructure analysis (SEM). To ensure Fe^{3+} content in thin films X-ray fluorescence spectrometry was used. The thickness of films was measured with profilometer.

2. Experimental

The first part of this work includes preparation of Fe-TiO_2 sol. Titanium tetraisopropoxide (TTIP, Aldrich, >97% purity) mixture with isopropanol was hydrolyzed in distilled water, followed by vigorous stirring that lead to peptization. Nitric acid (Lachner, 65%) was used as a catalyst for TiO_2 dispersion in media and Fe^{3+} ions were introduced with iron(III) nitrate nonahydrate (Sigma-Aldrich, >98% purity).

To produce an electrophoretic growth, Pt and Ti electrodes are placed parallel in Fe-TiO_2 sol with a distance 1 cm in between. A constant voltage of -20 V was applied by a DC System Power Supply Agilent Technologies N5751a between the electrodes and held for 1 – 4 min. The as-deposited thin films were first dried at 100 °C for 24 h and then annealed at 500 °C for 2 h.

The morphology of TiO2 films was characterized by scanning electron microscope FEI SEM "Quanta 200". The X-ray diffraction patterns (XRD) were recorded on a Philips X'Pert Pro MPD and Raman spectra by using spectrometer Spex Ramalog. Thickness of Fe-TiO_2 thin films was measured with Veeco Dektak-8 Stylus.

Measurements of photoactivity were carried out in two electrode cell with 0.1 M NaOH water solution as an electrolyte and Pt foil as reference electrode and counter electrode. The samples were exposed to light source (120 W mercury lamp) placed in distance of 25 cm from quartz window of cell. A filter BS3 was used in order to measure open-circuit potential only in visible light. The open-circuit voltage and photocurrent density were measured by potentiostat EG&G PAR Model 173.

3. Results and Discussion

The X-ray diffraction pattern is shown in Figure 1. Diffraction peaks at 25.3° 2θ and 36.9° 2θ are characteristic for anatase phase (average crystalline size 50 – 60 nm). However X-ray diffractogram also shows less intense lines that represents rutile phase at 27.5 ° 2θ and 36.1° 2θ. It is well known, that the mixture of both forms performs with higher photoactivity than pure anatase phase [8].

Figure 2 shows the Raman spectrum of annealed TiO_2 thin films. The bands at 150 cm^{-1} (E_g), 400 cm^{-1} (B_{1g}), 515 cm^{-1} (A_{1g}) and 633 cm^{-1} (E_g) confirm phase transition of electrophoreticaly as-deposited TiO_2 to anatase phase.

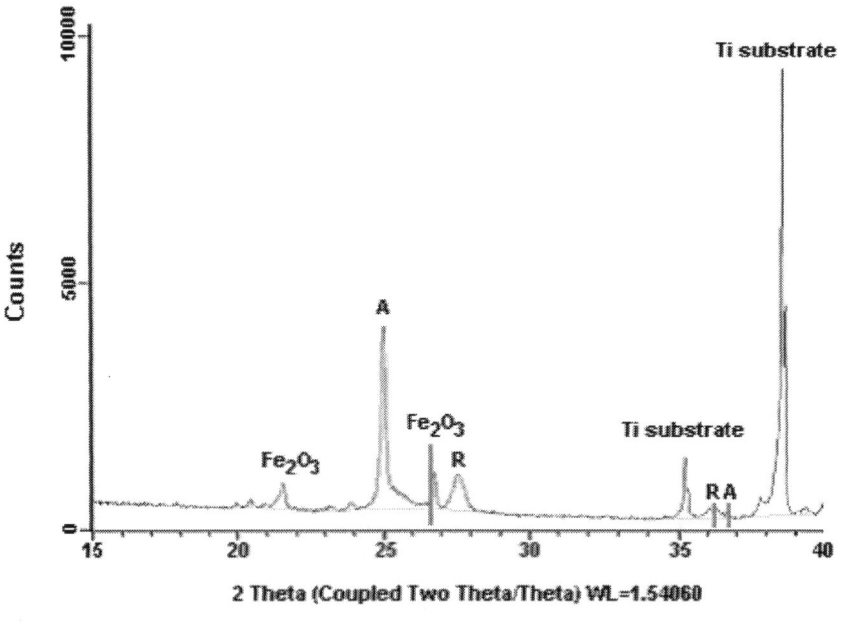

Figure 1. X-ray diffraction pattern of annealed TiO_2 thin film (EPD time –3 min) on Ti substrate (A – anatase, R- rutile)

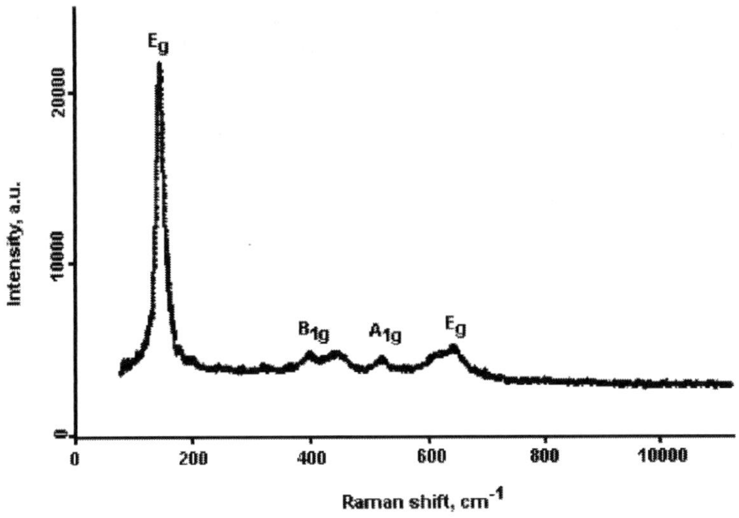

Figure 2. Raman spectrum of annealed TiO_2 thin film (EPD time –3 min) on Ti substrate.

Figure 3 shows the typical SEM images of TiO_2 thin films obtained after 1 min (A), 2 min (B) and 4 min (C) of electrophoresis.

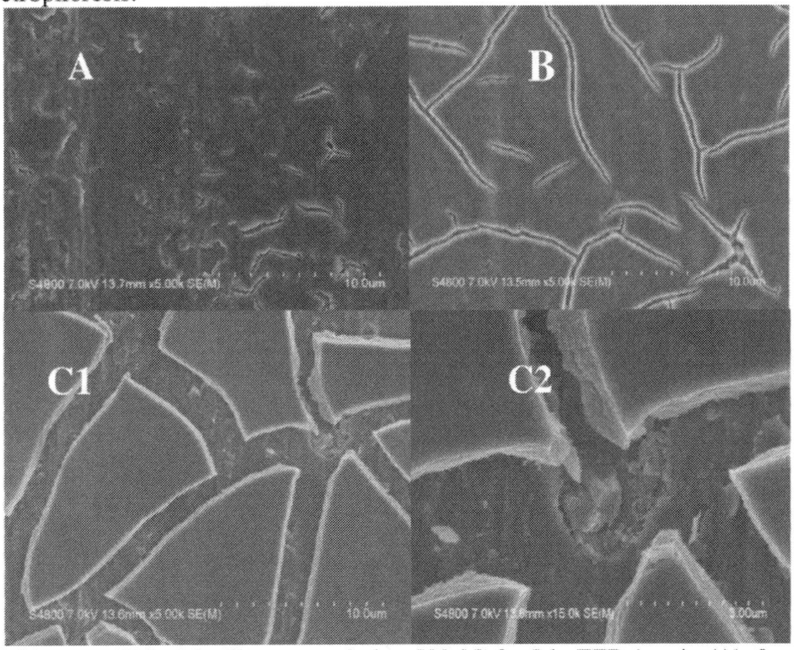

Figure 3. SEM images TiO_2 thin films annealed at 500 °C for 2 h. EFD 1 min (A), 2 min(B) and 3 min (C1 and C2).

Image A of thin film with average thickness of 1.9 μm shows a homogeneous layer of Fe-TiO_2 without significant defects. With the increase of deposition time the following samples seen in images B (3.5 μm) and C (8.1 μm) have developed cracks in thermal treatment that become wider with the increase of thin film thickness. The microstructure of cracks shown in image C2, confirms that the lower part of these cracks is just a deeper layer of Fe-TiO_2 and the splitting has not exposed Ti substrate.

Photoactivity expressed through open circuit potential changes is shown in Figure 4. The Fe-TiO_2 thin film obtained by 4 min electrophoretic deposition (EFD) exhibits the highest potential change both under the UV-VIS and VIS light radiation (0.065 V and 0.040 V, respectively). The same behavior was observed for thin films obtained by 3 min EFD (0.040 V and 0.030 V) and 2 min EFD (0.037 V and 0.030 V). The least activity both in UV-VIS and visible light was shown by Fe-TiO_2 film obtained by 1 min electrophoresis (0.025 V and 0.014 V, respectively). The measured current densities for all obtained thin films were in the range of 0.1 - 1 $\mu A/cm^2$.

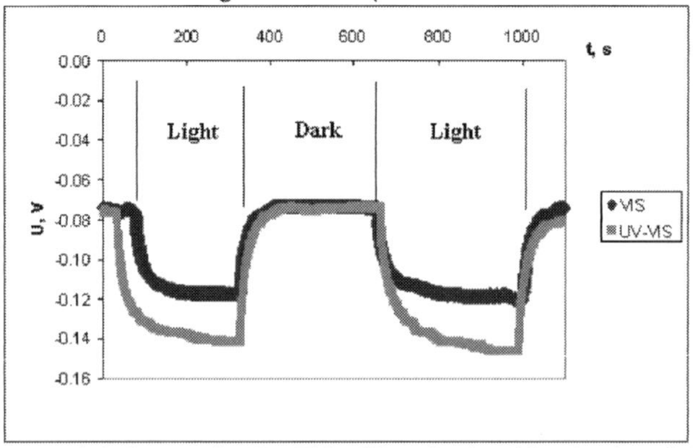

Figure 4. Open circuit potential change for Fe-TiO_2 film obtained by 4 min of electrophoresis.

The distinction of OCP results leads to conclude that active surface area of TiO_2 thin films has been expanded by increasing of the film thickness and surface cracking. Contrary, thinner and more homogenous Fe-TiO_2 thin films exhibit less photoactivity as it is shown for sample A (Fig 3).

Conclusions

In summary, photoactive Fe-TiO_2 thin films have been successfully obtained by a sol-electrophoretic deposition method on Ti substrate. XRD and Raman spectra confirmed that the crystallization of Fe-TiO_2 thin films annealed at 500 °C occurred mainly in the form of anatase phase. Obtained thin films demonstrate photoactivity under visible light radiation due to the doping with Fe^{3+} ions. Thickness and morphology of the Fe-TiO_2 thin films substantially have impact on their photoactivity. Increasing of the film thickness and surface area improves the photoactivity of Fe-TiO_2.

Acknowledments

This work was financially supported by European Regional Development Fund project No.2010/0243/2DP/2.1.1.1,0/10/APIA/VIAA/156. The authors thank J. Kosmaca and D. Erts, University of Latvia Institute of Chemical Physics, for their technical support for SEM observation. Also the authors thank J. Gabrusenoks for his support for Raman spectrum.

References

[1] Reyes-Coronado D, Rodríguez-Gattorno G, Espinosa-Pesqueira M E, Cab C, de Coss R and Oskam G 2008 *IPO Conf. Series: Nanotechnology* **19** 145605
[2] Zhang Y and Li Q 2013 *Solid state science* **16** 16
[3] Liepina I, Bajars G, Gabrusenoks J, Grinberga L, Kleperis J and Lusis A 2012 *IPO Conf. Series: Material Science and Engeneering* **38** 012059
[4] Lu X, Ma Y, Tian B and Zhang J 2011 *Solid state science* **13** 625
[5] Wantala K, Tipayarom D, Laokiat L, Grisdanurak N 2009 *Reaction Kinetics and Catalysis letters* **97** 249
[6] Zhao W, Fu W, Yang H, Tian C, Li M, Ding J, Zhang W, Zhou Z, Zhao H and Li Y 2011 *Nano-Micro letters* **3** 20
[7] Mohamed R M, Ordikhani F, Fray D J and Khomamizadeh F 2011 *Particuology* **9** 161
[8] Bakardjieva S, Šubrt J, Štengl V, Dianez M J and Sayagues M J 2005 *Applied Catalysis B: Environmental* **58** *193*

IOP Conf. Series: Materials Science and Engineering **49** (2013) 012061 doi:10.1088/1757-899X/49/1/012061

Annealing effect on CdS films: transition from glass to ITO

Natalia Maticiuc* **and Jaan Hiie**

Tallinn University of Technology, Ehitajate tee 5, Tallinn, Estonia

Abstract. Structural and optical properties of CdS films deposited on glass and ITO substrates and annealed in H_2, N_2 or H_2+N_2 atmospheres at 250 °C and 400 °C were compared. The optical thickness 405±10 nm of the CdS films was found not to be influenced by any annealing conditions or substrate. The as deposited films on both substrates had similar diffraction patterns with one narrow peak corresponding to the (111) cubic plane at 2θ 26.70° and 26.75°, respectively. Higher density of nucleation centres on the ITO surface generated denser CdS films. In the annealing process of CdS/ITO a slower shift of both (111) peak and lattice constant was observed in the direction of pure zinc blende modification. Similar values of band gap and transmittance were determined for as-deposited CdS while annealing at 400 °C decreased transmittance by 5% and 16% for CdS on glass and ITO, respectively. Transmittance dropped because of the destruction of hydroxide group in the CdS lattice, the formation of cadmium excess and the reduction of SnO_2 to black SnO in the ITO structure.

1. Introduction

CdTe- and CIGS- based photovoltaic solar cells in a superstrate configuration are made on glass substrates coated by transparent conductive oxide (TCO) like tin oxide (SnO_2) or indium tin oxide (ITO) over which CdS layer is deposited. Combined TCO–CdS window layers [1-4] and the effect of these bilayers on the final photovoltaic response [5] had been studied by many authors. Usually TCO-coated glass substrates with suitable photovoltaic quality are commercially available, while the optimization of CdS thin films remains in progress at laboratory scale.

As an optimization route for CdS thin film, annealing in different atmospheres at certain time-pressure-temperature conditions may be used. In [6] we studied the properties of CdS/glass thin films after H_2 annealing. A relationship between annealing conditions and desirable properties of CdS on glass was obtained: temperatures up to 400 °C had a weak influence on the morphological and optical properties; 1h annealing was enough to reach stable electrical properties; CdS films annealed at 250 - 400 °C had resistivity in the region of $2.3\cdot10^{-2} - 6.1\ \Omega\cdot cm$, respectively. It is interesting to verify if the knowledge gained about the properties of annealed CdS/glass may be applied for the annealed CdS/ITO/glass structures. Similar studies have shown the influence of annealing in air [7] or nitrogen [8] on CdS properties when deposited on different substrates such as glass, ITO, SnO_2 or other materials. However, information about CdS/ITO/glass bilayers annealed in H_2 atmosphere is scarce. This paper compares the optical, structural and electrical properties of CdS layers deposited on glass and ITO substrates and annealed in H_2, N_2 and H_2+N_2 gases.

*Corresponding author. Tel.: +372 620 3366
E-mail address: nataliamaticiuc@yahoo.com (N. Maticiuc)

2. Experiment

Polycrystalline CdS films were deposited by chemical bath deposition (CBD) on both uncoated soda-lime glass and commercial ITO coated glass substrates. The plates were accurately cleaned and immersed in the deposition solution. The temperature and agitation speed of the solution were 85 °C and 500 rpm, respectively. One deposition lasted for 30 minutes, but for thicker CdS films the process was repeated three times. After deposition the films were vacuum dried at 120 °C to remove most of the secondary phases of water, hydroxides and organic impurities. This drying was the last stage of preparation for the so-called as-deposited CdS layers. We choose the thermal treatment conditions based on our investigation on H_2 annealing of CdS/glass: normal pressure, 1 h duration at 250 °C and 400 °C. For comparison, annealing in N_2 and H_2+N_2 (H_2 concentration of was 5%) was also applied.

Crystallographic investigations were performed using the X-ray diffraction (XRD) technique in the Bragg-Brentano (θ-2θ) geometry by a Rigaku Ultima IV diffractometer with Cu-Kα radiation. Crystallite size, lattice constant and interplanar distance were computed by the PDXL software (Version 1.4.0.3) on the Rigaku system. The optical characteristics were evaluated at room temperature in the range of 200–2500 nm on the Jasco V-670 UV-VIS spectrophotometer equipped with an integrating sphere. Total optical transmission and reflection spectra were used to determine the band gap and optical thickness of CdS.

3. Results and discussions

Optical properties were analyzed on the basis of transmittance spectra and E_g values shown in figure 1a and table 1, respectively.

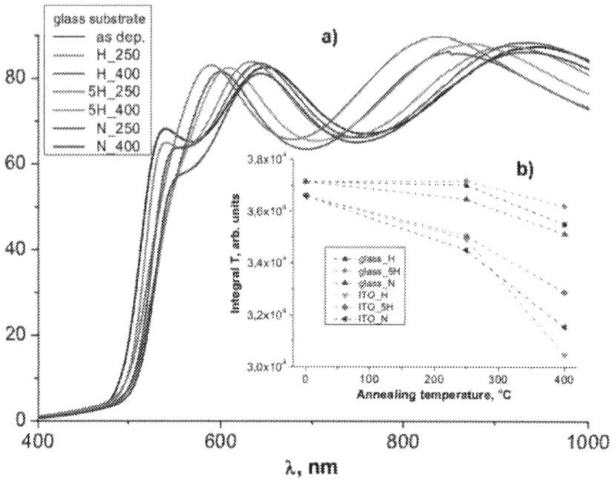

Figure 1. a) Optical transmittance spectra of the CdS films. b) Calculated integral transmittance of CdS/glass and CdS/ITO as a function of annealing temperature and ambient.

Table 1. Band gap (E_g), lattice constant (a), interplanar distance (d) and crystallite size (S) of the CdS films deposited on different substrates and annealed in H_2 (H), N_2 (N) and H_2+N_2 (5H) ambient.

	E_g, eV		**a, Å**		**d, Å**		**S, Å**	
Substrate →	*glass*	*ITO*	*glass*	*ITO*	*glass*	*ITO*	*glass*	*ITO*
as deposited	2.454	2.464	5.783	5.775	3.337	3.330	405.2	428.0
H-250 °C	2.412	2.373	5.803	5.793	3.346	3.341	428.3	457.0
H-400 °C	2.364	2.379	5.821	5.804	3.354	3.347	391.3	394.3
5H-250 °C	2.391	2.373	5.792	5.790	3.340	3.340	430.9	460.0
5H-400 °C	2.348	2.365	5.806	5.802	3.345	3.346	377.4	364.0
N-250 °C	2.377	2.376	5.797	5.787	3.344	3.338	420.0	472.0
N-400 °C	2.352	2.367	5.806	5.797	3.351	3.343	368.3	347.7

Almost similar changes of band gap were found for CdS on glass and ITO substrates, independent of the annealing atmosphere. When the annealing temperature increased, the band gap of CdS/glass gradually decreased for all three atmospheres from 2.45 eV (as-deposited CdS) to 2.35 eV (annealed at 400 °C). With ITO the E_g of CdS reached a constant value (2.375 eV) for N_2-containing ambients at both annealing temperatures (table 1).

We estimated the integral transmittance in the region of 500-1000 nm as the area under the T plot (figure 1b). With a glass substrate a similar trend of CdS transmittance was observed for all annealing gases - a slow decrease at 250 °C followed by an abrupt decrease at 400 °C (~5 %). However, the lowest transmittance was emphasized for a CdS sample annealed in N_2. The decreased transmitance of CdS could be explained by hydroxide group incorporated into the CdS lattice the destruction of which starts at 300 °C [6]. Independent of the annealing atmosphere, the hydroxide group incorporated into the CdS lattice leads to the formation of $CdS_{1-n}n(OH)_S$ solid alloy. When temperature is applied, H_2O evaporates and Cd precipitate [9] is formed (1, 2).

$$CdS_{1-n}n(OH)_S \xrightarrow{t} CdS_{1-2n}n(O_S)n(V_S) + nH_2O\uparrow \quad (1)$$

$$CdS_{1-2n}n(O_S)n(V_S) \xrightarrow{t} Cd_{1-m}S_{1-2n}n(O_S)n-m(V_S) + mCd\downarrow \quad (2)$$

When CdS/glass was annealed, the Cd precipitate (2) decreased the transmittance of CdS (figure 1b). However, in the presence of H_2 the CdO decomposes, as oxygen is eliminated by the release of H_2O, and CdS band gap increases [6]. Therefore, the CdS transmittance was higher after the H_2 annealing than that of N_2 one. This reveals an advantage of H_2 annealing for CdS films.

For a CdS film on ITO, the transmittance decreased faster up to 250 °C than with the glass case. Thus, the contribution of Cd precipitate (2) to the decomposition of SnO_2 should be taken into account (3, 4). Both products of reaction (3) resulted in the reduced transmittance of the CdS/ITO bilayer.

$$SnO_2 + Cd \rightarrow SnO + CdO \quad (3)$$

$$2SnO \rightarrow Sn + SnO_2 \quad (4)$$

Still, when CdS/ITO was annealed at 400 °C in H_2, the transmittance decreased by 16% compared to as-deposited CdS. A new reaction (5) may compete with (3) due to higher activity of the H_2 compared to the metallic Cd from CdS. These processes resulted in the blackening of the ITO substrate and an abrupt drop of CdS transmittance.

$$SnO_2 + H_2 \rightarrow SnO + H_2O\uparrow \quad (5)$$

The transmittance behavior of CdS on both glass and ITO substrates suggested that H_2+N_2 mixture instead of pure H_2 or N_2 gases is needed for annealing.

The structural properties of CdS were analyzed based on XRD measurement results. Similar XRD patterns with one narrow (111) diffraction peak of the cubic plane at 26.70° and 26.75° were found for CdS films on glass and ITO substrates, respectively. Figure 2 shows that both peaks are in the close proximity of the (002) hexagonal plane at 26.83°. Twice higher (111) peak of CdS/ITO than that of CdS/glass suggests a more oriented CdS film. This is attributed to the strong influence of the oriented crystalline surface of ITO. The high density of nucleation centers on ITO generates a denser CdS film, which might also explain the decreased transmittance of CdS on ITO (figure1b).

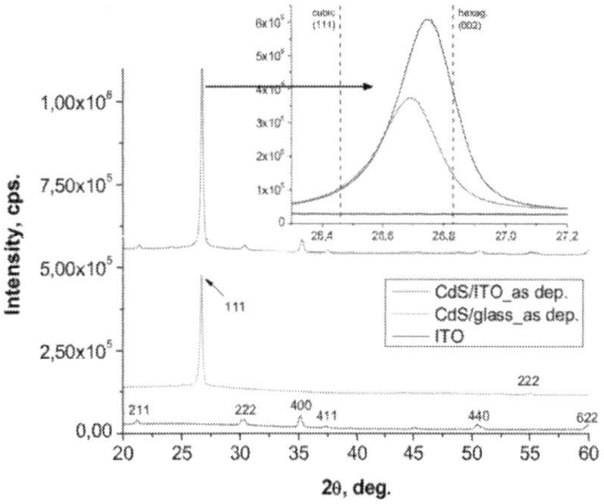

Figure 2. XRD pattern of as-deposited CdS on glass and ITO substrates.

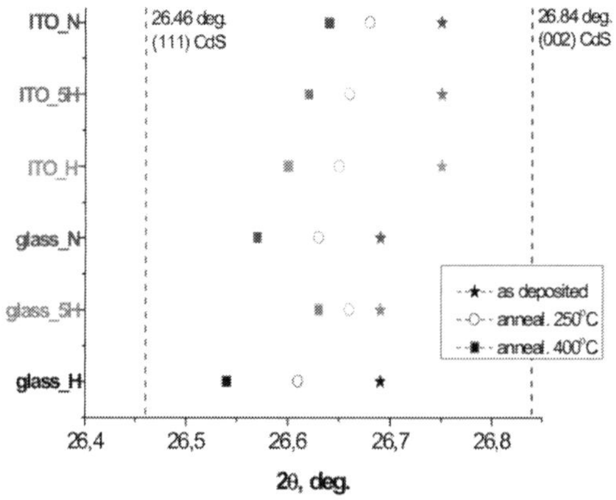

Figure 3. Influence of annealing conditions on the shift of the main (111) peak of CdS.

Interestingly for both substrates, the main (111) peak shifted in the direction of the zinc blende CdS plane at 26.46° in all three ambients when the annealing temperature increased (figure 3). However, a slower shift of the main cubic peak and lattice constant (table 1) in the direction of zinc blende modification of CdS was found in the case of the ITO substrate during annealing. The shift of this main peak infers that the stressed lattice of as-deposited CdS relaxed during annealing at 250 - 400 °C, after which a more pure cubic CdS phase was obtained. The comparison between the three annealing atmospheres in the case of glass substrate shows that the H_2 ambient contributed to a more cubic structure of CdS, while the H_2+N_2 ambient had a weak influence on the CdS lattice. In the case of the ITO substrate no significant differences were found when CdS was annealed at higher temperatures.

4. Conclusions

An abrupt reduction of transmittance was found for CdS/ITO annealed in H_2 at 400 °C. This might limit the applicability of CBD CdS for superstrate solar cells. We minimized the transmittance by annealing CdS/ITO in a mixture of H_2 + N_2. The oriented ITO contributes to denser CdS films. The typical shift of the main (111) peak is characteristic for CdS on both glass and ITO. Generally, the structural and optical properties of CdS films on glass and ITO are quite similar.

Acknowledgements

Support by DoRa, 7FP GA-2008-230861, TK114, ETF9142, AR10128, SF0140092s08.

References

[1] Oliva A I, Castro-Rodriguez R, Ceh O, Bartolo-Perez P, Caballero-Briones F, Sosa V 1999 *Appl. Surf. Sci.* **148** 42
[2] Schaffner J, Feldmeier E, Swirschuk A, Schimper H-J, Klein A, Jaegermann W 2011 *Thin Solid Films* **519** 7556
[3] Lee J-H 2007 *Thin Solid Films* **515** 6089
[4] Castro-Rodriguez R, Oliva A I, Sosa V, Caballero-Briones F 2000 *Appl. Surf. Sci.* **161** 340
[5] Romeo N, Bosio A, Canevari V, Terheggen M, Vaillant Roca L 2003 *Thin Solid Films* **431** 364
[6] Maticiuc N, Hiie J, Potlog T,Valdna V, Gavrilov A 2012 *Mater. Res. Soc. Symp. Proc.* **1324** 69
[7] Martinez M A, Guillen C, Herrero J 1999 *Appl. Surf. Sci.* **140** 182
[8] Metin H, Erat S, Durmus S, Ari M 2010 *Appl. Surf. Sci.* **256** 5076
[9] Moreno M S, Varela A, Otero-Diaz L C 1997 *Phys. Rev. B* **56(9)** 5186

IOP Conf. Series: Materials Science and Engineering **49** (2013) 012062 doi:10.1088/1757-899X/49/1/012062

Glucose oxidase as a biocatalytic enzyme-based bio-fuel cell using Nafion membrane limiting crossover

S Naidoo[1], Q Naidoo[2], H Blottnitz[1] and G Vaivars[3]

[1] University of Cape Town, Environmental and Process Systems Engineering, Chemical Engineering, Cape Town, South Africa
[2] HySA/Catalysis Competence Center, University of Cape Town, Cape Town, South Africa
[3] ISSP, Department of Chemistry, University of Latvia, Riga, Latvia

E-mail: Guntars.Vaivars@cfi.lu.lv

Abstract. A novel combination for an Enzyme-based Biofuel cell included a Nafion membrane as an ion transporter that maintained a working cell charge and inhibited membrane degradation. The prototype cell chamber used oxygen (O_2) in the cathode cell and glucose in the anode. The Nafion membrane stability studied here was evidently in the region of 0% loss of conductivity as the charge was constant and increased after the addition of glucose. The prototype cell chamber used NaCl in the cathode cell and glucose oxidase (GOx) in the anodic chamber was successfully studied for membrane stability showed in this study no evidence of poisoning from membrane leakage in a controlled pH environment. There was no crossover at the anaerobic operating ambient temperatures and under physiological pH 5 - 7 conditions. In this research we have successfully used a Nafion membrane together with GOx and under controlled conditions produced respectable power densities.

Introduction

Enzyme-based biofuel cells (EBBFC's) were first reported in 1964 using glucose oxidase (GOx) as the anodic catalyst and glucose as the fuel [1]. Developments to improve the enzyme based fuel cell have significantly advanced the technology since [2-5]. Biological fuel cells (BFC's) use Bio-catalysis where naturally occurring enzymes in a living organism that is able to transform organic compounds into sources of energy. The specialized function and applications include implantable devices, sensors, drug delivery, micro-chips, and portable power supplies [6-8]. Enzymes are isolated by laboratory validated techniques to perform these catalytic functions. Glucose oxidase alternatively known as Beta-D-glucose:oxygen 1-oxido-reductase, D-glucose-1-oxidase, Glucose aerodehydrogenase, Glucose oxyhydrase or GOD was used in this study. GOx catalyses the oxidation of glucose as follows:

Beta-D-glucose + $O_2 \rightarrow$ D-glucono-1,5-lactone + H_2O_2.................... (1)

GOx functioning depends on a cofactor, flavin adenine dinucleotide (FAD). FAD is commonly found as a component in biological oxidation-reduction (redox reactions). Redox reactions involve a reduction and oxidation process where gain or loss of electrons from a molecule occurs. In the GOx-catalysed redox reaction, FAD works as the initial electron acceptor and is reduced to $FADH_2$. The process involves the further processing of where $FADH_2$ is oxidized by the final electron acceptor, molecular oxygen (O_2), which can do so because it has a higher reduction potential. O_2 is then reduced to hydrogen peroxide (H_2O_2). Enzyme-based fuel cells feasibility has encouraged researchers to explore its potential further due to the increased production rates associated with enzymes that in turn provides an increased bio catalysis rate. BFC's can readily produce a source of electrons however there is a critical deficiency in transporting the electrons to the electrode for the active site. Advances involving the immobilization of enzymes at electrode surfaces using many different methods have significantly facilitated the electron transfer rate to the electrode making EBBFC's an even more attractive prospect. The immobilization, of the enzyme, Electrode Surface Enzyme Immobilization (ESEI) supported by electron carriers shuttling the electron to the electrodes was evidently in progress throughout the working period of the EBBFC. However previously direct electron transfer (DET) [9] between an enzyme and electrode has only been observed with several enzymes, such as cytochrome c, laccase, hydrogenase, and several peroxidases, including microperoxidases [10-13].

Although much of the early attention was devoted to the study of reactions occurring at anodes, biological molecules can also be utilised to catalyse the reduction of oxygen at the cathode as an alternative to the classical use of platinum. Microperoxidase [14] can be immobilised onto a gold cathode and along with a quinone modified cathode be utilized in a biofuel cell fuelled by NADH and H_2O_2. Similar cathodes were used along with apoglucose oxidase/quinone/flavin adenine dinucleotide phosphate modified anode [15] to construct a glucose/H_2O_2 fuel cell capable of generating 310 mV with a power output of 32 μW. The same anodes and cathodes could also be utilised in a biofuel cell containing two immiscible solvents with the anode being immersed in aqueous glucose and the cathode being immersed in cumene peroxide/ dichloromethane [16]. Chemical and physical methods of ESEI do exist and most of the EBBFC's reported thus far have been configured employed physically immobilized enzymes. It has become common practice to absorb the enzymes onto conductive particles such as carbon black or graphite powder. GOx was first adsorbed on synthetic graphite particles, and then the enzyme-adsorbed particles were suspended with 2-hexadecanone and ferrocene in a solvent of chloroform.

1. Experimental

The anode chamber contained Glucose Oxidase (Medical Grade) and glucose, 0.01% (m/v). The anode chamber contained neither facilitator nor mediator. The cathode chamber contained 10% (m/v) NaCl (Kimix). The electrodes used were carbon graphite rods (40mm length, 4mm diameter) 1 per chamber. Circuit connected using copper wire and chrome plated electrically conductive crocodile clip connections. The clip connections were not in contact with the solutions in either chamber. The experiments were conducted under ambient room temperature conditions and ranged from 20-25°C.

IR spectra were taken using Paragon 500 Infrared Spectrometer. pH influence on open circuit voltage (vs. E): 0.001 M HCl (Kimix batch no. 264) was monitored using Metrohm pH meter. Drop wise addition of NH4OH (Kimix) to maintain the optimum pH neutralizing the acidic properties of Nafion was employed. Electrochemical measurements have been performed using Metrohm Auto lab system.

2. Results and discussion

In this study a simplistic EBBFC construction of a glucose/O_2 biofuel cell and use of a of proton exchange membrane for adequate separation of anode and cathode compartments was tested. Obtained current density ranged from 0.6-0.5 $\mu A cm^{-2}$ at -0.30 V minimum. The OCV also varied and a minimum was recorded at -0.30 V. This was at physiological pH.

Enzyme based biofuel cell using GOx and glucose as a fuel source was conducted at pH as figure 1 represents the optimum functioning pH was approximately 5.5-6.0. The potential was found to be dependent on the pH. Using a combination of Graphite (Anode), Graphite (Cathode), membrane including Nafion use as the research focus area the following power output of 30-250 $\mu W/cm^2$ was obtained.

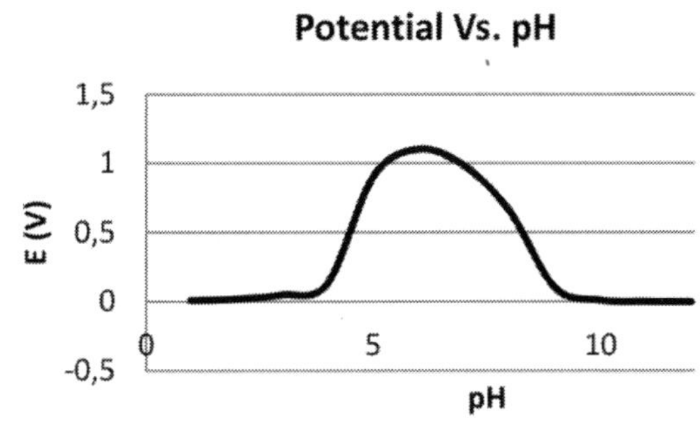

Figure 1. pH determination at the optimum potential.

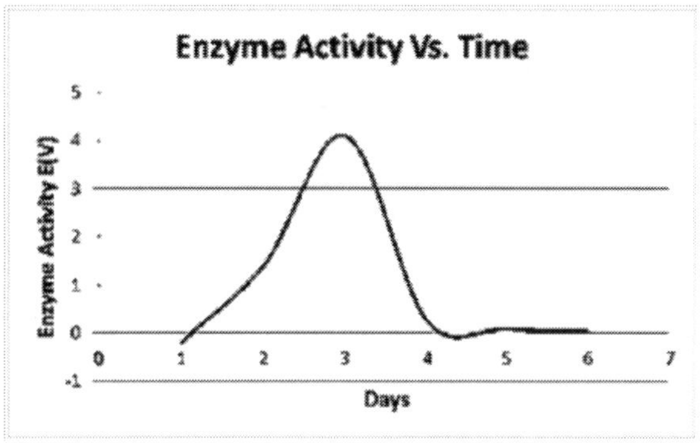

Figure 2. Enzyme activity over a period of 6 days.

The acidic properties of Nafion are not conducive for biofuel cell performance due to enzyme inactivation at low pH's; however, the use of NH_4OH to neutralize the sulphonic acids while maintaining the optimum pH of approximately 6.5 was able to ensure longer periods of operation.

The initial catalysis of glucose reactions were slow for the first 2 days, thereafter activity increased steadily. A peak output 250 $\mu W/cm^2$ was reached 72 hours after the glucose addition (figure 2-3). As this research study did not look at cathode modifications future enzyme based Biofuel cell use of Microperoxidase at cathode along with gold where the biofuel cell will be fueled by NADH and H_2O_2. Identification and continued pH sensitivity of anode and cathode reaction conditions will be tested also using acid/polyvinyl alcohol copolymers. A Biliruben Oxidase cathode will also be attempted in future projects. Other enzymes currently synthesized and sourced include alcohol dehydrogenase (ADH), amyloglucosidase (AGS), invertase (INV) and glucose dehydrogenase (GDH). These sources of oxidation reaction catalysis will be considered in future experiments to obtain the optimum combination of electrode type and absorbed immobilized enzyme.

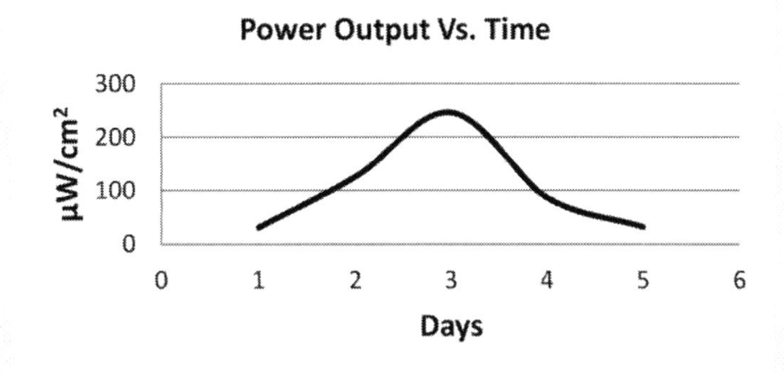

Figure 3. Power output of the MBFC determined over a period of 5 days.

3. Conclusions

Research communities are constantly seeking new sustainable energy sources and have identified BFC's and more specifically enzyme-based biofuel cells as a source of renewable and sustainable power. They are highly attractive for special miniaturized applications, such as implantable devices, sensors, physiological applications including drug delivery, microchips, and other portable power supplies. Several drawbacks, such as short lifetime and low power density, have limited enzyme-based biofuel cells from being used for practical applications. Recent developments in the newly emerging nano-biocatalysis appear to be promising because they provide some solutions in overcoming the present problems facing researches attempting to find cost effective alternative energy sources. Better understanding and further developments of nano-biocatalysis will fast track the improvement of biofuel cells, and once these variables have been identified and optimized to produce high performance biofuel cells the future role of BFC's will be significant in this dynamic energy market. In this research we have successfully used a Nafion membrane together with GOx and under controlled conditions produced respectable power densities.

References

[1] Yahiro A T, Lee S M and Kimble D O 1964 *Biochim Biophys Acta* **88** 375
[2] Bockris J O M and Srinivasan S. Fuel cells: their electrochemistry. New York: McGraw-Hill. 1969.
[3] Govil G and Saran A 1982 *J. Indian Chem. Soc.* **59** 1226
[4] Aston W J and Turner A P F 1984 *Biotechnol. Genet. Eng. Rev.* **1** 89
[5] Palmore G T R and Whitesides G M 1994 *ACS Symp. Ser.* **566** 71
[6] Katz E and Willner I 2003 *J. Am. Chem. Soc.* **125** 6803
[7] Barton S C, Gallaway J and Atanassov P 2004 *Chem. Rev.* **104** 4867
[8] Heller A 2004 *Phys. Chem. Chem. Phys.* **6** 209
[9] Jungbae K, Hongfei J and Ping W 2006 *Biotechnology Advances* 24 296
[10] Varfolomeev S D, Kurochkin I N and Yaropolov A I 1996 *Biosens Bioelectron* **11** 863
[11] Ghindilis A L, Atanasov P and Wilkins E 1997 *Electroanalysis* **9** 661
[12] Schuhmann W 2002 *Rev. Mol. Biotechnol.* **82** 425
[13] Freire R S, Pessoa C A, Mello L D and Kubota L T 2003 *J. Braz. Chem. Soc.* **14** 230
[14] Willner I, Arad G and Katz E 1998 *Bioelectrochem. Bioeng* **44** 209
[15] Willner I, Katz, E, Patolsky F and Buckmann A F 1998 *J. Chem. Soc., Perkin Trans.* **2** 1817
[16] Katz E, Willner I and Kotlyar A B 1999 *J.Electroanal. Chem.* **479** 64

Institute of Physics
Dirac House, Temple Back
Bristol BS1 6BE UK

CURRAN ASSOCIATES INC
proceedings.com